Journal of
*Composites Science*

# Discontinuous Fiber Composites

Edited by
**Tim A. Osswald**

Printed Edition of the Special Issue Published in
*Journal of Composites Science*

www.mdpi.com/journal/jcs

MDPI

# Discontinuous Fiber Composites

# Discontinuous Fiber Composites

Special Issue Editor

**Tim A. Osswald**

MDPI • Basel • Beijing • Wuhan • Barcelona • Belgrade

**MDPI**

*Special Issue Editor*
Tim A. Osswald
University of Wisconsin Madison
USA

*Editorial Office*
MDPI
St. Alban-Anlage 66
4052 Basel, Switzerland

This is a reprint of articles from the Special Issue published online in the open access journal *Journal of Composites Science* (ISSN 2504-477X) from 2017 to 2018 (available at: https://www.mdpi.com/journal/jcs/special_issues/Discontinuous_Fibers)

For citation purposes, cite each article independently as indicated on the article page online and as indicated below:

LastName, A.A.; LastName, B.B.; LastName, C.C. Article Title. *Journal Name* **Year**, *Article Number*, Page Range.

**ISBN 978-3-03897-491-8 (Pbk)**
**ISBN 978-3-03897-492-5 (PDF)**

Cover image courtesy of Courtesy Sebastian Goris.

# Contents

About the Special Issue Editor . . . . . . . . . . . . . . . . . . . . . . . . . . . . . . . . . . . . . . . vii

Preface to "Discontinuous Fiber Composites" . . . . . . . . . . . . . . . . . . . . . . . . . . . ix

Tim A. Osswald
Editorial for the Special Issue on Discontinuous Fiber Composites
Reprinted from: *J. Compos. Sci.* **2018**, *2*, 63, doi:10.3390/jcs2040063 . . . . . . . . . . . . . . . . . 1

Malte Schemmann, Juliane Lang, Anton Helfrich, Thomas Seelig and Thomas Böhlke
Cruciform Specimen Design for Biaxial Tensile Testing of SMC
Reprinted from: *J. Compos. Sci.* **2018**, *2*, 12, doi:10.3390/jcs2010012 . . . . . . . . . . . . . . . . . 4

Anna Trauth, Pascal Pinter and Kay André Weidenmann
Investigation of Quasi-Static and Dynamic Material Properties of a Structural Sheet Molding
Compound Combined with Acoustic Emission Damage Analysis
Reprinted from: *J. Compos. Sci.* **2017**, *1*, 18, doi:10.3390/jcs1020018 . . . . . . . . . . . . . . . . . 25

Yuyang Song, Umesh Gandhi, Takeshi Sekito, Uday K. Vaidya, Jim Hsu, Anthony Yang and
Tim Osswald
A Novel CAE Method for Compression Molding Simulation of Carbon Fiber-Reinforced
Thermoplastic Composite Sheet Materials
Reprinted from: *J. Compos. Sci.* **2018**, *2*, 33, doi:10.3390/jcs2020033 . . . . . . . . . . . . . . . . . 43

Katharina Albrecht, Tim Osswald, Erwin Baur, Thomas Meier, Sandro Wartzack and
Jörg Müssig
Fibre Length Reduction in Natural Fibre-Reinforced Polymers during Compounding and
Injection Moulding—Experiments Versus Numerical Prediction of Fibre Breakage
Reprinted from: *J. Compos. Sci.* **2018**, *2*, 20, doi:10.3390/jcs2020020 . . . . . . . . . . . . . . . . . 59

Mohammed Y. Abdellah, Hesham I. Fathi, Ayman M. M. Abdelhaleem and
Montasser Dewidar
Mechanical Properties and Wear Behavior of a Novel Composite of
Acrylonitrile–Butadiene–Styrene Strengthened by Short Basalt Fiber
Reprinted from: *J. Compos. Sci.* **2018**, *2*, 34, doi:10.3390/jcs2020034 . . . . . . . . . . . . . . . . . 76

Benedikt Fengler, Luise Kärger, Frank Henning and Andrew Hrymak
Multi-Objective Patch Optimization with Integrated Kinematic Draping Simulation for
Continuous–Discontinuous Fiber-Reinforced Composite Structures
Reprinted from: *J. Compos. Sci.* **2018**, *2*, 22, doi:10.3390/jcs2020022 . . . . . . . . . . . . . . . . . 88

Hongyu Chen and Donald G. Baird
Prediction of Young's Modulus for Injection Molded Long Fiber Reinforced Thermoplastics
Reprinted from: *J. Compos. Sci.* **2018**, *2*, 47, doi:10.3390/jcs2030047 . . . . . . . . . . . . . . . . . 107

Tom Mulholland, Sebastian Goris, Jake Boxleitner, Tim A. Osswald and Natalie Rudolph
Process-Induced Fiber Orientation in Fused Filament Fabrication
Reprinted from: *J. Compos. Sci.* **2018**, *2*, 45, doi:10.3390/jcs2030045 . . . . . . . . . . . . . . . . . 120

**Timothy Russell, David A. Jack, Blake Heller, and Douglas E. Smith**
Prediction of the Fiber Orientation State and the Resulting Structural and Thermal Properties of
Fiber Reinforced Additive Manufactured Composites Fabricated Using the Big Area Additive
Manufacturing Process
Reprinted from: *J. Compos. Sci.* **2018**, *2*, 26, doi:10.3390/jcs2020026 . . . . . . . . . . . . . . . . . **134**

**Zhaogui Wang and Douglas E. Smith**
Rheology Effects on Predicted Fiber Orientation and Elastic Properties in Large Scale Polymer
Composite Additive Manufacturing
Reprinted from: *J. Compos. Sci.* **2018**, *2*, 10, doi:10.3390/jcs2010010 . . . . . . . . . . . . . . . . . **152**

**Florian Wittemann, Robert Maertens, Alexander Bernath, Martin Hohberg, Luise Kärger and
Frank Henning**
Simulation of Reinforced Reactive Injection Molding with the Finite Volume Method
Reprinted from: *J. Compos. Sci.* **2018**, *2*, 5, doi:10.3390/jcs2010005 . . . . . . . . . . . . . . . . . **170**

**Christoph Kuhn, Ian Walter, Olaf Täger and Tim Osswald**
Simulative Prediction of Fiber-Matrix Separation in Rib Filling During Compression Molding
Using a Direct Fiber Simulation
Reprinted from: *J. Compos. Sci.* **2018**, *2*, 2, doi:10.3390/jcs2010002 . . . . . . . . . . . . . . . . . **186**

# About the Special Issue Editor

**Tim A. Osswald**, originally from Cúcuta, Colombia, is a Professor of Mechanical Engineering and Director of the Polymer Engineering Center at the University of Wisconsin–Madison. He holds honorary professorships at the Friedrich Alexander University in Erlangen, Germany, and the National University in Bogotá, Colombia. Osswald teaches polymer processing and designing with polymers and researches in the same areas, specifically in fiber orientation, fiber density, and fiber length distributions of discontinuous fiber reinforced composites. Professor Osswald's interests also include the history of polymers and polymer industry. Professor Osswald has published over 300 papers and the books Materials Science of Polymers for Engineers (Hanser, 1996, 2003, 2012), Polymer Processing Fundamentals (Hanser 1998), Injection Molding Handbook (Hanser, 2001, 2007) Compression Molding (Hanser, 2003), Polymer Processing Modeling and Simulation (Hanser 2006), International Plastics Handbook (Hanser 2006), Plastics Testing and Characterization (Hanser, 2008), Understanding Polymer Processing (2010, 2017), and Polymer Rheology (Hanser 2015). His books have been translated into Italian, German, Spanish, Japanese, Chinese, Korean, and Russian. Professor Osswald is also the series editor of Plastics Pocket Power (Hanser, 2001), which currently includes 6 books, is the Editor for the Americas of the *Journal of Polymer Engineering*, and the English language Editor for the *Journal of Plastics Technology*.

# Preface to "Discontinuous Fiber Composites"

Discontinuous fiber-reinforced polymer composites are special forms of composite materials that are appealing to the aerospace and automotive industries because they are easy to process into components and structures of complex shapes in an automated fashion, using compression and injection molding, as well as extrusion processes. The extrusion or mold filling process not only leads to fiber orientation but also results in fiber attrition and agglomeration, which all have a profound impact on the properties of the finished product.

The papers published here give the reader insight into what controls the micromechanical arrangement of the discontinuous fibers and how these influence the mechanical performance of the part, as well lead to a deeper understanding of the process–microstructure–property relationships. The papers present a balance between academic and industrial research, and clearly reflect the collaborative work that exists between the two communities, in a joint effort to solve existing problems.

**Tim A. Osswald**
*Special Issue Editor*

Journal of
*Composites Science*

MDPI

*Editorial*

# Editorial for the Special Issue on Discontinuous Fiber Composites

Tim A. Osswald

Polymer Engineering Center, Department of Mechanical Engineering, University of Wisconsin-Madison, Madison, WI 53706, USA; tosswald@wisc.edu

Received: 11 October 2018; Accepted: 16 October 2018; Published: 23 October 2018

The papers published in this special edition of the *Journal of Composites Science* will give the polymer engineer and scientist insight into what the existing challenges are in the discontinuous fiber composites field, and how these challenges are being addressed by the research community. The papers present a balance between academic and industrial research, and clearly reflect the collaborative work that exists between the two communities, in a joint effort to solve the existing problems.

Discontinuous fiber-reinforced composites are a special subcategory of composite materials that are used, due to the ability to process them into parts and structures of complex shape in an automated fashion via compression and injection molding, as well as extrusion processes. Discontinuous fiber-reinforced composites commonly consist of a thermoset or thermoplastic matrix material. Sheet molding compound (SMC) is the most prominent type of discontinuous fiber-reinforced thermoset and is processed using compression molding to manufacture automotive body panels and structures. Schemmann et a. [1] designed various test specimens to measure the mechanical behavior of SMC under biaxial loading, and Trauth et al. [2] studied the effect of volume fracture and SMC semi-finished charge manufacturing conditions on the mechanical properties of the molded parts. A variety of thermoplastic matrices and processes are also available for discontinuous fiber-reinforced composites. Thermoplastics, such as polyamide (PA) or polypropylene (PP), represent most matrix materials used for thermoplastic fiber reinforced parts due to their superior properties compared to other plastics. Glass fibers are frequently used for reinforcement due to their availability, low cost and high strength, although carbon fibers are also being implemented [3], as well as natural [4] and basalt [5] fibers. Discontinuous fiber-reinforced composites can be further classified as short fiber-reinforced thermoplastics (SFT) and long fiber-reinforced thermoplastics (LFT). The distinction between LFT and SFT is made by the average fiber aspect ratio (length to diameter). A fiber-filled material with an average aspect ratio of less than 100 is defined as short fiber-reinforced, while long fiber-reinforced composites have an average aspect ratio of more than 100. The performance and cost of LFT materials places them between continuous fiber-reinforced composites used for high performance applications and SFT compounds, because LFT can be processed economically with injection molding, while providing superior mechanical properties when compared to SFT. Often, products present a combination of continuous and discontinuous fiber composite systems [6], such as in the growing field of hybrid structures. Here, draped continuous fiber structures are over-molded with discontinuous fiber-filled resins, synergistically creating parts with superior mechanical properties.

The processing of discontinuous fiber-reinforced composites, such as mold filling in injection molding, or flow through the nozzle during fused filament fabrication, has a profound impact on the arrangement of the fibers within the finished part. As fiber filled polymers are shaped into the final part geometry, the fibers are not only oriented [2,3,7–11], but they are also broken down [4] and agglomerated [12], resulting in highly anisotropic products. While the anisotropy can be beneficial, when the fibers are aligned in the direction of highest stresses, most design processes of discontinuous fiber-reinforced composites parts do not consider all aspects of the process-microstructure-property relationship. Several papers in this issue relate mechanical properties to fiber microstructure [1–3,5,7,9,10] and the paper by Wang and Smith [10] relate the microstructure to rheological behavior. Not incorporating the effect of fiber microstructure can result in larger safety factors, which leads to higher weight than necessary, limiting the application of this class of materials. Although discontinuous fiber composites have been used for decades, the underlying physics that control fiber attrition, fiber matrix separation, and even fiber alignment are not fully understood. Without the ability to predict and fully control the fiber microstructure within the final product, the full potential of discontinuous fiber reinforced plastics for lightweight applications cannot be achieved.

Structural analysis can only provide accurate results if the fiber microstructure is estimated accurately. Today, a handful of software companies offer a wide range of tools to predict fiber microstructure which is helpful during the design process of discontinuous fiber-reinforced composites. One can say that these tools are constantly under development and continuously being improved. These tools are indispensable to industry, as parts made of composites for automotive and aerospace are required to go through a comprehensive numerical analysis and design process before being considered for production.

To fully exploit the full potential of discontinuous fiber filled polymer composite materials, it is crucial that the engineer controls and predicts the structural capabilities of fiber-reinforced molded parts, including the configuration of the fibers. Only by adequately incorporating the process-induced fiber microstructure in the design process will it be possible to achieve a reliable prediction of the performance of the molded part. The collection of papers in this issue may help advance technology and bring industry closer to understanding the underlying phenomena that control the microstructure development during processing of discontinuous fiber filled systems, and thus being able to confidently implement them into lightweight applications.

**Conflicts of Interest:** The authors declare no conflict of interest.

## References

1. Schemmann, M.; Lang, J.; Helfrich, A.; Seelig, T.; Böhlke, T. Cruciform Specimen Design for Biaxial Tensile Testing of SMC. *J. Compos. Sci.* **2018**, *2*, 12. [CrossRef]
2. Trauth, A.; Pinter, P.; Weidenmann, K.A. Investigation of Quasi-Static and Dynamic Material Properties of a Structural Sheet Molding Compound Combined with Acoustic Emission Damage Analysis. *J. Compos. Sci.* **2017**, *1*, 18. [CrossRef]
3. Song, Y.; Gandhi, U.; Sekito, T.; Vaidya, U.K.; Hsu, J.; Yang, A.; Osswald, T. A Novel CAE Method for Compression Molding Simulation of Carbon Fiber-Reinforced Thermoplastic Composite Sheet Materials. *J. Compos. Sci.* **2018**, *2*, 33. [CrossRef]
4. Albrecht, K.; Osswald, T.; Baur, E.; Meier, T.; Wartzack, S.; Müssig, J. Fibre Length Reduction in Natural Fibre-Reinforced Polymers during Compounding and Injection Moulding—Experiments Versus Numerical Prediction of Fibre Breakage. *J. Compos. Sci.* **2018**, *2*, 20. [CrossRef]
5. Abdellah, M.Y.; Fathi, H.I.; Abdelhaleem, A.M.M.; Dewidar, M. Mechanical Properties and Wear Behavior of a Novel Composite of Acrylonitrile–Butadiene–Styrene Strengthened by Short Basalt Fiber. *J. Compos. Sci.* **2018**, *2*, 34. [CrossRef]
6. Fengler, B.; Karger, L.; Henning, F.; Hrymak, A. Multi Objective Patch Optimization with Integrated Kinematic Draping Simulation for Continuous–Discontinuous Fiber-Reinforced Composite Structures. *J. Compos. Sci.* **2018**, *2*, 22. [CrossRef]
7. Chen, H.; Baird, D.G. Prediction of Young's Modulus for Injection Molded Long Fiber Reinforced Thermoplastics. *J. Compos. Sci.* **2018**, *2*, 47. [CrossRef]
8. Mulholland, T.; Goris, S.; Boxleitner, J.; Osswald, T.A.; Rudolph, N. Process-Induced Fiber Orientation in Fused Filament Fabrication. *J. Compos. Sci.* **2018**, *2*, 45. [CrossRef]
9. Russell, T.; Heller, B.; Jack, D.A.; Smith, D. Prediction of the Fiber Orientation State and the Resulting Structural and Thermal Properties of Fiber Reinforced Additive Manufactured Composites Fabricated Using the Big Area Additive Manufacturing Process. *J. Compos. Sci.* **2018**, *2*, 26. [CrossRef]

10. Wang, Z.; Smith, D.E. Rheology Effects on Predicted Fiber Orientation and Elastic Properties in Large Scale Polymer Composite Additive Manufacturing. *J. Compos. Sci.* **2018**, *2*, 10. [CrossRef]
11. Wittemann, F.; Maertens, R.; Bernath, A.; Hohberg, M.; Kärger, L.; Henning, F. Simulation of Reinforced Reactive Injection Molding with the Finite Volume Method. *J. Compos. Sci.* **2018**, *2*, 5. [CrossRef]
12. Kuhn, C.; Walter, I.; Täger, O.; Osswald, T. Simulative Prediction of Fiber-Matrix Separation in Rib Filling During Compression Molding Using a Direct Fiber Simulation. *J. Compos. Sci.* **2018**, *2*, 2. [CrossRef]

*Journal of*
*composites science*

MDPI

*Article*

# Cruciform Specimen Design for Biaxial Tensile Testing of SMC

Malte Schemmann [1], Juliane Lang [1], Anton Helfrich [2], Thomas Seelig [3] and Thomas Böhlke [1,*]

[1] Institute of Engineering Mechanics, Chair for Continuum Mechanics, Karlsruhe Insititute of Technology (KIT), Kaiserstr. 10, 76131 Karlsruhe, Germany; malte.schemmann@kit.edu (M.S.); juliane.lang@kit.edu (J.L.)
[2] Institute of Production Science, Manufacturing and Materials Technology, Karlsruhe Insititute of Technology (KIT), Kaiserstr. 12, 76131 Karlsruhe, Germany; anton.helfrich@kit.edu
[3] Institute of Mechanics, Karlsruhe Institute of Technology (KIT), Otto-Ammann-Platz 9, 76131 Karlsruhe, Germany; thomas.seelig@kit.edu
* Correspondence: thomas.boehlke@kit.edu; Tel.: +49-721-608-48852

Received: 30 December 2017; Accepted: 19 February 2018; Published: 1 March 2018

**Abstract:** This paper presents an investigation of different cruciform specimen designs for the characterization of sheet molding compound (SMC) under biaxial loading. The considered material is a discontinuous glass fiber reinforced thermoset. We define various (material-specific) requirements for an optimal specimen design. One key challenge represents the achievement of a high strain level in the center region of the cruciform specimen in order to observe damage, at the same time prevention of premature failure in the clamped specimen arms. Starting from the ISO norm for sheet metals, we introduce design variations, including two concepts to reinforce the specimens' arms. An experimental evaluation includes two different loading scenarios, uniaxial tension and equi-biaxial tension. The best fit in terms of the defined optimality criteria, is a specimen manufactured in a layup with unidirectional reinforcing outer layers where a gentle milling process exposed the pure SMC in the center region of the specimen. This specimen performed superior for all considered loading conditions, for instance, in the uniaxial loading scenario, the average strain in the center region reached 87% of the failure strain in a uniaxial tensile bone specimen.

**Keywords:** discontinuous fibers; sheet molding compound (SMC); biaxial tensile testing; cruciform specimen design; unidirectional reinforcements

## 1. Introduction

### 1.1. Motivation

With their high mass-specific strength and stiffness, fiber reinforced polymers receive considerable attention in mass-reduction strategies in the automotive and aerospace sector. The stiffest and strongest class of composites are continuous fiber reinforced composites. This work, however, focuses on discontinuous fiber reinforced composites in form of sheet molding compounds (SMCs). SMCs offer a large freedom of design and the low cycle times allow for mass production. Characteristics of SMC include brittleness, process-induced inhomogeneity and anisotropy, and comparably large microstructure dimensions.

The application of SMC is hindered by the lack of a precise understanding of its mechanical behavior. The focus of the present work lies in biaxial tensile testing of SMC, which is driven by two main factors. Firstly, the characterization under biaxial stress states covers a wide range of application loads on typical shell-like SMC structures. Secondly, as the implementation of virtual process chains and, especially, damage modeling increases in popularity, a detailed characterization of the damage behavior is essential to understand the complex damage phenomena and calibrate corresponding constitutive models. For instance, biaxial tensile testing allows for a validation of the anisotropic evolution of stiffness.

The cruciform specimen design significantly influences the characterization range of stress states as well as the precision of the characterized material properties. Optimization of specimen design is a material-dependent multi-objective task.

### 1.2. State-of-the-Art

In the past, many attempts have been made to find an appropriate cruciform specimen design, demonstrating that the specimen design is one of the most challenging aspects of the biaxial testing. Several authors have proposed specimen designs for specific applications and materials, varying the cut shape, the type of the thickness-reduced area and the type of slits in the specimen arms, which serve to reduce undesired lateral constraints on the strain in the center region of the specimen [1]. Deng et al. [2], Kuwabara et al. [3] and Makinde et al. [4] optimized specimens for metals. Demmerle and Boehler [5] compared different proposed specimen designs and chose the specimen of Kelly [6] with slits and a thickness reduced area to perform an optimization for isotropic materials. Boehler et al. [7] investigated anisotropic sheet metals with this specimen type. Hoferlin et al. [8] presented an alternative specimen design with slits and reinforced arms. Hannon and Tiernan [9] reviewed planar biaxial tensile test systems for sheet metals. ISO 16842 [10] is the first standardization in this field but applies to sheet metals only. Green et al. [11] proposed a sandwich design for aluminum sheet alloys in which the sample sheet is bonded by an adhesive between two face sheets, while leaving the center region free on both sides.

Other authors conducted research to find a suitable specimen for composites and polymers. Smits et al. [12] and Van Hemelrijck et al. [13] investigated the influence of parameters like the radius of the corner fillet and the thickness and the geometry of the biaxially loaded area by finite element simulations and experiments. This investigation led to a suitable specimen for fiber reinforced composite laminates with a reduced thickness area in the center region of the specimen, in combination with a fillet corner between arms. Makris et al. [14] and Makinde et al. [4] optimized the specimen shape by a numerical optimization technique with a parametric finite element model. Lamkanfi et al. [15] and Gower and Shaw [16] showed that geometrical discontinuities like the transition zone to the tapered thickness area have a major influence on the strain distribution leading to premature failure. Serna Moreno et al. [17] compared specimens with arms of different widths for chopped glass-reinforced polyester. They presented a specimen that was suitable to achieve failure in the center area in different loading cases but pointed out that there are still problems such as stress concentrations outside the center area. This glance into the literature shows that, despite the promising improvements already achieved, there are still many challenges ahead in the search for suitable cruciform specimen designs for composites.

### 1.3. Present Work

This work presents an investigation of different specimen designs for biaxial tensile testing of SMC. We aim for an optimal geometry for the characterization of damage evolution under a wide range of biaxial planar stress states. The outline is as follows: Section 2.2 presents the background of biaxial tensile testing and introduces the two loading scenarios we apply for the experimental evaluation of the specimen design. Section 3 begins with a definition of our specific specimen optimality criteria. In the following, we introduce four specimen designs and evaluate them experimentally and partly with finite element simulations. Section 4 discusses and compares the results in terms of the introduced optimality criteria.

## 2. Materials and Experiment

### 2.1. Materials and Manufacturing Process

The SMC considered here consists, specifically, of an unsaturated polyester polyurethane hybrid (UPPH) matrix system, reinforced with 23 vol % glass fibers. The SMC was manufactured at the Fraunhofer Institute of Chemical Technology (ICT) at Pfinztal, Germany. The detailed composition of the matrix system is listed in Table 1. The length of the fibers is 25.4 mm.

**Table 1.** Composition of the unsaturated polyester polyurethane hybrid (UPPH) resin [18].

| Component | Trade Name | Weight Fraction |
|---|---|---|
| UPPH resin | Daron ZW 14142 | 77% |
| Adherent and flow aids | BYK 9085 | 1.5% |
| Impregnation aid | BYK 9076 | 2.3% |
| Deaeration aid | BYK A-530 | 0.38% |
| Inhibitor | pBQ | 0.0023% |
| Peroxide | Trigonox 117 | 0.77% |
| Isocyanate | Lupranat M20R | 18% |

The SMC pre-impregnated fibers were manufactured on a belt conveyor system. The initial charge had 60% mold coverage (with exception of the specimens presented in Section 3.4) in a square mold with the dimensions 455 × 455 mm and consisted of three layers of pre-impregnated fibers. With regard to more details of the manufacturing process, we refer to [18]. All specimens were cut with a water jet cutter from the center region of the plate to minimize the influence of inhomogeneous fiber dispersion, e.g., investigated in [19].

### 2.2. Biaxial Tensile Experiments

#### 2.2.1. Fundamentals of Biaxial Tensile Testing

In this section, we briefly introduce our biaxial tensile testing device at the Institute of Engineering Mechanics (KIT) and some fundamentals on biaxial tensile testing. The biaxial tensile testing device shown in Figure 1a consists of four horizontally positioned elctro-mechanical actuators that are arranged perpendicular to each other. Each axis allows for a maximum load of 150 kN and is equipped with a load cell. The deformation of the specimen is measured by an integrated optical strain measurement system via the displacement of five points on the bottom of the specimen. The movement of these points is the input for the measurement of the strain load and midpoint control. The active midpoint control allows for bending-free load applications, even in the case of heterogeneous specimens. On the upper side of the specimen, a speckle pattern allows for full field strain measurements via digital image correlation. Due to the, in general, inhomogeneous stress and strain fields in the cruciform specimens the parameter identification is typically not as straightforward as, e.g., in uniaxial tensile tests. More details concerning inverse parameter identification are, e.g., covered in [20–25].

Figure 1b depicts an exemplary cruciform specimen. If we assume the specimen as homogeneous, the opposing forces on the specimen are equal and given by $F_1$ in positive and negative $e_1$-direction and $F_2$ in positive and negative $e_2$-direction. The ratio of these two forces is known as the loading ratio

$$\Gamma = \frac{F_2}{F_1}. \tag{1}$$

The loading ratio of $\Gamma = 0$ defines uniaxial tension or compression in the $e_1$-direction, whereas the loading ratio $\Gamma = 1$ defines equi-biaxial tension or compression. We define the area of interest $A^I$ as the measuring region which later can serve as input for the parameter identification. In all following contour plots, the area of interest is highlighted by a purple frame. For the cruciform specimen design, the area-averaged strain over the area of interest,

$$\bar{\varepsilon}_{ij}^{I} = \frac{1}{A^{I}} \int_{A^{I}} \varepsilon_{ij}(x) \, dA, \tag{2}$$

is a key quantity in the evaluation of the specimen geometries, with $\varepsilon_{ij} = \left(\partial u_i / \partial x_j + \partial u_j / \partial x_i\right) / 2$ being the components of the infinitesimal strain tensor defined from the displacement vector $u_i$.

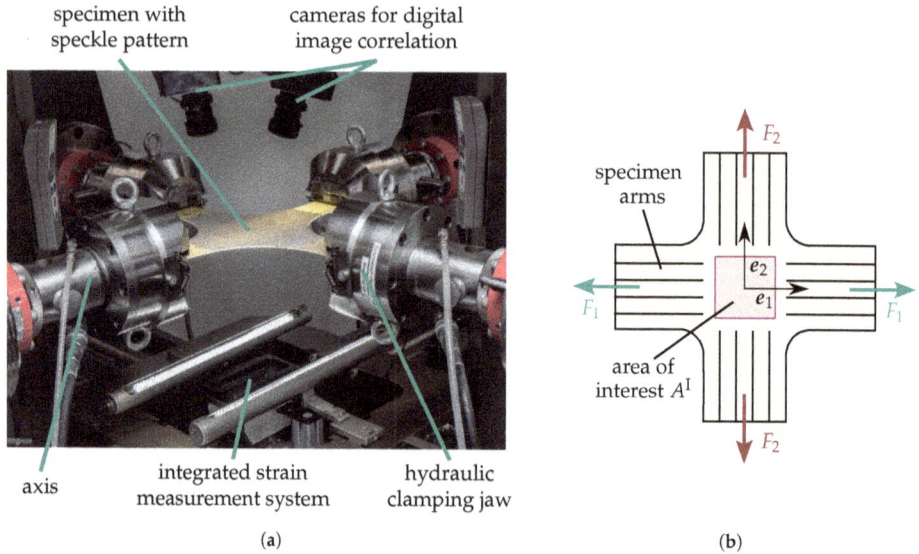

Figure 1. Biaxial tensile testing setup and an exemplary cruciform specimen. (a) Biaxial tensile testing device; (b) Cruciform specimen.

## 2.2.2. Experimental Procedures

For the experimental and simulative evaluation of the specimen designs, we focus on two loading scenarios. We avoid compressive loads as they may lead to buckling issues. The distinction of the mechanical phenomena like, for instance, elasticity, stiffness degradation (damage), plasticity, and viscoelasticity is simplified by a stepwise load application and waiting times between the load application steps.

The first procedure $P_{\Gamma=0}$ is based on cyclic uniaxial tension in $e_1$-direction with a stepwise load increase. Between these loading steps, we apply a constant uniaxial tension in the $e_2$-direction. This allows for an observation of stiffness degradation in and perpendicular to the main loading direction. Figure 2 shows a schematic stress and strain path for the loading procedure $P_{\Gamma=0}$. The tensile load is applied strain-controlled. Perpendicular to the respective loading direction, a force control ensures a low contact force.

The second procedure $P_{\Gamma=1}$ (see Figure 3) is designed such that the damage inducing load is equi-biaxial tension. Between the biaxial tension steps of increasing amplitude, uniaxial tension steps of small and constant amplitude are applied in the $e_1$-direction and subsequently in the $e_2$-direction in order to estimate the stiffness evolution in these directions. The strain and force control is analogous to the first procedure.

We chose these two procedures because we assume that the specimens which perform well—in the sense of allowing to detect an anisotropic damage evolution—at the extreme loading ratios $\Gamma = 0$ and $\Gamma = 1$ also perform well at the loading ratios in between.

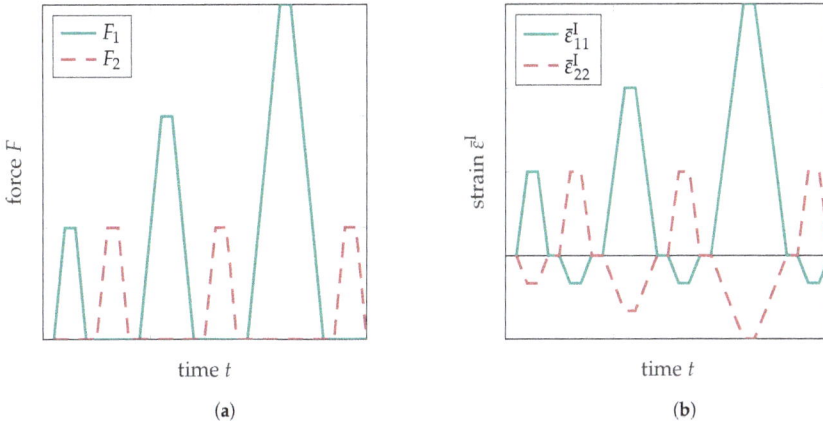

**Figure 2.** Schematic force and strain paths for experimental procedure $P_{T=0}$. (**a**) Force path for procedure $P_{T=0}$; (**b**) Strain path for procedure $P_{T=0}$.

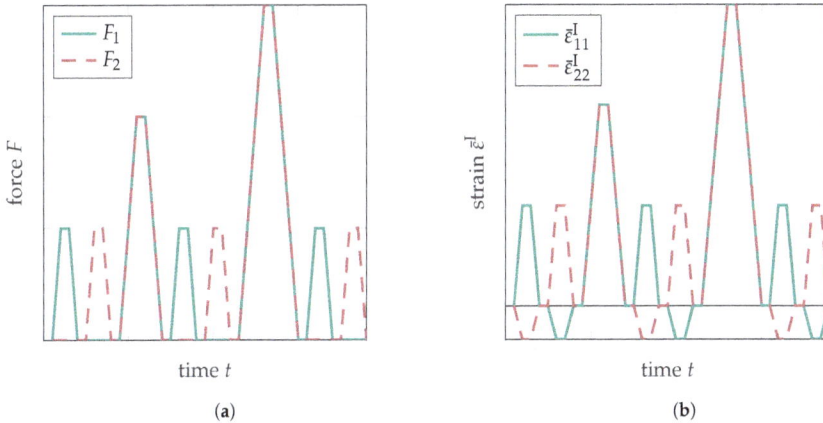

**Figure 3.** Schematic force and strain paths for experimental procedure $P_{T=1}$. (**a**) Force path for procedure $P_{T=1}$; (**b**) Strain path for procedure $P_{T=1}$.

## 3. Specimen Designs and Experimental Results

### 3.1. Specimen Requirements

Our goal for biaxial tensile tests is the robust measurement of the anisotropic evolution of damage and material strength of SMC under various planar stress states. Smits et al. [12] listed criteria for optimal cruciform specimens of unidirectional composites. We modified and extended this list by taking the specific properties of SMC into consideration. We believe that the optimality of a specimen is qualitatively determined by the following criteria:

(1) *Wide range of achievable stress states.* An optimal specimen geometry allows for all biaxial tension stress states, i.e., ratios between normal stresses. By coordinate transformation, all planar stress states (not in magnitude, but in relation to each other) are covered. The measurement of all stress states with the same geometry does not only ensure a good comparability in contrast to multiple specimen geometries, but also allows for a straightforward application of non-monotonic loading paths.

(2) *Damage dominantly in the area of interest.* Since it is our goal to inspect damage in the area of interest, we would like to avoid premature specimen failure in the arms, and thus analyze the material behavior at highest possible strains in the area of interest.

(3) *Homogeneity of stress state in the area of interest.* For the analysis of damage, it is desirable to reach a homogeneous stress state in the area of interest. This implies the demand to avoid stress concentrations.

(4) *Robust parameter identification.* The parameter identification must be a well-posed problem and robust with respect to noise of measured quantities (forces and strain field) [26]. A robust parameter identification is essential for reproducibility. The robustness of the parameter identification is, however, not considered in this paper.

(5) *Large area of interest.* The microstructural dimensions are in case of SMC, compared to other discontinuous fiber reinforced polymers, relatively large. The typical fiber roving length is 25.4 mm, whereas one roving is assembled of thousands of filaments. As the specimen size is limited, it is our goal to achieve a considerably large area of interest.

(6) *Low production effort.* In contrast to uniaxial tensile specimens, the load ratio is an additional parameter to be considered in the design of experiments. SMC is known to show significant scatter in experimental results. Additionally, the anisotropy and inhomogeneity must be considered in the design of experiments. The resulting high number of required experiments can better be coped with, if the economical effort for the specimen production is low.

*3.2. Unreinforced Specimen Arms*

3.2.1. Specimen Design

We introduce the first specimen design in line with the ISO 16842 norm [10] for biaxial tensile testing of sheet metals. Typical metals analyzed in biaxial tensile testing are aluminum and deep-drawing steels. These materials show, in contrast to SMC, large strain in the nonlinear (plastic) regime as well as work hardening. Figure 4 shows the corresponding design of a subsequently tested SMC specimen. The main advantages of water jet cutting include excellent heat removal and a minimum slit width of 1.2 mm. Aiming at a maximum surface area in the arms, we decided to introduce only three slits, which lead to a reduction of the arm surface area of 7.2%. In contrast to the normed design, we chose the slits to extend along the entire arms, in order to avoid a damaged area enforced by the starting points of water jet cutting.

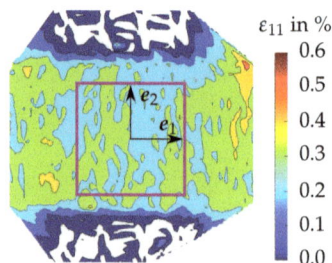

**Figure 4.** Image of the unreinforced cruciform specimen, thickness of the area of interest: 2 mm.

3.2.2. Results

For both the testing procedure $P_{\Gamma=0}$ and the testing procedure $P_{\Gamma=1}$, the specimens failed in the arms before a sufficiently high strain in the area of interest could be reached. Figure 5a shows the force over the strain in the $e_1$-direction for testing procedure $P_{\Gamma=0}$. The corresponding contour plot (see Figure 5b) shows that the strain in the arms is approximately equal to the strain in the area of interest, which violates optimality criterion 2. Figure 6a shows the mean values of the forces and the strains in the $e_1$-direction and the $e_2$-direction for testing procedure $P_{T=1}$. The contour plot (Figure 6b) shows that the normal strain $\varepsilon_{11}$ is significantly higher in the arms than in the area of interest. The plot only shows the areas of positive strains. The asymmetry of the image section is due to the recording area of the cameras. A contour plot of the area of interest at the instant shortly before failure and an image of the failed specimen for all specimen designs can be found in Appendix A.

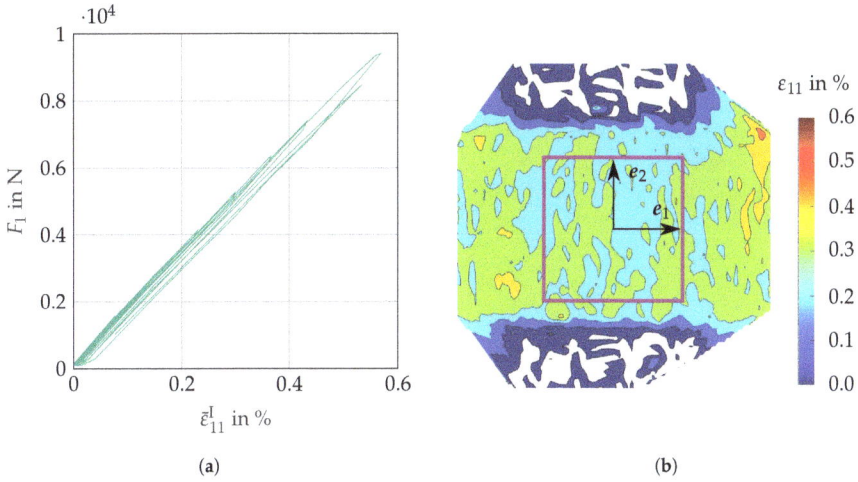

**Figure 5.** Unreinforced specimen arms with loading scenario $P_{T=0}$. (**a**) Force $F_1$ in $e_1$-direction over averaged normal strain $\bar{\varepsilon}^I_{11}$ in the area of interest; (**b**) Contour plot of the normal strain $\varepsilon_{11}$ at $\bar{\varepsilon}^I_{11} = 0.3\%$.

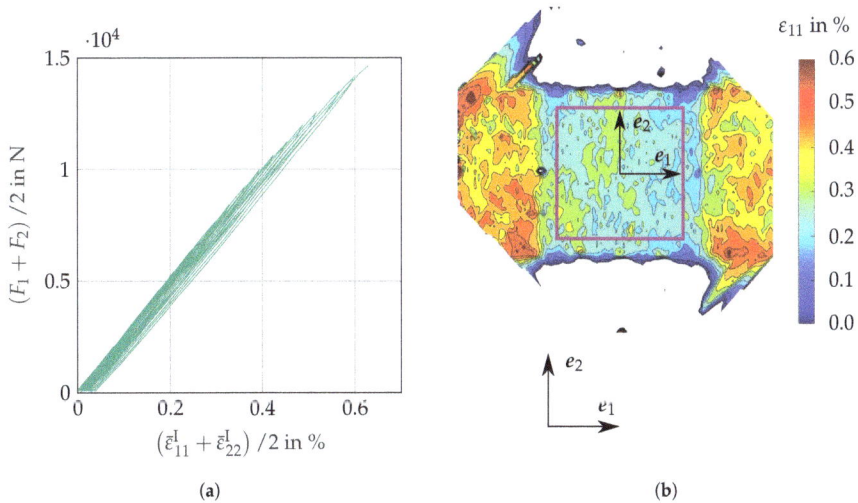

**Figure 6.** Unreinforced specimen arms with loading scenario $P_{T=1}$. (**a**) Force over strain, average values over the $e_1$- and the $e_2$-direction; (**b**) Contour plot of the normal strain $\varepsilon_{11}$ at $\bar{\varepsilon}^I_{11} = 0.3\%$.

### 3.3. Bonded Reinforcements on the Arms

#### 3.3.1. Specimen Design

To avoid premature failure in the arms, we reinforced the specimen by bonding strips on the arms as shown in Figure 7. The upper part of the picture shows the fixation of the reinforcing strips during the manufacturing process. The strips are cut from the same SMC plate as the specimen. We glued with epoxy adhesive, which has similar material properties as the resin material (UPPH). The preparation of the specimens requires many hours of manual work and, therefore, violates optimality criterion 6.

**Figure 7.** Image of the cruciform specimen with bonded reinforcements on the arms, thickness of the area of interest: 2 mm.

#### 3.3.2. Results

The bonded reinforcements significantly increase the maximum achievable strain. Here, the specimens also failed in the arms; however, a higher strain in $A^I$ can be reached compared to the unreinforced specimen. One reason for the premature failure in the arms was a partial detachment of the bonded reinforcing strips. The force strain diagram (see Figure 8a) demonstrates the increase of the maximum reachable strain for testing procedure $P_{\Gamma=0}$. Besides that, a stiffness degradation is already visible from the decrease of the slope of the force strain curve after each cycle. However, the contour plot (Figure 8b) shows that the strain in the arms is still approximately equal to the strain in the area of interest. Figure 9a shows the mean values of the forces and the strains in the $e_1$-direction and the $e_2$-direction for the testing procedure $P_{\Gamma=1}$. Here too the maximum strain is slightly higher compared to the unreinforced specimen. The contour plot (Figure 9b) shows that the normal strain in the arms is still significantly higher than the normal strain in the area of interest.

### 3.4. Continuous Fiber Reinforced Arms

#### 3.4.1. Particularities in Specimen Manufacturing

To overcome the disadvantages of bonded reinforcements on the arms, we considered the manufacturing process of SMC with unidirectional tapes [27]. The unidirectional tapes' constituents are UPPH resin (also in SMC) and 60 vol % carbon fibers. The plates were manufactured with 100% mold coverage to avoid in-mold slip of the layers during the compression molding. More details regarding the co-molding process can be found in [28]. Figure 10 shows the symmetric layup, where the middle layer consists of SMC and the outer layers are tapes with a fiber orientation perpendicular to each other.

Figures 10 and 11 depict the pockets machined into the continuous layers of the specimen. The milling process of these pockets is described in the following.

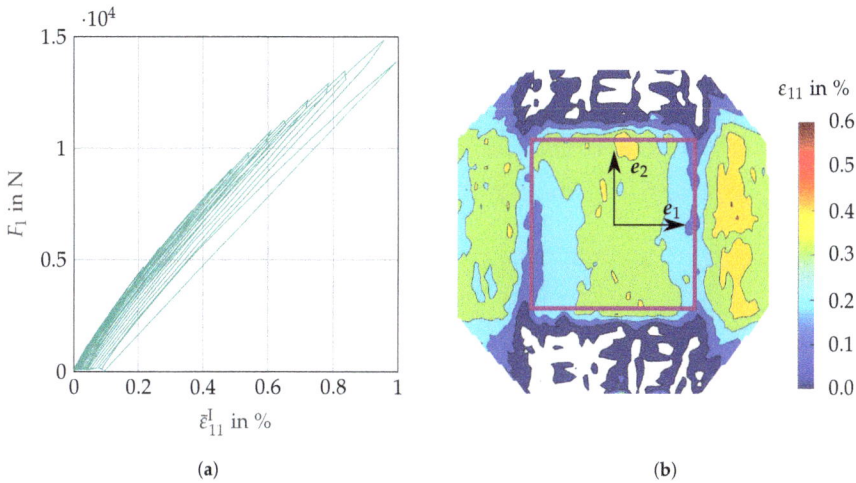

**Figure 8.** Bonded reinforcements on the arms with loading scenario $P_{T=0}$. (a) Force $F_1$ in $e_1$-direction over averaged normal strain in the area of interest $\bar{\varepsilon}^I_{11}$; (b) Contour plot of the normal strain $\varepsilon_{11}$ at $\bar{\varepsilon}^I_{11} = 0.3\%$.

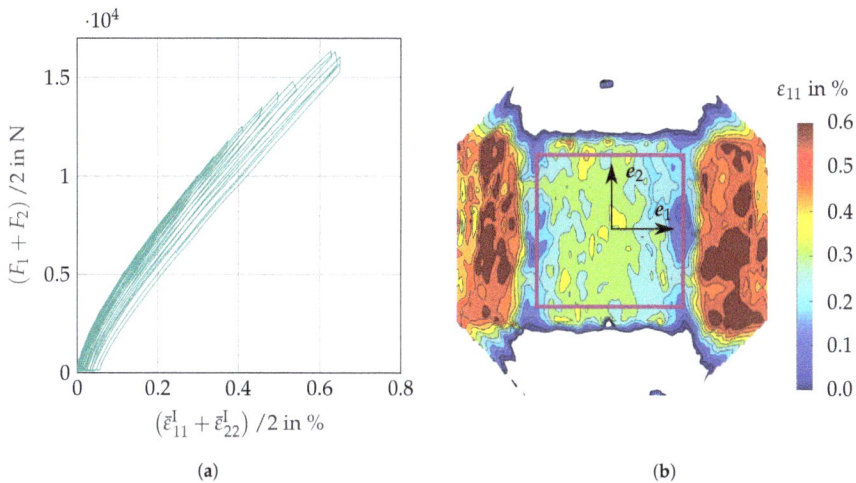

**Figure 9.** Bonded reinforcements on the arms with loading scenario $P_{T=1}$. (a) Force over strain, average values over the $e_1$- and the $e_2$-direction; (b) Contour plot of the normal strain $\varepsilon_{11}$ at $\bar{\varepsilon}^I_{11} = 0.3\%$.

For this process, a diamond coated end mill with eight cutting edges and a multi-cut geometry by Hufschmied GmbH (Bobingen, Germany) was used. The specimens were clamped into position with a line clamping on two sides of the pocket as to minimize the distance of the pocket center to either side of the clamping. This clamping strategy also allowed for a continuous process, i.e., no reclamping was necessary. We generated the tool path from a computer-aided design (CAD) model of the specimens with the Siemens NX computer-aided manufacturing (CAM) system. The values of the cutting speed $v_c$ and the feed per tooth $f_{PT}$ were experimentally qualified during preliminary tests with a similar material and the same tool. These parameters were set at $v_c = 66\,\mathrm{m/min}$ and $f_{PT} = 0.07\,\mathrm{mm}$, respectively. The main challenges in machining of the pockets were to prevent any kind of delamination, which is critical in plunging into the material, and to obtain a smooth, even surface without heat-induced damages or altered material properties. After the first cut with a depth of 1 mm, in some areas of the pocket, remnants of the continuous fibers were found. These remnants likely originate from a movement of the unidirectional tapes in $e_3$-direction during the compression molding process. Nevertheless, further machining steps were performed in 0.1 mm steps to ensure a pure SMC sector in the area of interest.

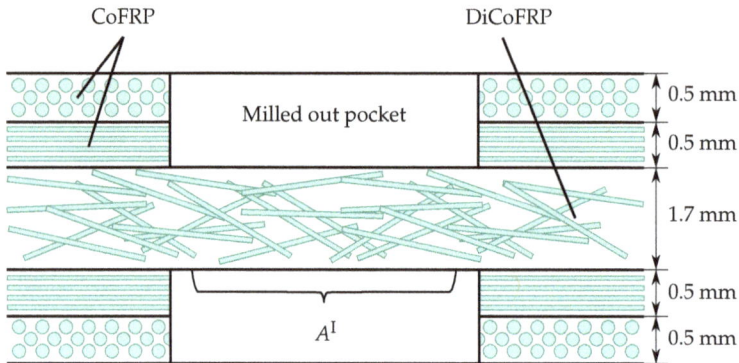

**Figure 10.** Cross section of continuous fiber reinforced SMC, with milled pockets in the area of interest.

### 3.4.2. Specimen Design

The novel manufacturing technique significantly increases the arms' stiffness and strength in the loading directions and, therefore, defines new constraints in the search for the optimal geometry. Here, we introduce two geometries that we consider to be a good compromise between the optimality criteria presented in Section 3.1. As there are difficulties in parametric shape optimization [29], we decided to perform a high number (thousands) of FE (finite element) simulations and thoroughly selected suitable solutions manually.

The first geometry, in the following called geometry 1, follows the design of the unreinforced specimen. Figure 11a shows an image of the design. The significant increase of arm strength allowed for incorporating six slits in each arm to reduce the peak stress at the slit's ends and reduce load transition in the slits perpendicular to the normal stress direction. The slits end slightly before the beginning of the milled out pockets.

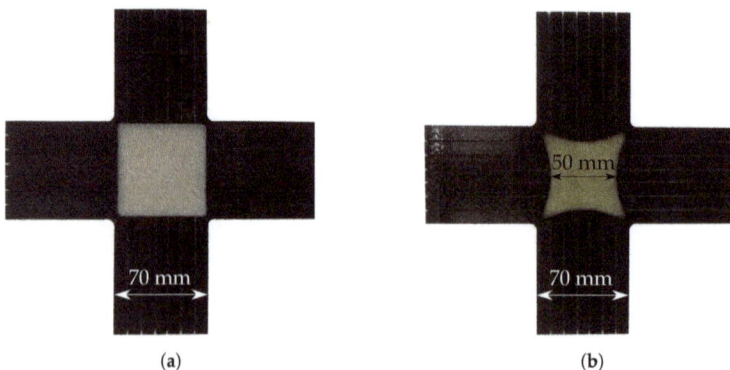

**Figure 11.** Images of specimens with continuous fiber reinforced arms, thickness of the area of interest: 1.7 mm. (**a**) Geometry 1; (**b**) Geometry 2.

Figure 12 depicts strain fields computed from finite element simulations on geometry 1. We took advantage of the specimen's symmetry and only simulated one fourth of the geometry with a fine discretization of shell elements. As we were only interested in trends of different geometry modifications and not precise strain fields, we assumed linear elastic isotropic material behavior with a different stiffness and element thickness in the pocket and the reinforced area. We would like to remark here that a precise quantitative comparison of designs should include the prediction of macroscopic cracks leading to macroscopic failure and an evaluation of the observability of the damage stage. This was not achievable within the scope of the presented research. The upper images show the strain distributions for a uniaxial loading scenario $P_{\Gamma=0}$ in the horizontal $e_1$-direction. The strain distribution in the area of interest shows only small fluctuations. The outer arms contribute more to the load transition into the specimen, as part of the outer load is transmitted into the tape-reinforced regions. The lower images visualize the strain distribution for an equi-biaxial tensile loading scenario $P_{\Gamma=1}$. The distribution of the normal strain $\varepsilon_{11}$ (Figure 12c) shows peaks at the end of the horizontal slits. Figure 12d depicts the distribution of the maximal principal strain $\varepsilon_1$. Due to the reduced stiffness in the area of interest, here, the strain is significantly higher than in the earlier discussed designs.

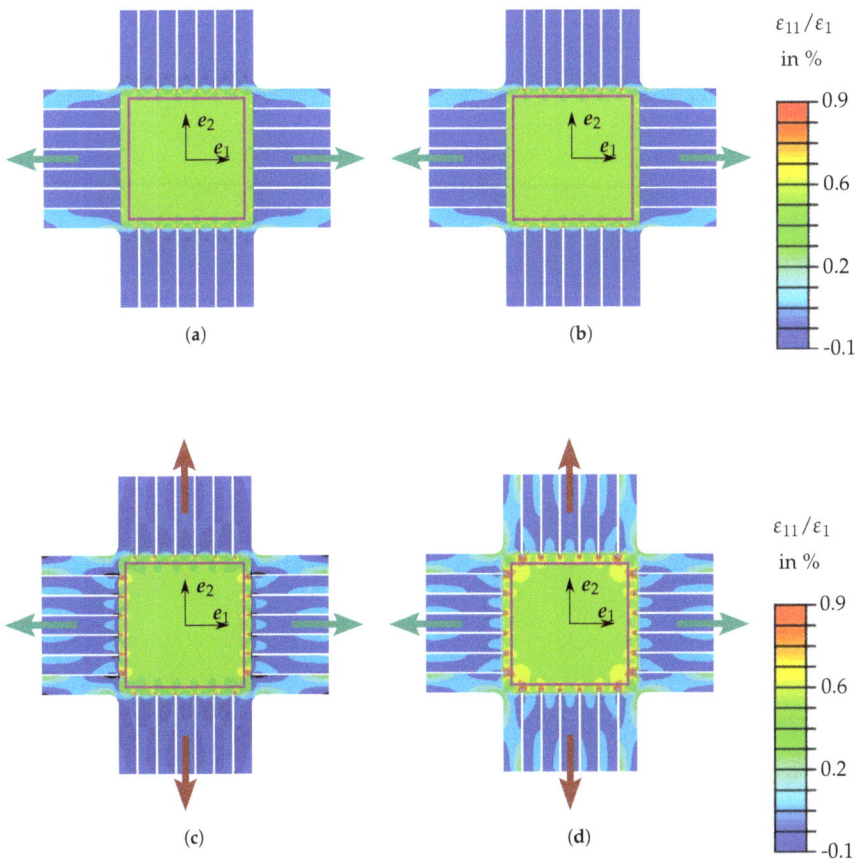

**Figure 12.** FE (finite element) results for geometry 1 specimen at a macroscopic strain of $\bar{\varepsilon}_{11}^{I} = 0.27\%$. The strain fields show the normal strain $\varepsilon_{11}$ and the largest principal strain $\varepsilon_1$. (**a**) $\varepsilon_{11}$ for uniaxial load in $e_1$-direction; (**b**) $\varepsilon_1$ for uniaxial load in $e_1$-direction; (**c**) $\varepsilon_{11}$ for equi-biaxial load; (**d**) $\varepsilon_1$ for equi-biaxial load.

Geometry 2 is different from geometry 1 in such a way that the milled out area has curved edges (see Figure 11). In analogy to uniaxial tapered bone specimens, our aim was to achieve an elevated stress level in the area of interest. Figure 13 shows the strain fields for uniaxial and biaxial load. For the uniaxial loading case, the stress distribution is more homogeneous; however, in front of the second outer slit end, there is a significant strain concentration. For the case of equi-biaxial loading, the strain concentrations, especially the principal strain $\varepsilon_1$ (see Figure 13d), shows significant peaks towards the end of the slits.

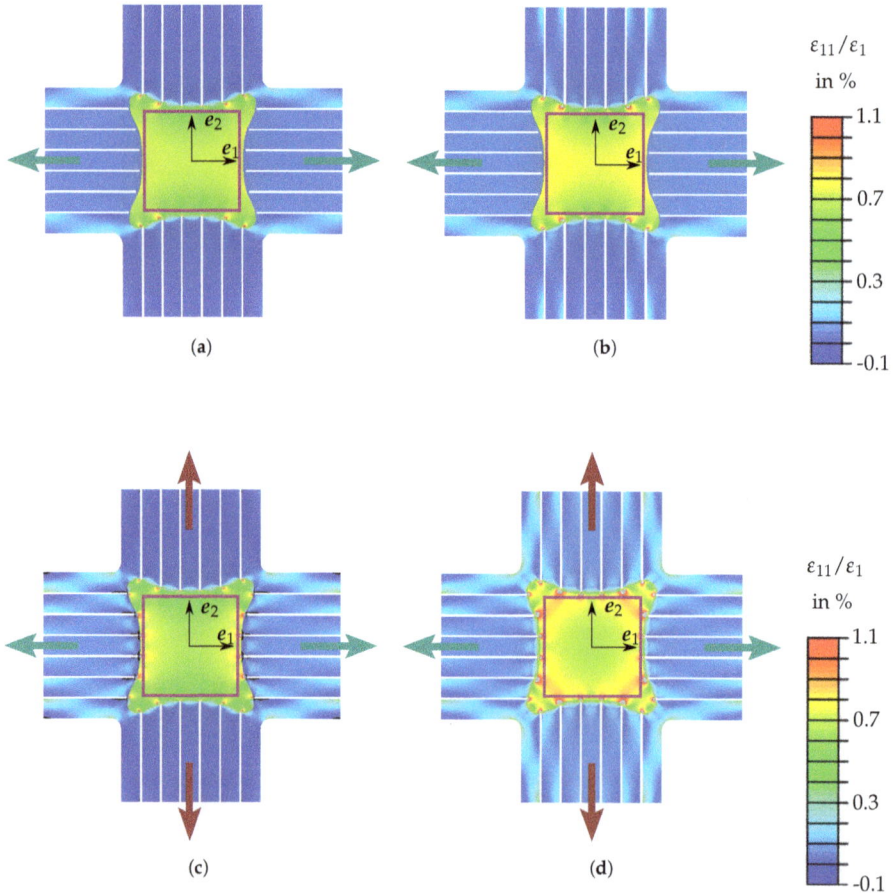

**Figure 13.** FE results for geometry 2 specimen at a macroscopic strain of $\bar{\varepsilon}_{11}^I = 0.27\%$. The strain fields show the normal strain $\varepsilon_{11}$ and the largest principal strain $\varepsilon_1$. (**a**) $\varepsilon_{11}$ for uniaxial load in $e_1$-direction; (**b**) $\varepsilon_1$ for uniaxial load in $e_1$-direction; (**c**) $\varepsilon_{11}$ for equi-biaxial load; (**d**) $\varepsilon_1$ for equi-biaxial load.

### 3.4.3. Results

The contour plots in Figures 14–17 show that the experimental strain distributions are similar to those predicted from the simulations. Figure 14b confirms the relatively homogeneous strain field of Figure 12a for geometry 1 and testing procedure $P_{\Gamma=0}$. For testing procedure $P_{\Gamma=1}$, the small strain peaks at the end of the slits mentioned in conjunction with Figure 12c are visible again (see Figure 15b). For both testing procedures, the strain in the area of interest is significantly higher than the strain in the arms. Failure occurs in the area of interest when a sufficiently high strain is reached. Figure 14a shows the force over the strain in the $e_1$-direction for the testing procedure $P_{\Gamma=0}$. This figure demonstrates that the continuous fiber reinforced layers significantly increase the maximum achievable strain. A stiffness degradation is clearly visible. Figure 15a shows that the reinforcement increases the maximum achievable strain also for testing procedure $P_{\Gamma=1}$.

Figures 16b and 17b show the contour plots of geometry 2. Here too, the strain in the area of interest is significantly higher than in the arms. The strain peaks at the end of the slits, as mentioned for Figure 13, are visible, particularly pronounced for testing procedure $P_{\Gamma=1}$. Failure occcurs in the area of interest, but is initiated at the end of the slits. The force strain diagrams (Figures 16a and 17a) show that also geometry 2 increases the maximum achievable strain, but slightly less than geometry 1.

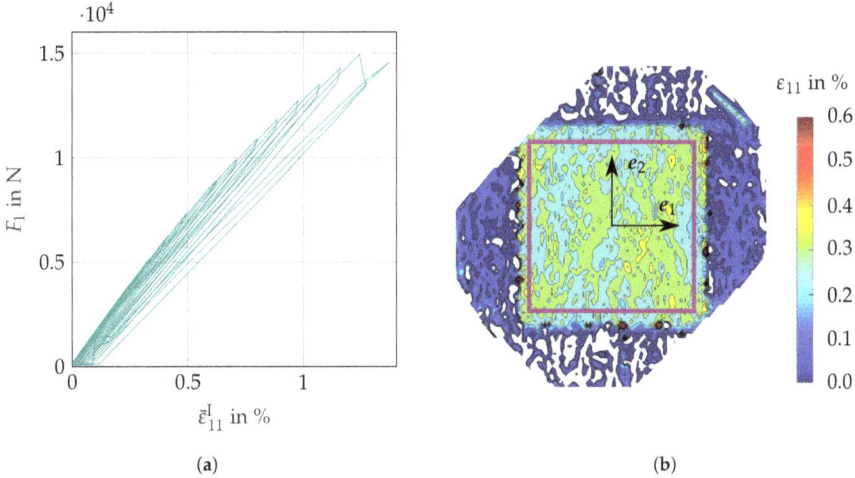

**Figure 14.** Geometry 1 with continuous fiber reinforced arms with loading scenario $P_{\Gamma=0}$. (**a**) Force $F_1$ in $e_1$-direction over averaged normal strain $\bar{\varepsilon}^I_{11}$ in the area of interest; (**b**) Contour plot of the normal strain $\varepsilon_{11}$ at $\bar{\varepsilon}^I_{11} = 0.3\%$.

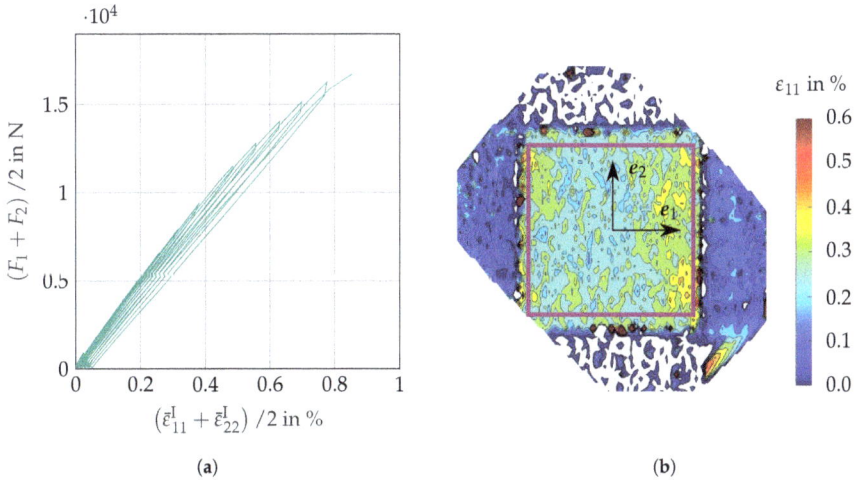

**Figure 15.** Geometry 1 with continuous fiber reinforced arms with loading scenario $P_{\Gamma=1}$. (**a**) Force over strain, average values over the $e_1$- and the $e_2$-direction; (**b**) Contour plot of the normal strain $\varepsilon_{11}$ at $\bar{\varepsilon}^I_{11} = 0.3\%$.

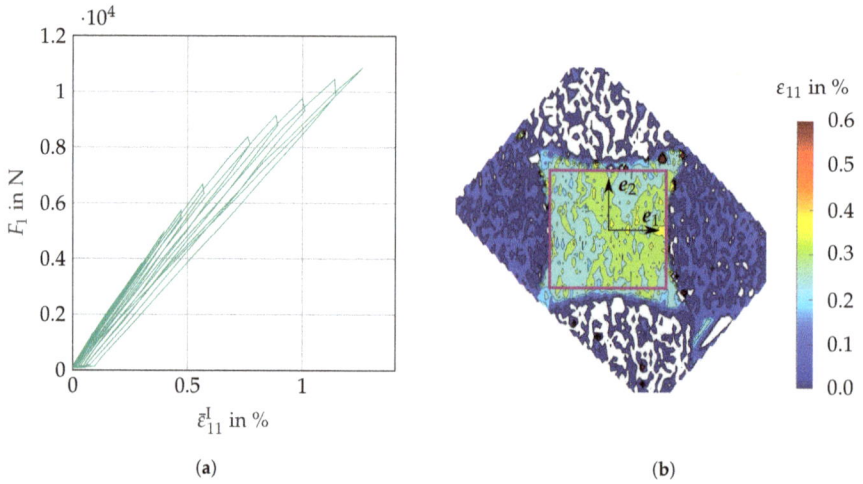

**Figure 16.** Geometry 2 with continuous fiber reinforced arms with loading scenario $P_{\Gamma=0}$. (a) Force $F_1$ in $e_1$-direction over averaged normal strain $\bar{\varepsilon}^I_{11}$ in the area of interest; (b) Contour plot of the normal strain $\varepsilon_{11}$ at $\bar{\varepsilon}^I_{11} = 0.3\%$.

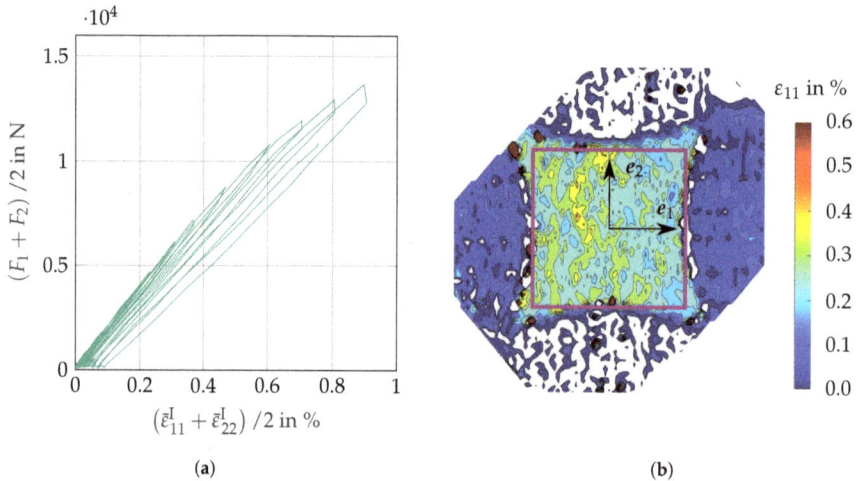

**Figure 17.** Geometry 2 with continuous fiber reinforced arms with loading scenario $P_{\Gamma=1}$. (a) Force over strain, average values over the $e_1$- and the $e_2$-direction; (b) Contour plot of the normal strain $\varepsilon_{11}$ at $\bar{\varepsilon}^I_{11} = 0.3\%$.

## 4. Discussion

Table 2 summarizes the maximum attained normal strains averaged over the area of interest max $\left(\bar{\varepsilon}^I_{11}\right)$ in loading scenario $P_{\Gamma=0}$ and max $\left(\bar{\varepsilon}^I_{11} + \bar{\varepsilon}^I_{22}\right)/2$ in loading scenario $P_{\Gamma=1}$ accordingly. These strains especially quantify the level of achievable strain in the area of interest and thus allow for evaluating the specimen with regard to criterion 2, since damage only occurs if sufficiently high strains are achieved (criterion 2 of the previously defined specimen requirements (Section 3.1)). The ranking of the specimens is equal for both loading scenarios. The unreinforced specimen fails at the lowest strains, bonded reinforcements on the arms lead to a significant improvement, the continuous fiber reinforced specimens perform better, whereas geometry 1 performs slightly better than geometry 2. In loading scenario $P_{\Gamma=0}$, the continuous reinforced geometry 1 reaches 87% of the failure strain measured in a uniaxial tensile bone specimen tested on the same machine. In parallel to the maximum strain reached in the area of interest, a steady reduction of the slope in the force–strain plots indicates the evolution of damage. A visual comparison of these plots indicates a good agreement with Table 2, such that the specimens that allow for the observation of higher strains also show a higher slope reduction in the force–strain plot. The interpretation of the results requires great caution, since only one experiment per specimen geometry and load case was performed.

**Table 2.** Maximum attained strain within one experiment for the different biaxial specimen designs and, for reference, the failure strain of a uniaxial bone specimen. Here, the operator $\max(\cdot)$ denotes the maximum value in the course of the experiment.

| Reinforcement Type | | None | Bonded SMC | Cont. Geom. 1 | Cont. Geom. 2 | Uniax. Bone |
|---|---|---|---|---|---|---|
| $\max\left(\bar{\varepsilon}_{11}^{I}\right)$ | $P_{\Gamma=0}$ | 0.57% | 1.00% | 1.37% | 1.26% | 1.57% |
| $\max\left(\bar{\varepsilon}_{11}^{I}+\bar{\varepsilon}_{22}^{I}\right)/2$ | $P_{\Gamma=1}$ | 0.63% | 0.65% | 0.85% | 0.76% | - |

Geometry 1 also performs equal or superior to the other specimens considering optimality criteria 1 and 3. Force $F_1$ is almost ideally transferred via a homogeneous normal stress in the $e_1$-direction, force $F_2$ governs the normal stress in the $e_2$-direction accordingly. Therefore, by a variation of the loading ratio $\Gamma$, we can achieve any planar tensile stress ratio. The strain state in the area of interest is rather homogeneous because of the high number of slits, that reduce the strain, as shown by DIC and FE simulations. Consequently, we may assume that (due to the slits) the disturbing effect of the arms on the stress distribution is weak and the stress field is likewise approximately homogeneous so that the stress state can be estimated from the forces divided by the material cross section in the area of interest. The thus obtained material parameters provide an estimate of the material behavior and can serve as a starting value for an inverse parameter identification (partly criterion 4).

When comparing geometry 1 and geometry 2 of the continuous fiber reinforced specimen, geometry 1 also shows advantages concerning the larger size of the area of interest (criterion 5). The manufacturing effort (criterion 6) is the lowest in the unreinforced specimen. The bonded specimen requires a large amount of manual work. The continuous fiber reinforced specimen preparation requires an additional manufacturing technique (milling), but is more time-efficient compared to the bonded specimen.

One experiment per design and load scenario is, of course, not statistically representative. We admit that our experimental investigation with only one specimen may not provide a precise quantitative design estimate; however, the experiments are (i) plausible within each experiment and in comparison of all experiments, and (ii) show a good agreement with the FE simulations (continuous fiber reinforced specimen). The experiments in this work, hence, provide a preliminary estimation of the specimens' suitability.

## 5. Conclusions

This work provided experimental as well as numerical results to support the selection of a suitable specimen design for the characterization of the damage behavior of SMC. The authors, however, did not aim to characterize the anisotropic damage behavior of SMC, as the number of experiments was significantly too low and the differentiation of damage, viscoelasticity, and plasticity in discontinuous fiber reinforced polymers [30] along with their parameter identification are still not fully understood. One result could be an understanding of the anisotropic stress–strain behavior of SMC. Geometry 1 of the continuous fiber reinforced specimen performed superior for our application, i.e., the optimality criteria and their prioritization we aimed for. The main advantages are the highest achievable strain in the area of interest among all specimen, the homogeneity of the strain in the area of interest, the large area of interest, and a moderate manufacturing effort.

**Acknowledgments:** We appreciate the support by David Bücheler (Fraunhofer Institute of Chemical Technology) for the UPPH SMC production, including the SMC with the unidirectional reinforcements. The research documented in this paper has been funded by the German Research Foundation (DFG) within the International Research Training Group "Integrated engineering of continuous-discontinuous long fiber reinforced polymer structures" (GRK 2078). The support by the German Research Foundation (DFG) is gratefully acknowledged.

**Author Contributions:** Malte Schemmann, Juliane Lang, Thomas Seelig, and Thomas Böhlke discussed the arm reinforcement strategy, the considered geometries, and load paths in detail. Anton Helfrich was responsible for the specimen manufacturing. Juliane Lang performed the experiments and the necessary post processing. All authors discussed the results. Malte Schemmann, Juliane Lang, and Anton Helfrich wrote the paper. All authors provided significant editing efforts towards the improvement of the paper.

**Conflicts of Interest:** The authors declare no conflict of interest.

## Appendix A. Strain Fields Shortly before Failure and Images of Failed Specimens

This appendix shows the strain fields in the area of interest in terms of the last captured image before specimen failure. For better comparison, all contour plots refer to the legend in Figure A1. Additionally, we present images of the failed specimens. The fractured areas are marked in green.

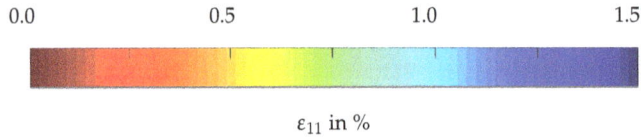

**Figure A1.** Legend for the following contour plots.

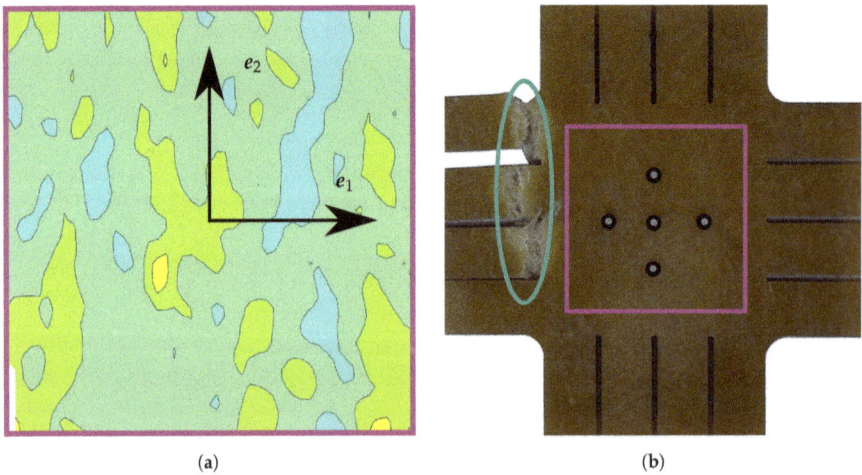

(a)                                   (b)

**Figure A2.** *Cont.*

(c)

(d)

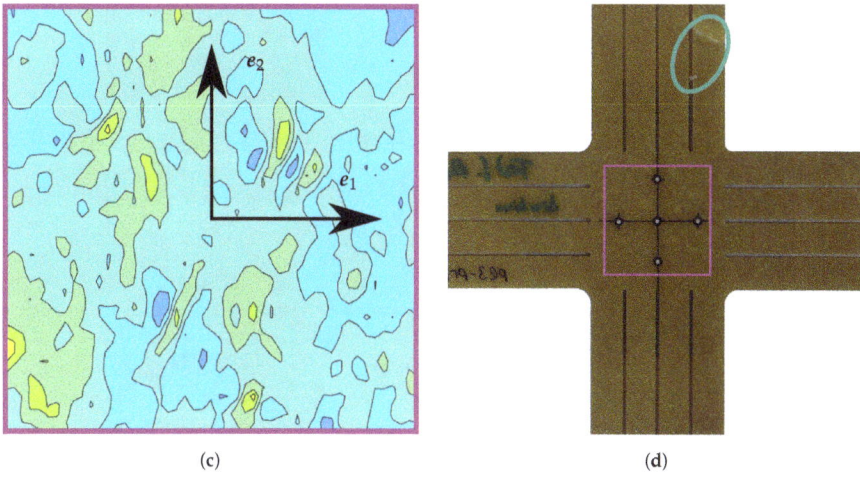

**Figure A2.** Unreinforced specimen arms. Last image of strain field before failure and failed specimen. (a) strain field $\varepsilon_{11}$ in $A^{I}$ for load case $P_{T=0}$; (b) failed specimen for load case $P_{T=0}$; (c) strain field $\varepsilon_{11}$ in $A^{I}$ for load case $P_{T=1}$; (d) failed specimen for load case $P_{T=1}$.

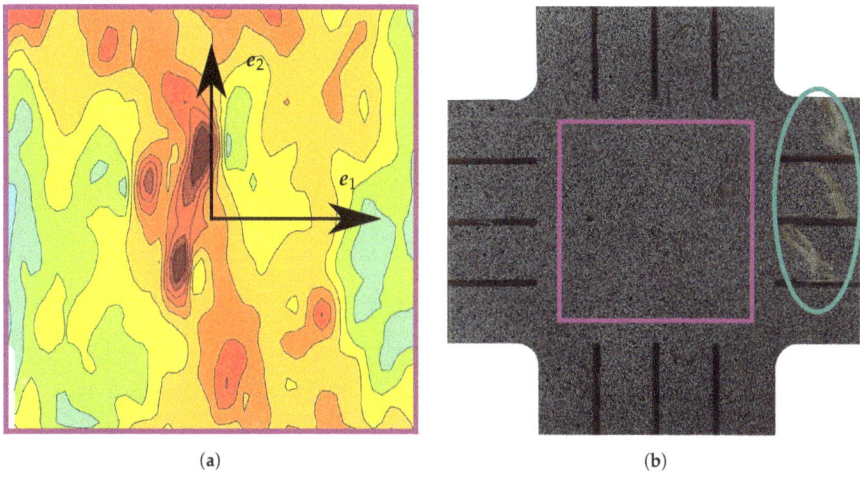

(a)

(b)

**Figure A3.** *Cont.*

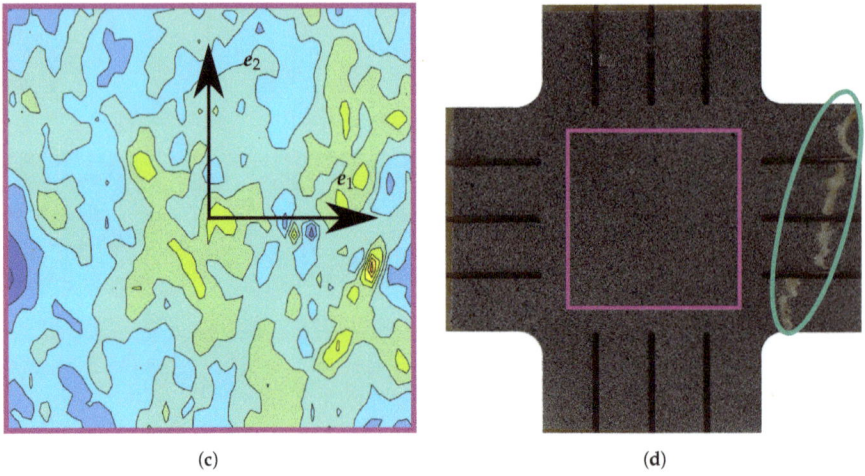

(c)                                              (d)

**Figure A3.** Bonded specimen arms. Last image of strain field before failure and failed specimen. (a) strain field $\varepsilon_{11}$ in $A^{\mathrm{I}}$ for load case $P_{\Gamma=0}$; (b) failed specimen for load case $P_{\Gamma=0}$; (c) strain field $\varepsilon_{11}$ in $A^{\mathrm{I}}$ for load case $P_{\Gamma=1}$; (d) failed specimen for load case $P_{\Gamma=1}$.

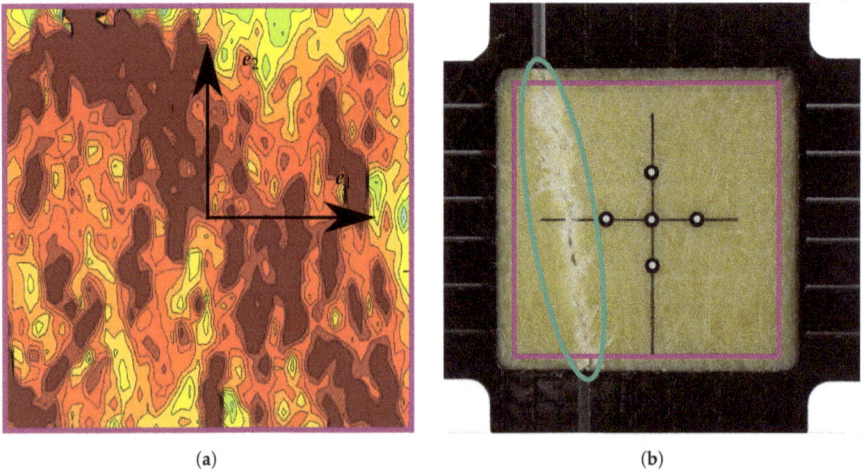

(a)                                              (b)

**Figure A4.** *Cont.*

(c)

(d)

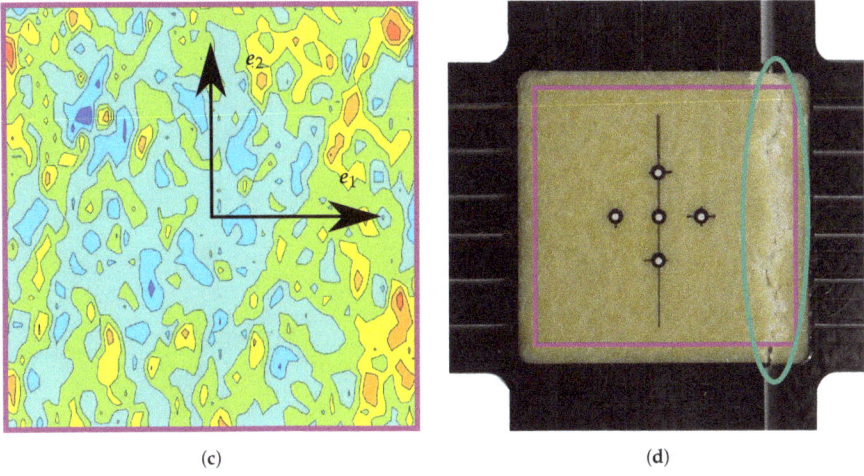

**Figure A4.** Continuous fiber specimen arms, geometry 1. Last image of strain field before failure and failed specimen. (**a**) strain field $\varepsilon_{11}$ in $A^{\mathrm{I}}$ for load case $P_{\mathrm{T}=0}$; (**b**) failed specimen for load case $P_{\mathrm{T}=0}$; (**c**) strain field $\varepsilon_{11}$ in $A^{\mathrm{I}}$ for load case $P_{\mathrm{T}=1}$; (**d**) failed specimen for load case $P_{\mathrm{T}=1}$.

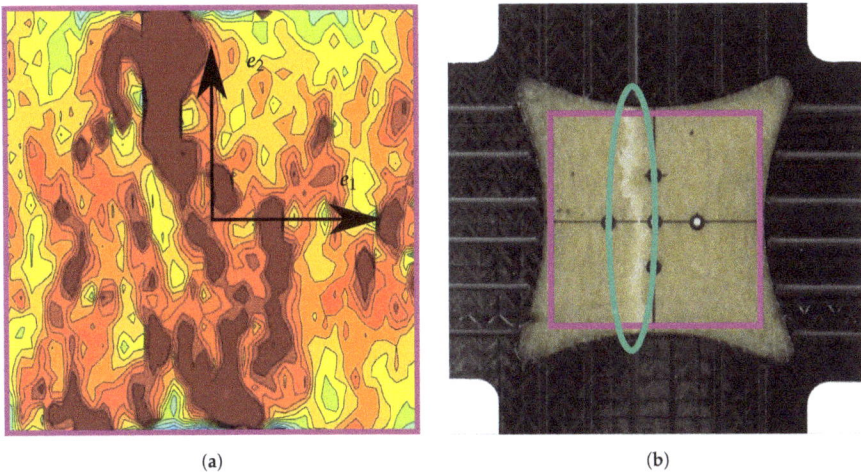

(a)

(b)

**Figure A5.** *Cont.*

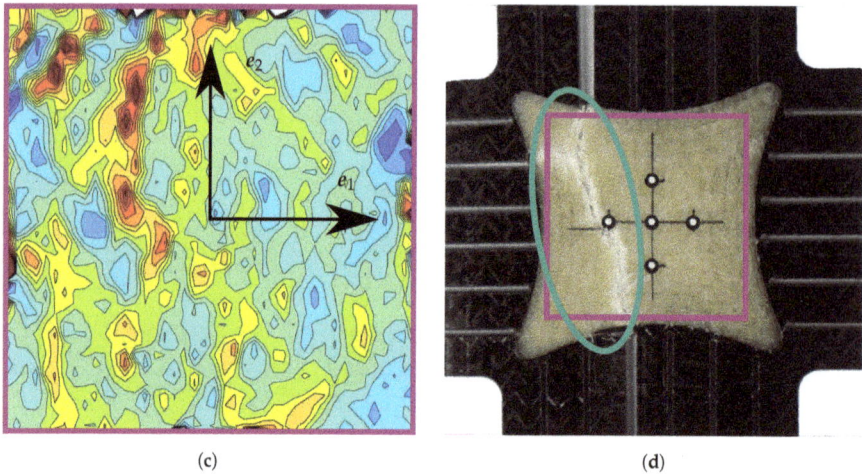

Figure A5. Continuous fiber specimen arms, geometry 2. Last image of strain field before failure and failed specimen. (**a**) strain field $\varepsilon_{11}$ in $A^I$ for load case $P_{T=0}$; (**b**) failed specimen for load case $P_{T=0}$; (**c**) strain field $\varepsilon_{11}$ in $A^I$ for load case $P_{T=1}$; (**d**) failed specimen for load case $P_{T=1}$.

## References

1. Ohtake, Y.; Rokugawa, S.; Masumoto, H. Geometry determination of cruciform-type specimen and biaxial tensile test of C/C composites. *Key Eng. Mater.* **1999**, *164–165*, 151–154.
2. Deng, N.; Kuwabara, T.; Korkolis, Y.P. Cruciform specimen design and verification for constitutive identification of anisotropic sheets. *Exp. Mech.* **2015**, *55*, 1005–1022.
3. Kuwabara, T.; Ikeda, S.; Kuroda, K. Measurement and analysis of differential work hardening in cold-rolled steel sheet under biaxial tension. *J. Mater. Process. Technol.* **1998**, *80–81*, 517–523.
4. Makinde, A.; Thibodeau, L.; Neale, K.W. Development of an apparatus for biaxial testing using cruciform specimens. *Exp. Mech.* **1992**, *32*, 138–144.
5. Demmerle, S.; Boehler, J.P. Optimal design of biaxial tensile cruciform specimens. *J. Mech. Phys. Solids* **1993**, *41*, 143–181.
6. Kelly, D.A. Problems in creep testing under biaxial stress systems. *J. Strain Anal.* **1976**, *11*, 1–6.
7. Boehler, J.P.; Demmerle, S.; Koss, S. A new direct biaxial testing machine for anisotropic materials. *Exp. Mech.* **1994**, *34*, 1–9.
8. Hoferlin, E.; Van Bael, A.; Van Houtte, P.; Steyaert, G.; De Maré, C. Design of a biaxial tensile test and its use for the validation of crystallographic yield loci. *Model. Simul. Mater. Sci. Eng.* **2000**, *8*, 423–433.
9. Hannon, A.; Tiernan, P. A review of planar biaxial tensile test systems for sheet metal. *J. Mater. Process. Technol.* **2008**, *198*, 1–13.
10. International Organization for Standardization. *Metallic Materials—Sheet and Strip—Biaxial Tensile Testing Method Using a Cruciform Test Piece*; ISO 16842; International Organization for Standardization: Geneva, Switzerland, 2014.
11. Green, D.E.; Neale, K.W.; MacEwen, S.R.; Makinde, A.; Perrin, R. Experimental investigation of the biaxial behaviour of an aluminum sheet. *Int. J. Plast.* **2004**, *20*, 1677–1706.
12. Smits, A.; Van Hemelrijck, D.; Philippidis, T.P.; Cardon, A. Design of a cruciform specimen for biaxial testing of fibre reinforced composite laminates. *Compos. Sci. Technol.* **2006**, *66*, 964–975.
13. Van Hemelrijck, D.; Ramault, C.; Markis, A.; Clarke, A.R.; Williamson, C.; Gower, M.; Shaw, R.; Mera, R.; Lamkanfi, E.; Van Paepegem, W. Biaxial testing of fibre reinforced composites. In Proceedings of the 16th International Conference on Composite Materials, Xi'an, China, 20–25 August 2007; pp. 1–8.
14. Makris, A.; Vandenbergh, T.; Ramault, C.; Van Hemelrijck, D.; Lamkanfi, E.; Van Paepegem, W. Shape optimisation of a biaxially loaded cruciform specimen. *Polym. Test.* **2010**, *29*, 216–223.

15. Lamkanfi, E.; Van Paepegem, W.; Degrieck, J.; Ramault, C.; Makris, A.; Van Hemelrijck, D. Strain distribution in cruciform specimens subjected to biaxial loading conditions. Part 2: Influence of geometrical discontinuities. *Polym. Test.* **2010**, *29*, 132–138.

16. Gower, M.R.L.; Shaw, R.M. Towards a planar cruciform specimen for biaxial characterisation of polymer matrix composites. *Appl. Mech. Mater.* **2010**, *24–25*, 115–120.

17. Serna Moreno, M.; Martínez Vicente, J.; López Cela, J. Failure strain and stress fields of a chopped glass-reinforced polyester under biaxial loading. *Compos. Struct.* **2013**, *103*, 27–33.

18. Hohberg, M.; Kärger, L.; Bücheler, D.; Henning, F. Rheological In-Mold Measurements and Characterizations of Sheet-Molding-Compound (SMC) Formulations with Different Constitution Properties by Using a Compressible Shell Model. *Int. Polym. Process.* **2017**, *32*, 659–668.

19. Asadi, A.; Miller, M.; Sultana, S.; Moon, R.J.; Kalaitzidou, K. Introducing cellulose nanocrystals in sheet molding compounds (SMC). *Compos. Part A Appl. Sci. Manuf.* **2016**, *88*, 206–215.

20. Mahnken, R.; Stein, E. A unified approach for parameter identification of inelastic material models in the frame of the finite element method. *Comput. Methods Appl. Mech. Eng.* **1996**, *136*, 225–258.

21. Mahnken, R.; Stein, E. Parameter identification for viscoplastic models based on analytical derivatives of a least-squares functional and stability investigations. *Int. J. Plast.* **1996**, *12*, 451–479.

22. Cooreman, S.; Lecompte, D.; Sol, H.; Vantomme, J.; Debruyne, D. Identification of mechanical material behavior through inverse modeling and DIC. *Exp. Mech.* **2007**, *48*, 421–433.

23. Lecompte, D.; Smits, A.; Sol, H.; Vantomme, J.; Van Hemelrijck, D. Mixed numerical-experimental technique for orthotropic parameter identification using biaxial tensile tests on cruciform specimens. *Int. J. Solids Struct.* **2007**, *44*, 1643–1656.

24. Schemmann, M.; Brylka, B.; Müller, V.; Kehrer, L.; Böhlke, T. Mean field homogenization of discontinous fiber reinforced polymers and parameter identification of biaxial tensile tests through inverse modeling. In Proceedings of the 20th International Conference on Composite Materials, Copenhagen, Denmark, 19–24 July 2015.

25. Schemmann, M.; Gajek, S.; Böhlke, T. Biaxial tensile tests and microstructure-based inverse parameter identification of inhomogeneous SMC composites. In *Advances in Mechanics of Materials and Structural Analysis: In Honor of Reinhold Kienzler*; Altenbach, H., Jablonski, F., Müller, W.H., Naumenko, K., Schneider, P., Eds.; Springer International Publishing: Cham, Switzerland, 2018; Volume 80, pp. 329–342.

26. Hartmann, S.; Gilbert, R.R. Identifiability of material parameters in solid mechanics. *Arch. Appl. Mech.* **2017**, *1*, 1–24.

27. Bücheler, D.; Trauth, A.; Damm, A.; Böhlke, T.; Henning, F.; Kärger, L.; Seelig, T.; Weidenmann, K.A. Processing of continuous-discontinuous-fiber-reinforced thermosets. In Proceedings of the SAMPE Europe Conference 2017, Leinfelden-Echterdingen, Germany, 14–16 November 2017; pp. 1–8.

28. Bücheler, D. Locally Continuous-Fiber Reinforced Sheet Molding Compound. Ph.D. Thesis, Karlsruhe Institute of Technology, Karlsruhe, Germany, 2017.

29. Bauer, J.; Priesnitz, K.; Schemmann, M.; Brylka, B.; Böhlke, T. Parametric shape optimization of biaxial tensile specimen. *Proc. Appl. Math. Mech.* **2016**, *16*, 159–160.

30. Brylka, B.; Schemmann, M.; Wood, J.; Böhlke, T. DMA based characterization of stiffness reduction in long fiber reinforced polypropylene. *Polym. Test.* **2018**, *66*, 296–302.

*Journal of*
*composites science*

MDPI

*Article*

# Investigation of Quasi-Static and Dynamic Material Properties of a Structural Sheet Molding Compound Combined with Acoustic Emission Damage Analysis

**Anna Trauth \*, Pascal Pinter and Kay André Weidenmann**

Institute for Applied Materials, Karlsruhe Institute of Technology, 76131 Karlsruhe, Germany;
pascal.pinter@kit.edu (P.P.); kay.weidenmann@kit.edu (K.A.W.)
\* Correspondence: anna.trauth@kit.edu; Tel.: +49-721-608-46596

Received: 28 October 2017; Accepted: 11 December 2017; Published: 14 December 2017

**Abstract:** Sheet molding compounds (SMC) are discontinuously fiber-reinforced thermosets, attractive to the automotive industry due to their outstanding specific strength and stiffness, combined with a cost efficient manufacturing process. Increasingly important for structural components, a structural SMC-based improved resin formulation featuring no fillers is investigated in this study. The influence of fiber volume content, fiber length, and manufacturing induced fiber orientation on quasi-static and dynamic mechanical properties of vinylester-based SMC is characterized. Stiffness and strength increased with increasing fiber volume content for tensile, compression, and flexural loadings. Fiber length distribution did not significantly influence the mechanical properties of the material. The movement of the conveyor belt leads to an anisotropic fiber orientation and orientation-dependent mechanical properties. Acoustic emission coupled with machine learning algorithms enabled the investigation of the damage mechanisms of this discontinuous glass fiber SMC. The acoustic emission analysis was validated with micro computed tomography of damaged specimens. The dominant failure mechanisms of the SMC exposed to bending loading were matrix cracking and interface failure.

**Keywords:** mechanical characterization; sheet molding compound; fiber volume content; fiber length; fiber orientation; micro computed tomography (μCT) analysis; acoustic emission; failure mechanisms

---

## 1. Introduction

Discontinuous fiber-reinforced polymeric composites are attractive materials, especially in the automotive industry, due to their high specific stiffness and strengths combined with low materials and manufacturing costs.

Among these materials, sheet molding compounds (SMC) stand out due to the many favorable aspects of their production, providing the ability to manufacture structures at very high productivity rates combined with good part reproducibility, cost efficiency, and surface quality, as well as the ability to manufacture complex part geometries. SMCs are thermosetting resin-based fiber-reinforced semi-finished materials. To a large extent, SMC is also a generic term describing this type of compound combined with the process to convert it into a composite part, which is usually based on compression molding [1]. In 2016, SMCs, in combination with bulk molding compounds (BMC), were the most produced glass fiber-reinforced composites [2], highlighting the importance of this material class for numerous technical applications. SMC have played an important role in different technical sectors since the 1960s. However, compared to SMC materials developed in the past, a novel class of SMC materials, called structural SMC, is rapidly advancing and different car manufacturers have already successfully included structural SMC components into their vehicle concepts [3–5]. The resin formulations for structural SMC aim to manufacture a material that can fulfill more stringent demands on its mechanical properties. Much research has been done to determine the influence of resin formulations and fiber content on the mechanical properties of standard SMC. This contribution aims to identify the influence of fiber volume content and fiber length on mechanical material properties of a structural unfilled SMC. The following literature review focuses on some advanced and structural SMC materials, so all of them feature a significant amount of fillers.

Boylan et al. [6] investigated the mechanical material properties of soft and hard glass fiber-reinforced SMC, featuring a fiber volume content of 21%. The material was based on an unsaturated polyester resin, whereas the resin formulation contained a significant amount of calcium carbonate ($CaCO_3$) as filler. Their results indicated that tensile strength and stiffness strongly depend on the fiber type. Longer fibers tended to increase mechanical performance. This study also demonstrated the anisotropic mechanical material properties for SMC sheets due to material flow during compression molding. The study did not consider different fiber volume contents but only the mixture of soft and hard glass fiber featuring different fiber lengths.

Considering material anisotropy described by Boylan et al. [6], similar results were found by Lamanna et al. [7]. This research group investigated the mechanical properties of polyester resin-based glass fiber-reinforced SMC with calcium carbonate with a nominal wt/wt/wt ratio of 39:27:34. The investigated material was considered for structural components in the automotive industry. The focus in the study was a broad material characterization for one specific material. No variation of fiber volume content or fiber length was considered. The observed specific SMC exhibited substantial in-plane anisotropy in terms of tensile stiffness and strength.

Oldenbo et al. [8] investigated the mechanical properties of SMC material developed for automotive exterior body panels containing hollow glass micro-spheres and thermoplastic toughening additives, compared to a conventional standard SMC containing $CaCO_3$ as filler. The hollow glass spheres reduced the density of the investigated material, but Young's modulus and compressive strength also decreased, which was explained by the replacement of stiffer $CaCO_3$ fillers by hollow glass spheres.

These three examples show the investigations and developments in the field of SMC. Nevertheless, the SMC recipes investigated contain fillers to compensate for the resin shrinkage during molding for a superior surface quality, and none of these studies focused on unfilled SMC. To further improve the mechanical properties of structural SMC, resin formulations without fillers must be developed, as reducing the filler content allows for increasing the glass fiber content. Increased glass fiber content is a crucial factor for the improved mechanical performance, allowing for the structural use of SMC. In this regard, the usual outstanding surface quality in comparison to filled SMC is a secondary aspect. Scientific publications on filler-free structural SMC are rare. To the best of the authors' knowledge, the mechanical material properties of unfilled SMC have rarely, if at all, been previously considered.

The International Research Training group on the "Integrated engineering of continuous-discontinuous long fiber reinforced polymer structures" (GRK 2078) focused on the development of a structural hybrid continuous-discontinuous SMC, whereas the vinylester-based discontinuous SMC considered within this study offers the possibility of being hybridized in a one shot compression molding process [9]. Since the material flow of the discontinuous SMC can lead to misalignment of the continuous locally placed tapes as well as fiber misalignment [10], the sheets in the study were manufactured with a 100% mound coverage. Due to the material movement on the conveyor belt during manufacturing of the semi-finished sheets, the fibers tend to orient in the manufacturing direction. This study was the first step in investigating the anisotropic material properties resulting from the manufacturing of semi-finished sheets, which did not flow during compression molding. The study also aimed to prove it was possible to manufacture SMC sheets with 100% mold coverage. To further improve this novel hybrid material class that has superior mechanical properties, a profound understanding of the two individual components, the discontinuous glass fiber SMC and the continuous carbon fiber SMC, is necessary.

Thus, our study is the next step to better understand the material behavior of an unfilled structural discontinuous glass fiber SMC, based on vinylester with a focus on the influence of fiber volume content and fiber length on mechanical quasi-static and dynamic material properties. We also wanted to evaluate the anisotropy introduced due to the manufacturing of SMC semi-finished sheets.

The locally reinforced structural SMC is of major importance as flow molding may displace continuous fiber inlays [10]. This can be effectively prevented by high mold coverage and, for this purpose, the semi-finished material considered within this study was compression molded with 100% mound coverage to ensure no additional flow of the material in the mold. Hence, the composition and the potential processing impacts on the material properties are different between conventional, filled SMC, and structural SMC, even at the early stage of the manufacturing of the semi-finished materials. This influence is unique to structural SMC with respect to the hybridization option and has not been investigated to date.

To further improve the simulation tools and optimize applications of structural SMC components, a profound understanding of damage and failure of SMC is of major importance. Hence, this paper analyzed the damage mechanisms of discontinuous glass fiber SMC. In general, the damage to a fiber-reinforced material is a complex phenomenon based on various microscopic failure types [11]. The goal of this study was to gain a profound understanding of damage and failure mechanisms of SMC by combining in situ acoustic emission (AE) with micro computed tomography (μCT) observation of damaged specimens. The acoustic emission technique has already been proven to be a suitable technique to characterize damage of materials. If a material is mechanically loaded, it stores elastic energy. The release of this strain energy in the form of a stress wave, which travels through the material by creating cracks or defects, can be measured with the appropriate sensors [12]. For this purpose, piezoelectric sensors were attached to the surface. These sensors detect the waves and produce a voltage output [13]. Different research groups indicated a correlation between failure type and AE signal for glass fiber-reinforced materials.

For chopped glass fiber-reinforced polypropylene, Barré et al. showed that low amplitude signals can be linked to matrix cracking, whereas higher amplitude signals were linked to fiber failure [14]. The fracture behavior of SMC has also been successfully investigated using AE [15].

In addition to the amplitude of AE signals, other signal properties, such as signal duration, energy distribution, and counts/duration ration, are suitable aspects used to investigate composite failure [16,17]. A problem with AE analysis is the large amount of signals resulting from a structural test, since each failure event creates thousands of hits, to which numerous AE parameters can be associated. In addition, filtering of background noise and signals from non-structural sources is necessary [18]. To address this, recognition methods were proven to be a promising technique to reduce this type of uncertainty pattern. The proposed method was based on the formation of different clusters to separate AE signals. The formation of clusters was based on pattern recognition algorithms [19]. Machine learning and pattern recognition techniques are suitable tools to characterize the failure of fiber-reinforced composites by means of acoustic emission. [20–23].

To enhance these studies, this paper addressed the application of pattern recognition techniques to investigate the failure of discontinuous glass fiber-reinforced SMC exposed to flexural loadings. The objective was to identify and better understand the failure mechanisms to provide data for failure and damage simulation, which is a crucial factor for the use of structural SMC. Furthermore, this study aimed to prove the suitability of combining machine learning algorithms with the acoustic emission technique to identify the failure mechanisms of discontinuous fiber reinforced materials.

This research focused on structure–property relationships of an unfilled structural SMC, and namely the influence of fiber volume content and fiber length on mechanical material properties, as well as on anisotropic material properties resulting from the movement of the conveyor belt. An additional objective was the investigation of failure and damage mechanisms. To address this purpose, an in situ acoustic emission (AE) analysis was combined with a μCT observation ofpost-damaged specimens. This study also aimed to prove the suitability of machine learning algorithms to process AE data and to identify different failure mechanisms for SMC materials.

## 2. Materials

### 2.1. Material Manufacturing

The material considered within this study is a discontinuous glass fiber SMC based on a vinylester resin (Altac XP810X supplied by Alyancis (former DSM), Schaffhausen, Switzerland). To optimize structural properties no fillers were added, and only a low amount of flow additives (BYK 9085 supplied by BYK, Wesel, Germany), peroxide (Trigonox 17 supplied by Akzonobel, Amsterdam, The Netherlands), and an MgO thickener (Luvatol EK 100 KM supplied by Lehmann&Voss&Co., Hamburg, Germany) were mixed into the resin. A flat conveyor belt (type HM-LB-800 by Schmidt & Heinzmann, Bruchsal, Germany) was used to manufacture the semi-finished sheets. The length of the reinforcing glass fiber (Multistar 272 by Johns Manville, Denver, CO, USA) was set to either 25.4 mm (1 inch) or to a mixture of 25.4 mm (1 inch) and 50.8 mm (2 inch) with a 6:5 ratio, and different fiber volume contents (FVCs) ranging from 17 to 31 volume percent (vol %) were used. After maturation at 30 °C for several days, the semi-finished SMC sheets were cut into plies, stacked, and compression molded into plaques at approximately 150 °C, with a maximum force of 1600 kN, and a 92-second mound closing time.

Due to material movement on the conveyor belt during manufacturing of the semi-finished sheets, the fibers tend to orient in the manufacturing direction. To evaluate this anisotropy introduced due to manufacturing, the sheets were compression molded with 100% mound coverage to ensure no flow of the material in the mound. The dimensions of the compression molded plaques were 250 mm × 800 mm. Depending on fiber volume content, the sheets had thicknesses between 2 mm and 2.7 mm.

### 2.2. Specimen Preparation

Specimens for quasi-static and dynamic characterization were extracted using a water jet cutting technique by GENTHNER SchneidTechnik GmbH & Co. KG in Straubenhardt, Germany. The water-jet beam had a width of 0.8 mm with and cutting pressure of 360 N/mm$^2$. GMA Garnet™ 80 mesh was used as the abrasive. Before mechanical testing, the specimens were stored at room temperature (22 °C) and at a relative humidity of approximately 50% for several days. Mechanical testing was performed under the same conditions

## 3. Methods

### 3.1. Determination of Fiber Volume Content

To determine the real fiber volume content of the sheets, a thermogravimetric analysis (TGA) was completed with a Leco TGA701 (LECO Corporation, St. Joseph, MI, USA). Heating rate was 37 °C/min up to 550 °C. This temperature stayed constant for 2 h. Nine rectangular specimens per sheet, having a section of approximately 100 mm$^2$ each, were used to determine the real fiber volume content of at least three different sheets for each configuration.

### 3.2. Mechanical Characterization

3.2.1. Tensile Testing

The quasi-static tensile properties were measured with a ZwickRoell ZMART.PRO universal testing machine (Zwick Roell Group, Ulm, Germany) with a maximum capacity of 200 kN. Longitudinal displacement was measured with a tactile extensometer. Force and displacement data were acquired at 10 Hz. The tests were performed following the DIN EN ISO 527-4 standard [24], but a slightly different specimen type was considered within this study. The specimens had dog-bone geometry with a gauge length of 60 mm × 20 mm. The clamping distance was 150 mm. Testing velocity was 2 mm/min leading to a test duration of approximately one minute per specimen. The tensile modulus was calculated in the strain range of $\varepsilon_t$ = 0.05–0.25%. Additionally, the tensile strength, the maximum tensile stress sustained by the specimen during the test, was evaluated. For the six material configurations, the five samples that failed in the measurement region for each orientation (0°, 45°, 90°, and 135° with respect to the longitudinal axes of the plaques) were analyzed.

### 3.2.2. Flexural Testing

Three-point bending tests were performed according to DIN EN ISO 14125 [25] on a ZwickRoell Z2.5 testing machine with a 2.5-kN load cell. Deflection of the specimen was measured with a tactile displacement transducer. Force and deflection data were acquired at 10 Hz. The rectangular specimen had a width of 15 mm and a thickness to span ratio of 1:16.

According to [25], the testing velocity $v$ was determined from a target strain rate of $\dot{\varepsilon}$ = 0.001/min. To consider the different thicknesses of the specimens, the testing velocity was set individually for each of the six configurations. Bending stress $\sigma_f$ and surface strain $\varepsilon_f$ were calculated according [25].

Based on these values, the flexural modulus was calculated in the strain range of $\varepsilon_f$ = 0.05–0.25%. Additionally, the bending strength, the maximum bending stress sustained by the specimen during the test, was determined. At least four specimens of each configuration in 0°, 45°, 90°, and 135°, with respect to the longer side of the plaque, were used for testing.

### 3.2.3. Compression Testing

Compression tests were conducted on a ZMART.PRO 100 kN universal testing machine according to DIN EN ISO 14126 [26]. The machine was equipped with a Hydraulic Composite Compression Fixture (HCCF) by Zwick clamping unit. The rectangular specimens had a length of 110 mm and a width of 10 mm. The measurement length was 12 mm. Deformation was detected by two clip-on sensors, one at each side of the specimen to additionally measure bending of the specimens. Testing velocity was 1 mm/min and the compression modulus was calculated over the strain range of $\varepsilon_c$ = 0.05–0.25%. The compression strength was determined as the maximum compression stress sustained by the specimen during the test.

For each material configuration, five samples in the 0°, 45°, 90°, and 135° direction, that failed in the measurement region and did bend during loading, were analyzed. Bending was controlled by evaluating the bending factor according to [26].

### 3.2.4. Charpy Impact Testing

To assess impact strength, Charpy impact tests were conducted on rectangular specimens with a length of 50 mm and a width of 15 mm according to DIN EN ISO 179 [27]. The specimen, which was supported near its ends as a horizontal beam, was impacted by a swinging striker (impact energy of 5 J) and the Charpy impact strength $a_c$ was calculated according to EN ISO 179. At least 14 specimens per configuration were tested. The non-instrumented impact tests allowed only determining the Charpy impact strength as the force-deflection response was not recorded.

### 3.3. Acoustic Emission (AE) Analysis

### 3.3.1. Data Acquisition

To assign the mechanisms of damage of the material, in situ acoustic emission was used. When in service, SMC components are most frequently loaded by bending loads. Due to this, the mechanisms of failure of discontinuous glass fiber SMC exposed to three point bending were examined. The tests were completed on an INSTRON E3000 universal testing machine (Instron, Norwood, MA, USA) with a load cell capacity of 3 kN. The testing velocity was 2.6 mm/min. The span of the two lower supports was 60 mm. The crosshead displacement was measured to determine the displacement of the specimen. Two broadband B-1025 piezoelectric transducers (DigitalWave Corporation, Centennial, CO, USA), attached to the lower surface of the specimen, captured the AE signals. The two transducers were linked to a preamplifier type AEP3 (Vallen System GmbH, Icking, Germany). The distance between the two sensors was 40 mm (Figure 1).

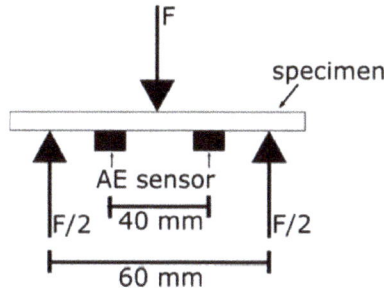

**Figure 1.** Testing setup for three-point bending with in-situ acoustic emission analysis.

A coupling agent, treacle, ensured contact between transducers and specimen. The AE signals were captured with a sampling rate of 10 MHz and an AMSY-4 system (Vallen System GmbH). To reduce the captured signals arising from testing machine or background, the AE acquisition threshold was set to 23 dB. The duration discrimination time was 200 µs and the rearm time was set to 0.4 ms. The signal duration of one hit was the time from first exceeding the threshold until it is not crossed for the duration discrimination time.

### 3.3.2. Data Processing

Data processing included amplifying the sensors' output by 49 dB and a filtering of the signal with a 10 kHz high pass filter. Force and displacement signals, continuously recorded by the load frame with a sampling rate of 100 Hz, were fed to the AMSY-4 system, enabling a correlation of the captured AE signal with the load-displacement curve resulting from bending of the specimen.

### 3.3.3. Data Clustering

Data clustering of acquired signals was largely based on the method introduced by Sause et al. [20]. First, different features of the signal that were the most promising to find natural clusters within the dataset were chosen (Table 1).

**Table 1.** Initial feature space.

| Peak Frequency (Hz) | $f_{peak}$ |
|---|---|
| Frequency Centroid (Hz) | $f_{centroid} = \dfrac{\int f \cdot U(f) df}{\int U(f) df}$ |
| Weighted Peak Frequency (Hz) | $< f_{peak} > = \sqrt{f_{peak} \cdot f_{centroid}}$ |
| Amplitude (dB) in Time Domain | A |
| Partial Power 1 … 6 (%) | $P_{P1} \ldots P_{P6} = \dfrac{\int_{f_1}^{f_2} U^2(f) df}{\int_{0kHz}^{1200kHz} U^2(f) df}$ |
| | $P_{P1} : f_1 = 0$ kHz, $f_2 = 150$ kHz |
| | $P_{P2} : f_1 = 150$ kHz, $f_2 = 300$ kHz |
| | $P_{P3} : f_1 = 300$ kHz, $f_2 = 450$ kHz |
| - | $P_{P4} : f_1 = 450$ kHz, $f_2 = 600$ kHz |
| | $P_{P5} : f_1 = 600$ kHz, $f_2 = 900$ kHz |
| | $P_{P6} : f_1 = 900$ kHz, $f_2 = 1200$ kHz |

### 3.3.4. Signal Processing

Signal processing was completed in MATLAB (Version R2016b, The MathWorks, Inc., Natick, MA, USA). The acoustic emission data was imported with an import tool implemented by Vallen. Signal processing started with a Fourier transformation of the recorded signals by the MATLABbuilt-in fast Fourier transformation with 2048 data points. With a cropped signal using the hat function, edge-effects results were due to the sudden drop or rise of the time signal at the borders. To reduce this effect, the time window was multiplied with a hamming function of the same length. Finally, all features in the frequency domain described in Table 1 could be derived.

To find a suitable feature space for the following clustering, all possible combinations of the 10 preselected features listed in Table 1 were compared. The minimum number of features $Q_{min} = 2$ and a maximum of $Q_{max} = 6$ for a cluster were set as restrictions for the algorithm. This led to 837 different combinations. All combinations with the number of clusters ranging from $P_{min} = 2$ to $P_{max} = 6$, were considered, and in total there were 4185 clustering processes to solve. The clustering was based on a Gaussian mixture model with P components.

The validation of each configuration was based on an algorithm developed by Günter et al. [28], which is a combination of different validation algorithms. In this study, different validation methods were considered, and the methods introduced by Davies et al. [29], Rousseeuw et al. [30], and Calinski et al. [31] were applied. The different methods were combined as proposed by Sause et al. [20], and the procedure was performed twice: once to find the optimum number of clusters for each feature combination for the number of clusters for all combinations of features defined, then a second time, for all combinations of features, considering the optimum number of clusters that was determined in the previous step. A similar voting scheme, as proposed by Sause et al. [20], was considered, with 25 points being the best configuration, 24 for the second, and 23 for the third. A feature combination, recognized as the best solution by all three cluster validation methods, attained up to 75 points in the best case scenario. The objective of this method was to find an optimum configuration of features with a number of clusters for the present dataset.

### 3.4. Micro Computed Tomography

To assign natural clusters from AE to a certain failure mechanism, μCT scans were used to obtain insight into the damaged samples. Therefore, samples were scanned in an Yxlon-CT precision computed tomography system (Yxlon International Ct GmbH, Hattingen, Germany) containing an open micro-focus X-ray transmission tube with tungsten target and a $2048 \times 2048$ pixel flat panel detector from Perkin Elmer (Waltham, MA, USA). The acceleration voltage was 100 kV and the tube current 0.05 mA. The scans were acquired with a focus–object distance of 64.13 mm and a focus–detector distance of 999.79 mm, leading to a voxel size of 12.83 μm.

## 4. Results

### 4.1. Fiber Volume Content and Fiber Length

Table 2 lists the measured fiber volume content (FVC) and fiber length of the six different considered configurations.

**Table 2.** Real fiber volume content and fiber length distribution of the six different sheet molding compounds (SMC) materials considered within this study.

| Configuration | 1 | 2 | 3 | 4 | 5 | 6 |
|---|---|---|---|---|---|---|
| Measured Fiber Volume Content (FVC) (%) | $17 \pm 2$ | $25 \pm 2$ | $31 \pm 2$ | $17 \pm 2$ | $25 \pm 2$ | $31 \pm 2$ |
| Fiber Length (mm) | 25.4 | 25.4 | 25.4 | 25.4 and 50.8 (6:5) | 25.4 and 50.8 (6:5) | 25.4 and 50.8 (6:5) |

*4.2. Mechanical Material Properties*

The following section addresses the mechanical properties of the considered SMC materials. The figures within this section show the mechanical properties for specimens extracted in different directions with respect to the movement of the conveyor belt.

4.2.1. Tensile, Compression, and Flexural Stiffness

Figures 2–4 show the elastic moduli resulting from tension, compression, and flexural loading of the specimens, which were extracted in different directions with respect to the manufacturing direction, where 0° is the direction of the movement of the conveyor belt. In general, the investigated SMC materials showed a slight anisotropic material behavior for tensile compression and flexural loading due to the orientation of the fibers on the conveyer belt.

The elastic tensile modulus increased with increasing fiber volume content from 7.3 GPa to 12.5 GPa, from lowest to highest fiber volume content in the manufacturing direction (Figure 2). The modulus perpendicular to the manufacturing direction increased from 7.3 GPa to 11 GPa for materials with fibers only 25.4 mm long. For materials with two different fiber lengths, the tensile modulus ranged between 8.2 GPa and 13.6 GPa in the manufacturing direction, and from 7 GPa to 12.1 GPa perpendicular to it. Thus, only a slight increase in tensile stiffness occurred for materials consisting of longer fibers. The higher the FVC of the material, the higher the observed anisotropy.

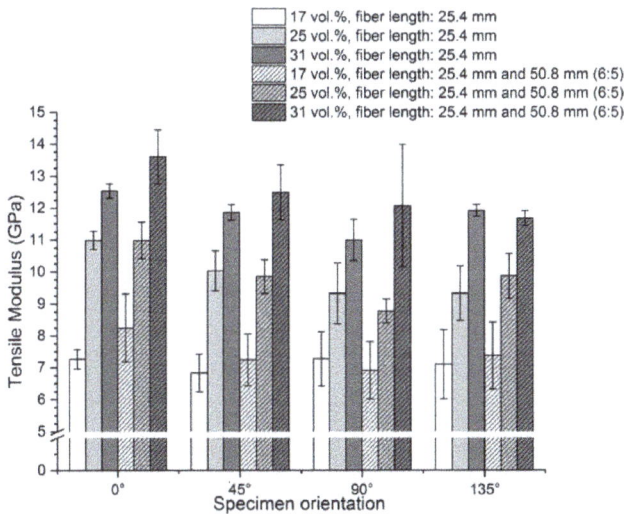

**Figure 2.** Tensile modulus of different SMC materials in different directions with respect to the movement of the conveyor belt.

Compression stiffness (Figure 3) showed a slightly anisotropic trend, though anisotropy was less severe than the tensile loads. The compression modulus increased from 6.8 GPa to 10 GPa for specimens with 25.4 mm long fibers extracted in the manufacturing direction, and from 6.4 GPa to 8.2 GPa perpendicular to the manufacturing direction. An increase in fiber length did not significantly increase the compression. The considered SMC materials showed lower stiffness due to compression loads compared to tensile loads.

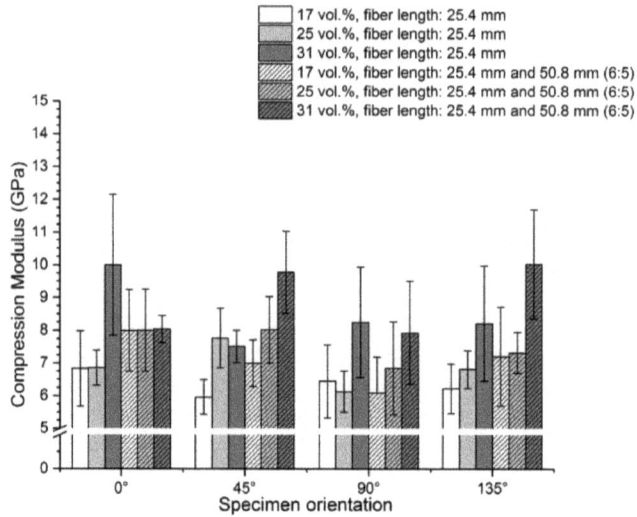

**Figure 3.** Compression modulus of different SMC materials in different directions with respect to the movement of the conveyor belt.

If the material was exposed to flexural loads (Figure 4), only those materials with the highest FVC showed a significant increase in flexural stiffness. The flexural modulus ranged between 7.7 GPa and 12.2 GPa in the manufacturing direction, for materials with 25.4 mm long fibers. Perpendicular to the manufacturing direction, the values ranged from 7.1 GPa to 10.6 GPa. Longer fibers were found not to significantly influence the flexural stiffness of the material. For SMC with 25.4 mm and 50.8 mm long fibers, the flexural modulus ranged from 7.5 GPa to 13.4 GPa in the manufacturing direction, and from 7.1 GPa to 11.3 GPa perpendicular to the movement of the conveyer belt.

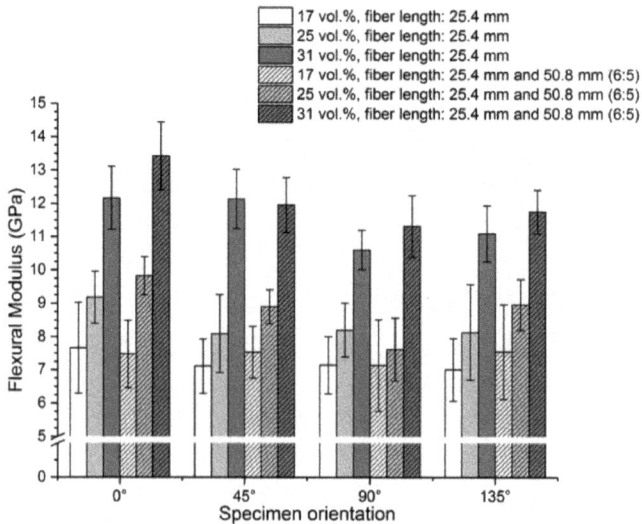

**Figure 4.** Flexural modulus of different SMC materials in different directions with respect to the movement of the conveyor belt.

### 4.2.2. Tensile, Compression, and Flexural Strength

Figures 5–7 depict the tensile, compression, and flexural strengths of the considered SMC material configurations. Tensile strength (Figure 5) increased significantly with increasing FVC, from 64 MPa to 150 MPa for the specimens extracted in the manufacturing direction of SMC with fibers only 25.4 mm long. Due to the movement of the conveyor belt during manufacturing, the strength exhibited an anisotropic trend for materials with higher FVC. The tensile strength perpendicular to the manufacturing direction ranged from 67 MPa to 116 MPa, and was slightly lower than in the direction of manufacturing. A variation in fiber length only influenced the tensile strength for materials with higher FVCs, ranging from 71 MPa to 165 MPa in the manufacturing direction, and from 62 MPa to 133 MPa perpendicular to the conveyor movement direction for SMC containing 25.4 mm and 50.8 mm long fibers.

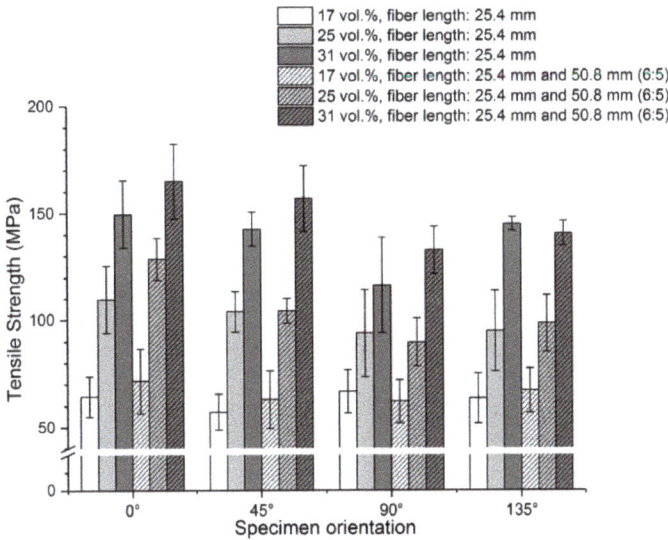

**Figure 5.** Tensile strength of different SMC materials in different directions with respect to the movement of the conveyor belt.

The considered SMC materials generally showed higher values for compression than tensile strength (Figure 6), ranging from 149 MPa to 196 MPa in the manufacturing direction, and from 132 MPa to 202 MPa for specimens perpendicular to conveyor belt motion with 25.4 mm long fibers. If the material also consisted of 50.8 mm long fibers, the compression strength varied from 157 MPa to 194 MPa in the manufacturing direction, and from 139 MPa to 207 MPa perpendicular to it. An increase in fiber length only slightly changed the compression strength, as the measured values were highly scattered.

The flexural strength (Figure 7) of the material was directionally dependent on FVC. The flexural strength ranged from 157 MPa to 277 MPa for materials with 25.4 mm long fibers in the manufacturing direction, and from 156 MPa to 245 MPa perpendicular to it. Materials with longer fibers showed a slight increase in flexural strength, ranging from 164 MPa to 303 MPa in the manufacturing direction, and from 141 MPa to 261 MPa perpendicular to it.

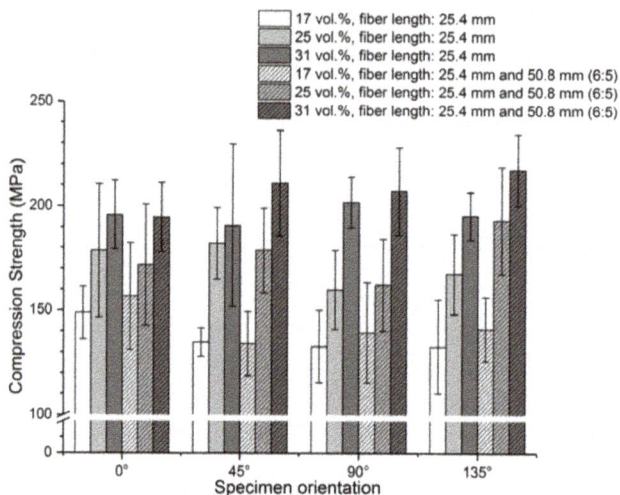

**Figure 6.** Compression strength of different SMC materials in different directions with respect to the movement of the conveyor belt.

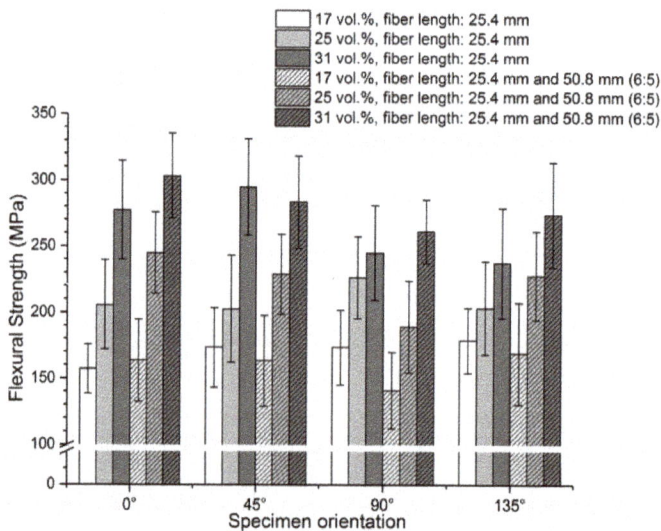

**Figure 7.** Flexural strength of different SMC materials in different directions with respect to the movement of the conveyor belt.

### 4.2.3. Charpy Impact Properties

Specific to Charpy impact properties, an increase in FVC resulted in increased energy absorption capabilities (Figure 8). For the SMC material with 25.4 mm fibers, the energy absorption capability increased from 51 kJ/m$^2$ to 91 kJ/m$^2$ from the lowest to highest FVC, for specimens extracted in manufacturing direction (Figure 8). Perpendicular to the manufacturing direction, absorbed energy increased from 46 kJ/m$^2$ to 79 kJ/m$^2$. The higher the FVC, the more significant the resulting anisotropy. No significant influence of fiber length was found on Charpy impact properties. Energy absorption capability increased from 58 kJ/m$^2$ to 95 kJ/m$^2$ in the manufacturing direction, and from 42 kJ/m$^2$ to 83 kJ/m$^2$ in the perpendicular direction for SMC consisting of 25.4 mm and 50.8 mm fibers.

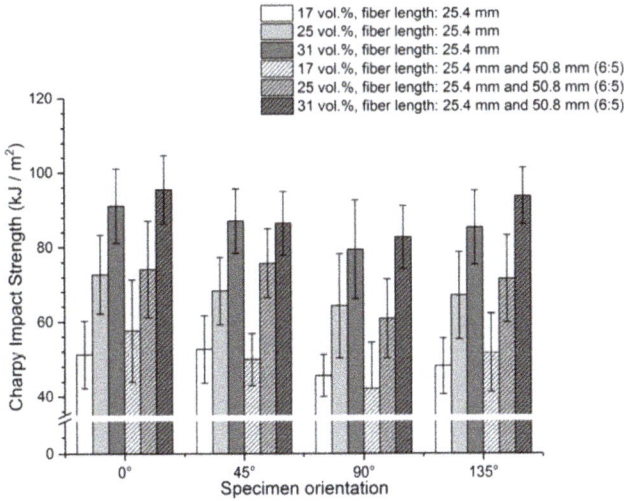

**Figure 8.** Charpy impact strength of different SMC materials considered in different directions with respect to the movement of the conveyor belt.

### 4.3. Acoustic Emission Analysis and Damage Evolution

The algorithm applied in this study leads to two natural clusters for the discontinuous glass fiber reinforced SMC (Figure 9). The best clustering results were obtained by considering the weighted peak frequency and partial power three, with the highest rating possible, and 75 points.

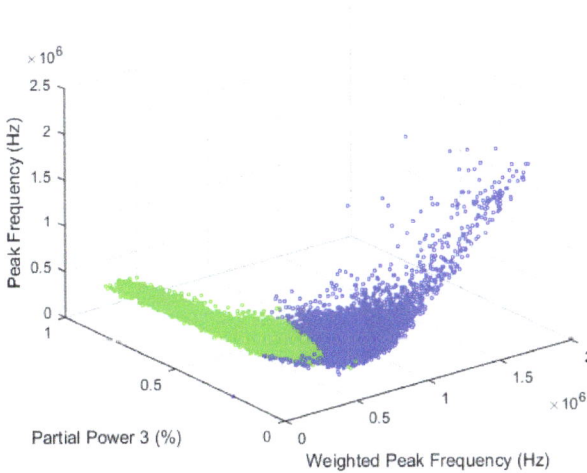

**Figure 9.** Results of clustering algorithms and the clusters in feature space.

Considering the accumulated energy of each cluster for the discontinuous glass fiber SMC specimen, cluster one rapidly increased in energy, with a small drop in force at a displacement of approximately 3.7 mm. From this point, the slope of the force-deflection curve decreased slightly, and did not show any further linear increase. This indicates the first appearance of damage inside the material. Cluster two also showed a growth at this point, but not as significant as cluster one (Figure 10). The discontinuous sample showed an almost steady rise of cluster one and two, without sharp jumps in both of the curves. Nevertheless, cluster one increased its energy much faster than cluster two. Computed tomography of the damaged discontinuous glass fiber SMC specimen showed two important failure mechanisms: matrix cracking and interface failure (Figure 11). Since signal processing of the AE data leads to two different clusters, each failure mechanism can be linked with one of these clusters. Since matrix cracking most likely starts and ends at a fiber-matrix interface, these two failure mechanisms cannot be considered separately, but significantly influence each other.

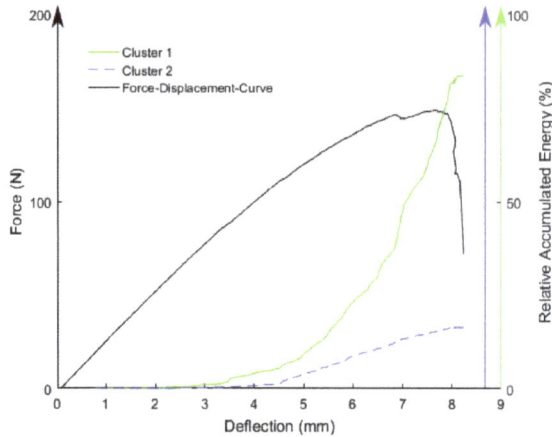

**Figure 10.** Relative accumulated energy of cluster one and cluster two and the force-deflection response.

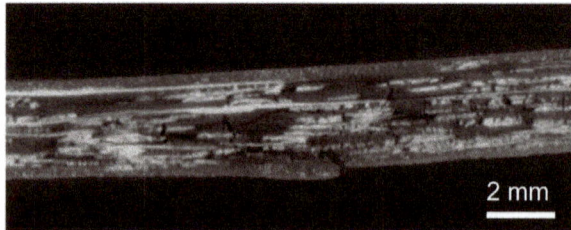

**Figure 11.** Micro computed tomography (μCT)-scan of damaged discontinuous glass fiber SMC specimen exposed to bending loading.

## 5. Discussion

For this study, vinylester-based unfilled structural SMC sheets were successfully manufactured with 100% mold coverage. Due to the mixing of the individual resin components in a vacuum atmosphere, entrapped air was minimized, leading to mechanical material properties comparable to and even better than standard SMC material investigated by different research groups [6–8]. This study aimed to determine the influence of fiber volume content, fiber length, and fiber orientation on the mechanical properties of a glass fiber-reinforced SMC based on a vinylester resin. The six investigated SMC materials showed increasing mechanical properties with increasing FVC in the investigated range, from approximately 17 vol % to 31 vol %. This remained the same for quasi-static and for dynamic loading.

*J. Compos. Sci.* **2017**, *1*, 18

The observed material behavior aligns well with the material behavior of chopped fiber reinforced polymers. No significant difference was found between tensile and flexural stiffness for the considered materials in this study, but the observed compression moduli were smaller for all six material configurations. One possible reason for the lower compression moduli can be micro-buckling, which is frequently present when fiber-reinforced polymers are exposed to compressive loading. For the different materials, the strength was observed to increase from tensile to compression to flexural loads. An increase in fiber length did not significantly influence the mechanical properties for the materials considered within this study. From the authors' point of view, this result cannot be generalized, since, contrary to the findings, the results observed by Boylan et al. [6] showed an increase in tensile strength of SMC as fiber length increased, whereas the tensile modulus did not change with different fiber lengths. This contradiction can be explained by different specimen types considered in the different studies.

To evaluate the results within this study, the specimen size and thus the representative volume differed for the different test configurations, although the testing was based on standardization, and even larger specimens than proposed by the standard were considered for tensile testing. Marissen et al. [32] indicated that specimen geometry and specimen size have an influence on measured mechanical material properties. This must be considered when evaluating the measured mechanical properties and anisotropy. According to the study of Marissen et al. [28], the width of the specimen largely affects the mechanical properties, since a specimen with a small width contains many cut fibers, which decreases the average fiber length. Thus, the effect of reinforcement is less efficient. Considering specimen size, the representative volume is very small, thus the longer fibers did not significantly influence the microstructure of the considered materials. Although the considered specimen volume for tensile loads was relatively big, only a small volume was loaded during compression, increasing the scatter of the measured properties. The difference in specimen size, and thus the considered volume loaded during mechanical testing, could also be a reason for the lower compression stiffness compared to the tensile stiffness of the material.

For the Charpy impact tests, the specimens were exposed to very localized loadings. Understandably, the longer fibers did not influence the impact properties due to this localization of loading. To obtain a better understanding of the influence of fiber length on the impact properties of SMC, another test, for example penetration tests with larger specimens, would be more suitable. Additionally, instrumented impact tests would allow for the investigation of the entire force-deflection response to gain a deeper insight into the elastic energy and energy absorption capability.

The SMC sheet, from which specimens were extracted for this study, did not flow during molding, but the material nevertheless showed anisotropic material properties, which resulted only from the movement of the conveyor belt. Although all considered loading cases within this study showed anisotropic mechanical material properties, the anisotropy was most severe for tensile loadings.

Anisotropic material properties of SMC were also reported by Lamanna et al. [7], but since this study did not provide any details on mold coverage and material properties for different orientations, a quantitative comparison is not possible.

The SMC material observed by Oldenbo et al. [8] featuring 21 vol % glass fibers, 20 vol % fillers, and 18 vol % of hollow glass spheres, only showed a slight anisotropy with a ratio of 1.21 for the elastic modulus measured in longitudinal (0°) and transverse (90°) directions. Tensile strength showed a ratio of 1.16. No details were provided on mold coverage during molding within this study, making it difficult to compare the findings by Oldenbo et al. [8] directly with the findings of this study. However, we assumed that a two-dimensional flow occurred, since the specimens for mechanical testing were extracted from a square mold. The ratio of longitudinal to transverse tensile modulus and tensile strength of the SMC investigated within this study, featuring 25 vol % of 1 inch long fibers, were equal to 1.18 and 1.16, respectively. The anisotropy of the SMC material, featuring the same fiber volume content but a mixture of one and two inch long fibers, was more pronounced, with a ratio of 1.25 (tensile modulus) and 1.44 (tensile strength). We concluded that only the movement of the conveyor belt led to the anisotropic material properties of the material.

The specimens investigated by Boylan et al. [6] were extracted from rectangular sheets, featuring either one or two inch long fibers and a fiber volume content of 21%, whereas the stack of semi-finished material was exposed to either one- or two-dimensional flow during compression molding. The observed ratio of tensile modulus was 1.85 for SMC featuring only one-inch fibers and 1.55 for SMC with only two inch long fibers. This is an opposite trend compared to the findings of this study. With a ratio of 2.8, the tensile strength of the material investigated by Boylan et al. [6] showed a significantly more pronounced anisotropy than the resulting mechanical properties, due to fiber orientation resulting only from the movement of the conveyor belt, as observed within this study. In general, we concluded that material flow during the manufacturing of the semi-finished sheets or during compression molding more significantly affects the material's strength than the elastic properties. Entrapped air, which normally causes problems in the manufacturing process of SMC [6] and is one reason for using a flow molding process, did not cause significant problems while processing the material investigated within this study.

The SMC material investigated by Oldendo et al. [8] featured a fiber volume content of 18–20% with approximately the same amount of hollow glass spheres and fillers. The material featured a density of 1.57 $g/cm^3$. The material investigated within this study had a lower density (density of approximately 1.48 $g/cm^3$ for SMC with approximately 25 vol % of glass fibers) for higher fiber volume contents, since no fillers were added to the resin formulation. With comparable tensile and compression properties, the introduced material offers an increased lightweight potential, since the presented structural SMC enables the manufacturing of a material featuring higher fiber volume contents combined with a lower density than standard SMC materials, as the fillers can be replaced by load carrying fibers.

This contribution also focused on the identification of failure mechanisms, which are present during bending loadings of glass fiber SMC. The μCT investigation of the damaged specimen showed two dominant failure mechanisms resulting from bending loading: matrix cracking and interface failure. These findings align with the results of different research groups, that investigated the failure of sheet molding compounds [33,34]. Fiber breakage was not observed as a failure mechanism for the investigated SMC material. The acoustic emission analysis performed in this contribution, being coupled with machine learning algorithms, led to the identification of two natural signal clusters. In general, different clusters represent different failure mechanisms. To describe the damage of a discontinuous glass fiber SMC, the two clusters were attributed to the two observed failure mechanisms. Machine learning tools for AE are still in an early stage and are being steadily developed. New feature spaces could help offer a sharper line between the clusters seen in Figure 9, and hence to minimize erroneously clustered hits. Various material systems could also use an adjustment of the feature space, such as the range of the frequencies for integration of partial powers. Nevertheless, machine learning algorithms represent a helpful tool to analyze the large volume of data created in AE analysis, which may be a first step toward a better understanding of the damage behavior and evolution of discontinuous composites, as it allows for continuous observation and identification of the damage evolution and the underlying mechanisms.

## 6. Conclusions

This paper investigated the influence of fiber length, fiber volume content, and fiber orientation on quasi-static and dynamic material properties of discontinuous glass fiber-reinforced vinylester-based SMC. Six different materials, featuring different fiber volume contents and two different fiber length configurations, were considered. The acoustic emission technique and data processing by means of machine learning algorithms were used to determine damage mechanisms of this material.

The results showed that manufacturing SMC sheets with a 100% mold coverage is possible. An increase in fiber volume content from 17 vol % to 31 vol % led to increased stiffness and strength

*J. Compos. Sci.* **2017**, *1*, 18

for tension, compression, and flexural loads for the considered unfilled SMC material. This behavior aligns well with the mechanical behavior of fiber-reinforced polymers featuring different fiber volume contents in general. The combination of one and two inch long fibers did not significantly influence the quasi-static and dynamic material properties for the considered materials. This result is possibly due to the small effective loaded regions of the specimen. Fiber orientation, which occurred only due to the movement of the conveyer belt during the manufacturing of the semi-finished materials, led to slightly anisotropic quasi-static and dynamic material properties, whereas the anisotropy was less severe than for SMC sheets which flew during compression molding. The omission of fillers led to materials with high fiber volume contents and lower densities, compared to standard SMC materials. Hence, the material considered within this study has high potential for lightweight applications. Machine learning algorithms are appropriate to cluster AE signals. Comparison of the results of the post-damaged μCT analysis and acoustic emission analysis, performed during loading of the specimen, pointed out that damage of discontinuous vinylester SMC was based on matrix cracking and interface failure. These findings are in good agreement with literature.

**Acknowledgments:** The research documented in this manuscript has been funded by the German Research Foundation (DFG) within the International Research Training Group "Integrated engineering of continuous-discontinuous long fiber reinforced polymer structures" (GRK 2078). The authors also kindly acknowledge the Fraunhofer ICT in Pfinztal, Germany for manufacturing the SMC sheets with a special thanks to David Bücheler and Leopold Giersch for his support to carry out parts of the mechanical testing. Furthermore, the authors would like to thank Markus Sause from Experimental Physics II in Augsburg for his suggestions in data processing as well as Florentin Pottmeyer and Stefan Dietrich from IAM-WK for valuable discussions.

**Author Contributions:** Anna Trauth conceived, designed and performed the experiments; Pascal Pinter performed the μCT scans; Anna Trauth, Pascal Pinter and Kay André Weidenmann analyzed the data; Pascal Pinter contributed machine learning and analysis tools; Anna Trauth and Pascal Pinter wrote the paper.

**Conflicts of Interest:** The authors declare no conflict of interest.

## References

1. Orgéas, L.; Dumont, P.J.J.; Nicolais, L. Sheet Molding Compounds. In *Wiley Encyclopedia of Composites*, 2nd ed.; Nicolais, L., Borzacchiello, A., Eds.; John Wiley & Sons, Inc.: Hoboken, NJ, USA, 2011.
2. Witten, E.; Kraus, T.; Kühnle, M. *Composites Market Report 2016: Market Developments, Trends, Outlook and Challenges*; AVK: Frankfurt, Germany, 2016.
3. Reimer, U.; Esswein, G.; Derek, H. Quo Vadis—SMC im Automobilbau. *Ku Kunststoffe Online-Archive* **2000**, *90*, 86–90.
4. Bruderick, M.; Denton, D.; Shinedling, M.; Kiesel, M. Application of Carbon Fiber SMC for the Dodge Viper. Available online: http://www.quantumcomposites.com/pdf/papers/viper-spe-paper.pdf (accessed on 21 October 2017).
5. Gardiner, G. Is the BMW 7 Series the Future of Autocomposites? Available online: http://www.compositesworld.com/articles/is-the-bmw-7-series-the-future-of-autocomposites (accessed on 21 October 2017).
6. Boylan, S.; Castro, J.M. Effect of Reinforcement Type and Length on Physical Properties, Surface Quality, and Cycle Time for Sheet Molding Compound (SMC) Compression Molded Parts. *J. Appl. Polym. Sci.* **2003**, *90*, 2557–2571. [CrossRef]
7. Lamanna, G.; Ceprano, A. Mechanical Characterization of Sheet Moulding Composites for the Automotive Industry. *Open Mater. Sci. J.* **2014**, *8*, 108–113. [CrossRef]
8. Oldenbo, M.; Fernberg, S.; Berglund, L. Mechanical Behaviour of 5SMC6 Composites with Toughening and Low Density Additives. *Compos. Part A Appl. Sci. Manuf.* **2003**, *34*, 875–885. [CrossRef]
9. Trauth, K.A. *Mechanical Properties of Continuously Discontinuously Fibre Reinforced Hybrid Sheet Moulding Compounds*; Euro Hybrid: Kaiserslautern, Germany, 2016.
10. Bücheler, D.; Henning, F. Hybrid Resin Improves Position and Alignment of Continuously Reinforced Prepreg during Compression Co-Molding with Sheet Molding Compound. In Proceedings of the 17th European Conference on Composite Materials, Munich, Germany, 26–30 June 2016.
11. Puck, A. *Festigkeitsanalyse von Faser-Matrix-Laminaten: Modelle für Die Praxis*; Carl Hanser Verlag: Munich, Germany, 1996.

12. Muravin, B. Acoustic Emission Science and Technology. Available online: http://www.muravin.com (accessed on 21 October 2017).
13. Jackson, C.N.; Moore, P.O.; Sherlock, C.N. *Nondestructive Testing Handbook, Acoustic Emission Testing*, 3rd ed.; ASNT: Columbus, OH, USA, 2005.
14. Barré, S.; Benzeggagh, M.L. On the Use of Acoustic Emission to Investigate Damage Mechanisms in Glass-Fibre-Reinforced Polypropylene. *Compos. Sci. Technol.* **1994**, *52*, 369–376. [CrossRef]
15. Jiao, G.Q.; Zheng, S.T.; Suzuki, M.; Iwamoto, M. Damage Evaluation of Sheet Moulding Compound Composite by Acoustic Emission. *Theor. Appl. Fract. Mech.* **1990**, *14*, 135–140. [CrossRef]
16. Alander, P.; Lassila, L.V.J.; Tezvergil, A.; Vallittu, P.K. Acoustic Emission Analysis of Fiber-Reinforced Composite in Flexural Testing. *Dent. Mater.* **2004**, *20*, 305–312. [CrossRef]
17. Huguet, S.; Godin, N.; Gaertner, R.; Salmon, L.; Villard, D. Use of Acoustic Emission to Identify Damage Modes in Glass Fibre Reinforced Polyester. *Compos. Sci. Technol.* **2002**, *62*, 1433–1444. [CrossRef]
18. Santulli, C. Matrix Cracking by Acoustic Emission in Polymer Composites and Counts/Duration Ratio. Available online: http://www.ndt.net/search/docs.php3?showform=off&id=13678 (accessed on 21 October 2017).
19. Dahmene, F.; Yaacoubi, S.; Mountassir, M.E.L. Acoustic Emission of Composites Structures: Story, Success, and Challenges. In Proceedings of the 2015 ICU International Congress on Ultrasonics, Metz, France, 10–14 May 2015.
20. Sause, M.; Gribov, A.; Unwin, A.R.; Horn, S. Pattern Recognition Approach to Identify Natural Clusters of Acoustic Emission Signals. *Pattern Recognit. Lett.* **2012**, *33*, 17–23. [CrossRef]
21. De Oliveira, R.; Marques, A.T. Health Monitoring of FRP Using Acoustic Emission and Artificial Neural Networks. *Smart Struct.* **2008**, *86*, 367–373. [CrossRef]
22. Mccrory, J.P.; Al-Jumaili, S.K.; Crivelli, D.; Pearson, M.R.; Eaton, M.J.; Featherston, C.A.; Guagliano, M.; Holford, K.M.; Pullin, R. Damage Classification in Carbon Fibre Composites Using Acoustic Emission: A Comparison of Three Techniques. *Compos. Part B Eng.* **2015**, *68*, 424–430. [CrossRef]
23. Sause, M.; Horn, S.; Klug, M.; Scholler, J. Anwendung von Mustererkennungsverfahren zur Schadensanalyse in Faserverstärkten Kunststoffen. In Proceedings of the 17th Colloquium Acoustic Emission, Bad Schandau, Germany, 24–25 September 2009.
24. Deutsches Institut für Normung e.V. *Determination of Tensile Properties of Plastics–Part 4: Test Conditions for Isotropic and Orthotropic Fibre-Reinforced Plastic Composites (ISO 527-4: 1997)*, German version EN ISO 527-4; Deutsches Institut für Normung e.V.: Berlin, Germany, 1997.
25. Deutsches Institut für Normung e.V. *Fibre-Reinforced Plastic Composites–Determination of Flexural Properties (ISO 14125: 1998 + Cor.1: 2001 + Amd.1: 2011)*, German version EN ISO 14125:1998 + AC: 2002 + A1; Deutsches Institut für Normung e.V.: Berlin, Germany, 2011.
26. Deutsches Institut für Normung e.V. *Fibre Reinforced Plastic Composites–Determination of Compressive Properties in the In-Plane Direction (ISO 14126: 1999)*, German version EN ISO 14126; Deutsches Institut für Normung e.V.: Berlin, Germany, 1999.
27. Deutsches Institut für Normung e.V. *Plastics–Determination of Charpy Impact Properties–Part 1: Non-Instrumented Impact Test (ISO 179-1: 2010)*, German version EN ISO 179-1; Deutsches Institut für Normung e.V.: Berlin, Germany, 2010.
28. Günter, S.; Bunke, H. Validation Indices for Graph Clustering. *Pattern Recognit. Lett.* **2003**, *24*, 1107–1113. [CrossRef]
29. Davies, D.L.; Bouldin, D. A Cluster Separation Measure. *IEEE Trans. Pattern Anal. Mach. Intell.* **1979**, *1*, 224–227. [CrossRef] [PubMed]
30. Rousseeuw, P.J. Silhouettes: A Graphical Aid to the Interpretation and Validation of Cluster Analysis. *J. Comput. Appl. Math.* **1987**, *20*, 53–65. [CrossRef]
31. Caliński, T.; Ja, H. A Dendrite Method for Cluster Analysis. *Commun. Stat.* **1974**, *3*, 1–27. [CrossRef]
32. Marissen, R.; Linsen, J. Variability of the Flexural Strength of Sheet Moulding Compounds. *Compos. Sci. Technol.* **1999**, *59*, 2093–2100. [CrossRef]

*J. Compos. Sci.* **2017**, *1*, 18

33. Fitoussi, J.; Guo, G.; Baptiste, D. A Statistical Micromechanical Model of Anisotropic Damage for S.M.C. Composites. *Compos. Sci. Technol.* **1998**, *58*, 759–763. [CrossRef]
34. Sayari, M.; Baptiste, D.; Bocquetand, M.; Fitoussi, J. Characterization and Damage Evolution of S.M.C Composites Materials by Ultrasonic Method. In Proceedings of the 12th International Conference on Composite Materials, Paris, France, 1999; Available online: http://www.iccm-central.org/Proceedings/ICCM12proceedings/site/papers/pap1129.pdf (accessed on 13 December 2017).

*Journal of*
*composites science*

MDPI

*Article*

# A Novel CAE Method for Compression Molding Simulation of Carbon Fiber-Reinforced Thermoplastic Composite Sheet Materials

Yuyang Song [1,*], Umesh Gandhi [1], Takeshi Sekito [2], Uday K. Vaidya [3], Jim Hsu [4], Anthony Yang [4] and Tim Osswald [5]

[1]  Toyota Research Institute North America, Ann Arbor, MI 48105, USA; umesh.gandhi@toyota.com
[2]  Material Creation & Analysis Department, Toyota Motor Corporation, Toyota 471-8572, Japan; takeshi_sekito@mail.toyota.co.jp
[3]  Department of Mechanical, Aerospace and Biomedical Engineering, University of Tennessee, Knoxville, TN 37996, USA; uvaidya@utk.edu
[4]  Moldex3d Northern America Inc., Farmington Hills, MI 48331, USA; jimhsu@moldex3d.com (J.H.); anthonyyang@moldex3d.com (A.Y.)
[5]  Polymer Engineering Center, University of Wisconsin-Madison, Madison, WI 53706, USA; tosswald@wisc.edu
*  Correspondence: Yuyang.song@toyota.com

Received: 20 April 2018; Accepted: 30 May 2018; Published: 1 June 2018

**Abstract:** Its high-specific strength and stiffness with lower cost make discontinuous fiber-reinforced thermoplastic (FRT) materials an ideal choice for lightweight applications in the automotive industry. Compression molding is one of the preferred manufacturing processes for such materials as it offers the opportunity to maintain a longer fiber length and higher volume production. In the past, we have demonstrated that compression molding of FRT in bulk form can be simulated by treating melt flow as a continuum using the conservation of mass and momentum equations. However, the compression molding of such materials in sheet form using a similar approach does not work well. The assumption of melt flow as a continuum does not hold for such deformation processes. To address this challenge, we have developed a novel simulation approach. First, the draping of the sheet was simulated as a structural deformation using the explicit finite element approach. Next, the draped shape was compressed using fluid mechanics equations. The proposed method was verified by building a physical part and comparing the predicted fiber orientation and warpage measurements performed on the physical parts. The developed method and tools are expected to help in expediting the development of FRT parts, which will help achieve lightweight targets in the automotive industry.

**Keywords:** compression molding; sheet material; computer-aided engineering (CAE); draping; recycled carbon fibers

## 1. Introduction

Fiber-reinforced composites offer exciting new possibilities for weight reduction due to their excellent mechanical properties, economical fuel consumption, and lower carbon footprint. For typical automotive components where large volume, low cost, and fast cycle time are desired, chopped reinforcing fibers with thermoplastic resin are increasing in popularity [1–3]. Usually, such composite materials are made using injection molding, compression molding, or resin transfer molding processes [4,5]. Of these processes, compression molding offers the highest potential to maintain longer fibers and hence better mechanical properties [6–8]; therefore, compression molding is the preferred method by which to manufacture fiber-reinforced thermoplastic components in the automotive industry.

There are typically two different initial formats for the fiber-reinforced charge used in the compression molding process, i.e., (1) bulk charge, as seen in Figure 1, and (2) sheet charge, as per Figure 2. Typically, the bulk charge is made using a single/twin screw low shear plasticator; the desired size of the bulk coming out from the plasticator is collected and placed in the die for compression. Sheet charge manufacturing is more complex. Chopped fibers are dispersed to form mats or spread on the resin sheets and are then partially consolidated. The sheet charge is preferred because it starts with longer fibers and, since it covers most of the mold cavity surface to begin with, the material flow required to fill the mold is shorter; therefore, there is a better possibility of maintaining a longer fiber length.

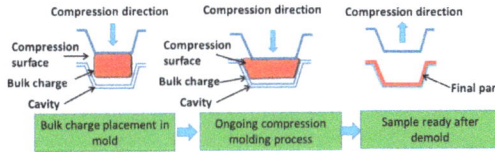

**Figure 1.** Typical compression molding process using bulk charge.

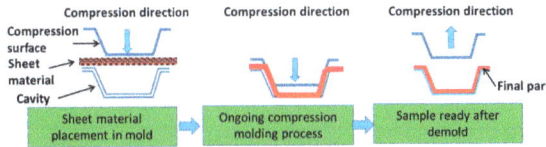

**Figure 2.** Typical compression molding process using sheet material.

As the use of fiber-reinforced polymer components in the industry is increasing, interest in using numerical simulation—often called computer-aided engineering (CAE)—to accelerate the development process is also increasing. There have been numerous research activities on the CAE simulation of fiber-reinforced polymeric resin materials in the past 30 years. Early on, Oswald [9,10] and Advani [11,12] developed the basic formulation for the compression molding simulation for fiber-reinforced polymeric materials. As a result of their work, commercial simulation software called Cadpress [13] was developed. Cadpress used simplified Barone–Caulk [14] flow geometry, which was suitable for sheet moulding compound (SMC) (discontinuous fiber + thermoset resin). Wang et al. [15] investigated compression molding simulation of a multilayer composite with continuous fibers and developed a method to estimate the draped shape and temperature distribution of the laminates. Kikuchi [16] studied the compression molding of woven-fabric thermoplastic composite laminates using finite element analysis, and developed an algorithm to optimize temperature distribution to minimize warpage. Three-dimensional thermo-viscoplastic analysis was conducted by Kim et al. [17], where the rheological characteristic of the SMC sheet material was modeled and the effects of dwelling time, and mold temperature on mold filling and curing were investigated using finite element analysis (FEA) analysis.

In summary, all the previous studies on the simulation of sheet materials for the compression molding process focused on SMC, i.e., discontinuous thermoset materials, continuous fiber laminate structures, or woven fabric with thermoset or thermoplastic resins. The majority of research activities have concentrated only on the thermoforming simulation. Thermoset materials flow well due to low viscosity and therefore were preferred. However, thermoset materials take a longer time to cure and hence the cycle time is longer. Therefore, thermoset materials are preferred for low-volume parts. Since we are interested in automotive applications that require a higher volume, thermoplastic materials that do not require curing and can be formed in a very short cycle time are desired. The molten thermoplastic materials are highly viscous and therefore the flow geometry in filling the mold cavity is different when compared to thermoset materials. Consequently, the fiber orientation and fiber lengths are also different. A major challenge in CAE simulation of the discontinuous long fiber thermoplastic parts is the ability to simulate compression molding to predict the fiber orientation and warpage of the final part. For continuous fibers, the fiber orientation can be estimated by using the draping process simulation. Recently, CAE methods to simulate the compression molding process of fiber-reinforced thermoplastic composites using the bulk charge format have been developed and validated with physical tests by Song et al. [18]. However, the compression molding simulation approach developed for bulk materials does not work for the compression molding of sheet materials. This is because, when heated fiber-reinforced thermoplastic sheets are placed in the mold cavity, the sheets have very little stiffness and hence drape easily in the mold cavity under very little compressive force. This draping process involves significant rigid body motion of the sheets. The fluid mechanics-based approach used for compression molding of bulk material [19,20] is therefore not adequate to simulate the large rigid body motions that occur during the draping process.

In this paper, a novel CAE method was proposed to address this challenge. The compression molding of sheet materials was simulated in two steps. First, the draping of the heated sheet was simulated using an explicit finite element approach. Such an explicit finite element approach, used for metal sheet stamping simulation, has been proven to be effective for large structural deformation as well as rigid body motions. Next, the draped shape of the fiber-reinforced thermoplastic sheet was used as a prepreg in the compression molding. This is similar to the compression molding of a bulk charge with a given shape, which is an already developed method in our previous work [18]. Furthermore, the proposed approach was demonstrated on a complex 3D shape part. The compression molding simulation was carried out using the proposed two-step approach and the fiber orientation and warpage of the finished part were estimated. To verify the proposed approach, actual parts were made and the measured warpage and fiber orientation results were compared with the predictions. Finally, the effect of the draping distance for the compression molding process was discussed and the ideal draping stroke for the simulation of the given part were estimated. Material properties required to support the simulation, i.e., high temperature stiffness and strength for the draping simulation and material flows and other temperature properties, were measured experimentally.

## 2. Challenges and Approaches

### 2.1. Sheet Material Compression Molding Issues

Detailed steps for compression molding of sheet materials are presented in Figure 3. There are four main steps. Step 1: Sheet material is heated above its melt temperature at 270 °C. Step 2: The heated sheet material is draped in the heated mold cavity, which includes a slight compression in the mold. Step 3: Compressing the sheet in the mold and holding at high pressure until solidification. Step 4: Solidified finished part is ejected from the mold and cooled.

From the literature review provided in the first section, it was observed that most commercial software can undertake draping analysis for a continuous unidirectional fiber-reinforced thermoset composite, mainly for wrinkling and thickness analysis. For the compression of discontinuous thermoplastic composite materials, a method to simulate the mold filling process, which can estimate fiber orientation and fiber length, is not available. This is mainly because fiber-reinforced thermoplastic sheets need to be draped onto the mold cavity and then compressed; a single computational tool to address both in the same step is not available. This is because of elastic–plastic structural behavior during draping and the flow behavior of resin melt during compression; the integration of structural and fluid flow simulation for a continuum in one physical system are difficult. Therefore, we propose a two-step approach, as shown in Figure 4. The entire compression molding of sheet material is divided into two steps. In the first step, the draping analysis is conducted using stamping capability using an explicit finite element analysis (LS-DYNA V971)Livermore Software Technology Corporation, Livermore, CA, USA). In the second step, the draped part is treated as a prepreg for the compression molding process using mold fill analysis (Moldex3D, R14) CoreTech System Co., Ltd., Hsinchu County, Taiwan). A translator to convert the deformed shape from LS-DYNA to the initial shape for compression molding was developed.

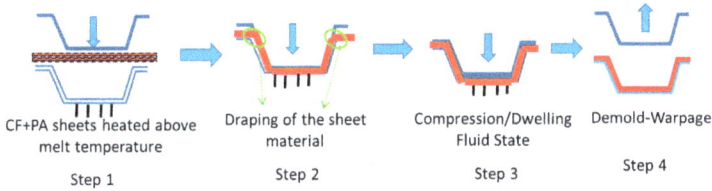

**Figure 3.** Challenges and issues when compression molding sheet materials.

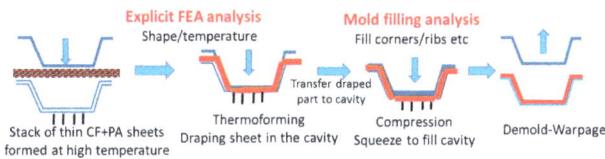

**Figure 4.** Approaches used for compression molding of sheet materials.

*2.2. Measurement of Material Properties of the Sheet Material*

2.2.1. Sheet Material Properties Measured for Draping Analysis

For accurate draping analysis, temperature-dependent material properties such as elastic modulus and Poisson's ratio are needed for the sheets. The sheet materials are made from a water slurry process, with 35% carbon fiber. The initial resin was made of polyamide 6 (PA6) fibers. Figure 5a shows the procedures used to measure the mechanical properties of the sheet material. As shown, the two layers of fluffy sheet materials are stacked together under a hot press and consolidated to form the composite plate. The tensile bar samples are cut from the consolidated plate. Tests were performed following the ASTMD638-02 standard [21], and the samples were tested in a temperature chamber to obtain the mechanical properties at different temperatures. The modulus of the material was obtained from the tensile test. The Poisson's ratio was also measured as per the ASTMD638 standard [21] using the optical imaging system. Table 1 shows the measured mechanical properties. The increase of the Poisson's ratio in the end is due to the polymeric phase change at around 180 °C, and crystallization happens again. The modulus and Poisson's ratio results are the average of three repeated tests.

**Table 1.** Temperature-dependent stress–strain curve for the sheet material.

| Temperature (C) | Modulus (Mpa) | Poisson's Ratio |
|---|---|---|
| 23 | 14,091 | 0.339 |
| 80 | 10,444 | 0.239 |
| 130 | 10,118 | 0.237 |
| 180 | 8728 | 0.179 |
| 200 | 7380 | 0.403 |

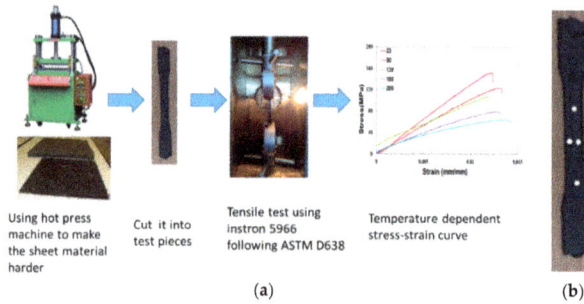

| Using hot press machine to make the sheet material harder | Cut it into test pieces | Tensile test using instron 5966 following ASTM D638 | Temperature dependent stress-strain curve |

(a)　　　　　　　　　　　　　　　　(b)

**Figure 5.** Temperature dependent young's modulus and Poisson's ratio measurement procedures. (a) Young's modulus measurement; (b) Poisson's ratio measurement.

## 2.2.2. Sheet Material Properties Measured for Compression Molding Analysis

To accurately simulate the compression molding process, actual material properties such as pressure–volume–temperature (PVT), viscosity, and thermal properties are also needed for this process [22]. The sheet material properties were measured by following the corresponding ASTM standards [23–26]. After the physical measurement, all the measurement results were converted to the moldex3D material format. These measured composite properties were then imported into the compression molding software Moldex3D's material database and were applied to the process simulation for the part we designed, as Figure 6 shows.

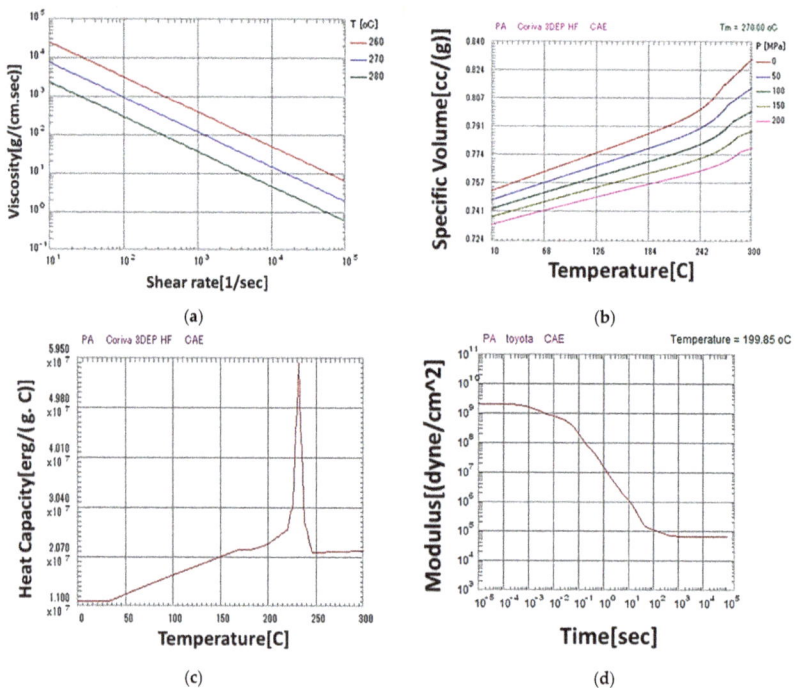

(a)　　　　　　　　　　　　　　　　(b)

(c)　　　　　　　　　　　　　　　　(d)

**Figure 6.** Measured charge material properties converted to the Moldex3D format. (a) Viscosity curve; (b) PVT curve; (c) Heat capacity curve; (d) Viscoelasticity curve.

## 3. CAE Simulation

### 3.1. Overall CAE Simulation Concept and Steps

Based on the potential needs for the light weighting of automotive components, a three-cavity tool was designed and the sample geometry is presented in Figure 7.

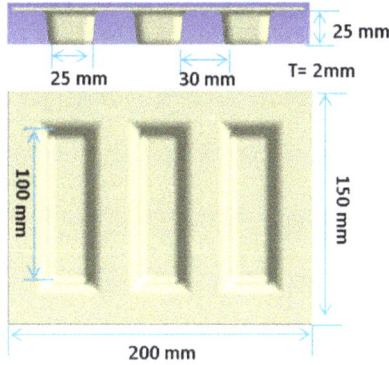

**Figure 7.** The shape and geometry of the designed part.

After the initial design of the part, the question was how to use the CAE software to predict the forming process from sheet materials. Another question of interest was how to predict the warpage and microstructural details such as fiber orientation, fiber length, etc. of the part post-compression molding. As proposed in Section 2, the integrated simulation process combining LS-DYNA and Moldex3D was used for this part.

Figure 8 shows the overall CAE simulation workflow for the part. The first step was the draping analysis in LS-DYNA. After the draping analysis, the draped part was transferred to the Moldex3D for compression molding process analysis. The details of each simulation process will be discussed separately in the following sections.

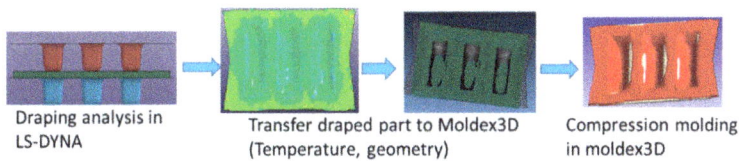

Draping analysis in LS-DYNA

Transfer draped part to Moldex3D (Temperature, geometry)

Compression molding in moldex3D

**Figure 8.** CAE steps for the compression molding of sheet materials.

### 3.2. Draping Analysis with LS-DYNA

For the first step of the simulation, LS-DYNA was used to predict the draping behavior of the sheet materials. A thermoforming module from LSTC was used in the simulation. The initial sheet material was modeled using solid elements, as is shown in Figure 9. In the model, the total number of solid elements in the model was 15,000, the total number of shell elements in the model was 25,896, and the total number of nodes was 49,280.

The basic model information and boundary condition are shown in Figure 9. During the thermoforming process, thermal properties of the sheet material and the tool were defined, and the mechanical properties of the sheets were also defined. Due to the draping/tearing of the sheet material at the corners, the adaptive remeshing technique was used in the simulation to reduce the instability or any numerical convergence issues [27,28]. Figure 10 shows the displacement contour of the draped part. Then, the shape of the draped part was exported to Moldex3D for the next simulation step.

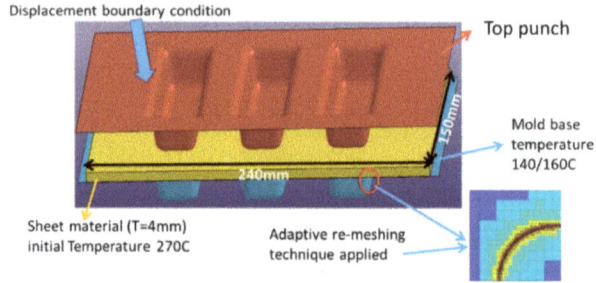

**Figure 9.** Draping model information in LS-DYNA.

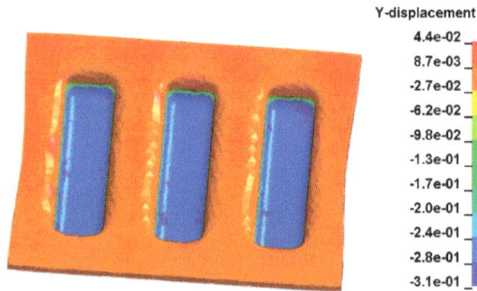

**Figure 10.** Displacement contours of the draped part (unit: m).

### 3.3. Compression Molding Analysis with Moldex3D

Based on the cavity design from Figure 6, a compression molding 3D model was built in Moldex3D, as shown in Figure 11. The element type used in the model was tetra and hex mesh; the total number of elements was 5,944,826 and the total number of nodes was 2,942,618. There were 11 elements through thickness. The green surface on the top is the compression surface, the pink area in the middle is the compression zone, and the gray area on the bottom is the cavity. Figure 12 shows the draped part as the charge layover the cavity. The red area on the top is assigned as the charge; the yellow area on the bottom is assigned as the cavity.

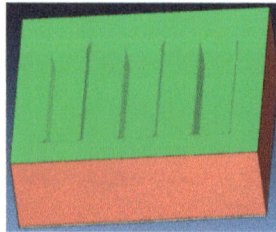

**Figure 11.** Compression molding model built in Moldex3D.

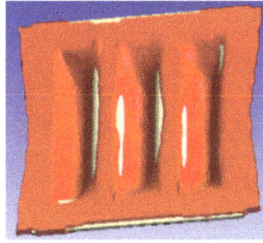

**Figure 12.** The draped part shown as charge layover the cavity.

The temperature information and deformed shape were taken into account in the compression molding simulation in Moldex3D. The governing equations of the fluid mechanics that describe the transient and non-isothermal compression flow motion are as follows:

$$\frac{\partial \rho}{\partial t} + \nabla \cdot \rho \mathbf{u} = 0 \tag{1}$$

$$\frac{\partial}{\partial t}(\rho \mathbf{u}) + \nabla \cdot (\rho \mathbf{u}\mathbf{u} - \sigma) = \rho \mathbf{g} \tag{2}$$

$$\sigma = -p\mathbf{I} + \eta\left(\nabla \mathbf{u} + \nabla \mathbf{u}^T\right) \tag{3}$$

$$\rho C_P \left(\frac{\partial T}{\partial t} + \mathbf{u} \cdot \nabla T\right) = \nabla \cdot (k\nabla T) + \eta \dot{\gamma}^2 \tag{4}$$

where $\rho$ is the density; $\mathbf{u}$ is the velocity vector; $t$ is the time; $\sigma$ is the total stress tensor; g is the acceleration vector of gravity; $p$ is the pressure; $\eta$ is the viscosity; $C_p$ is the specific heat; $T$ is the temperature; $k$ is the thermal conductivity; and $\dot{\gamma}$ is the shear rate. Tait's $p$VT model is to express a thermodynamic state relationship, where the volume of the material is a function of the temperature/pressure.

The power law model is used to describe complex viscosity behaviors, including the dramatic viscosity change in a lower shear rate range. Details of the numerical implementation are available elsewhere [29] and so are not repeated here. After the part is ejected from the mold, a free thermal shrinkage occurs due to the temperature difference. The equilibrium equation representing the stress is:

$$\nabla \sigma = \mathbf{0}. \tag{5}$$

The relationship between stress and strain is:

$$\sigma = \mathbf{C}(\varepsilon - \alpha_{CLTE}\Delta T) \text{ with } \varepsilon = \frac{1}{2}\left(\nabla \mathbf{U} + \nabla \mathbf{U}^T\right) \tag{6}$$

where $\sigma$ is the stress; $\mathbf{C}$ is a fourth order tensor and a function of the relaxation modulus $E$; $\alpha_{CLTE}$ is the coefficient of the linear thermal expansion tensor; $\varepsilon$ is the strain tensor, which is transferred by the volume shrinkage during molding; and $\mathbf{U}$ is the displacement vector, respectively. In order to consider the viscoelastic behavior after molding, a master function for the relaxation modulus is described using the generalized maxwell model, which is a function of time and temperature with temperature shift factor $a_T$ described below:

$$E = E(T_{\text{ref}}, time/a_T) \text{ with } a_T = 10^{\frac{-C_1(T-T_{\text{ref}})}{C_2+T-T_{\text{ref}}}}, \tag{7}$$

where $C_1$, $C_2$ are constant; and $T_{\text{ref}}$ is the reference temperature.

The compression molding process simulations were used in Moldex3D, version R14 [13]. The iARD–RPR (the improved anisotropic rotary diffusion model combined with the retarding principal rate model) was used in the simulation with $Ci = 0.01$. The other parameters used in the iARD–RPR model were $Cm = 0.005$, $Ci = 0.1$, and the fiber matrix interaction alpha factor was 0.7. The processing conditions in CAE are the same as the one used to make the physical part. The material properties used in the simulation are listed in Table 2 and Figure 6.

**Table 2.** Material properties used in the simulation.

| Items | Carbon Fiber | PA6 |
|---|---|---|
| Density (g/cc) | 1.78 | 1.13 |
| Fiber Weight Percentage | 35% | N/A |
| Young's Modulus E1 (Mpa) | 230,000 | 2400 |
| Young's Modulus E2 (Mpa) | 23,000 | 2400 |
| Poisson's Ratio | 0.26 | 0.42 |
| Fiber Aspect Ratio | 400 | N/A |
| Fiber CLTE at fiber direction (1/K) | $1 \times 10^{-6}$ | N/A |
| Fiber CLTE at transverse direction (1/K) | $1 \times 10^{-5}$ | N/A |
| Polymer CLTE (1/K) | N/A | $8.3 \times 10^{-5}$ |

## 4. Experimental Procedure

### 4.1. Carbon Fiber Sheet Material Manufacturing

The first step in the production of fiber-reinforced thermoplastic composites is the combining of the fiber reinforcement with the matrix resin to form the sheet material. One technique that has been demonstrated to produce a high-quality self-supporting preform is the wet-lay paper making process [30–32]. In this process, the reinforcing fibers and thermoplastic polymer fibers are dispersed in water along with bonding agents. This creates a slurry that is cast onto a moving forming wire and then dewatered, leaving a nonwoven mat of thermoplastic and reinforcing fibers. The carbon fibers, which are made from recycled carbon fiber composites, were supplied by Carbon Conversion (Lake City, SC, USA). The carbon fiber mats have a random fiber orientation, and an average fiber length of about 12 mm, and diameter about 6–7 µm. The resin was PA6 fibers with an average length of about 10 mm, and diameter of 20 µm. Each sheet with a thickness of about 0.2 mm was stacked to form a 2 mm nominal thickness sheet. The mats were fluffy and had in-plane random fiber orientation. The size of the mats we used in this study were 355 mm × 254 mm. Figure 13 shows the steps used in the water slurry process to form the sheet materials.

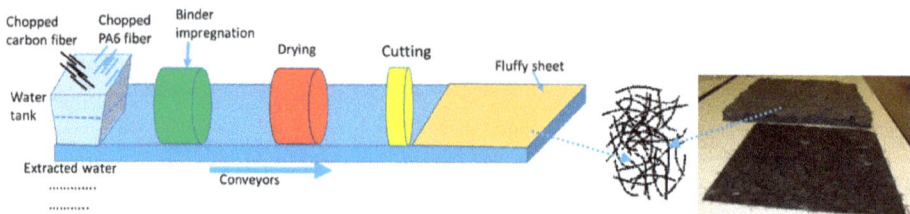

**Figure 13.** Sheet material manufacturing process using water slurry method.

### 4.2. Actual Part Manufacturing

Three cavity parts were built using carbon fiber (CF) (35%) + PA6 sheet material. The basic procedure for the compression molding of the part is shown in Figure 14. After the sheet material was prepared from the water slurry process, it was moved to the infrared oven, and heated to 343 °C before it was moved to the press.

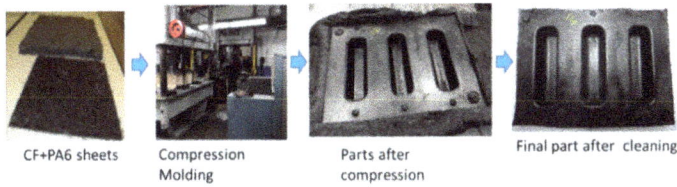

Figure 14. Three-cavity tool and the part made using sheet material.

Table 3 shows the processing conditions of the manufacturing process. After the sample is demolded and cleaned, it is ready for warpage and fiber orientation measurement. Figure 15 shows the warpage measurement method used for the three-cavity tool. The FARO-Arm (FARO Technologies, Inc., Lake Mary, FL, USA) [33,34], which is a state-of-the-art non-contact scanning device for warpage measurement, was used to collect the point cloud of the final part.

The scanned point cloud was then superimposed with the original computer-aided design (CAD) data of the part, as shown. From the superimposed contour, the warpage information was obtained. Computed tomography (CT) scanning and volume graphics (VGSTUDIO MAX, Volume Graphics GmbH, Heidelberg, Germany) analysis [13,35] were used for measuring the fiber orientation. For fiber orientation tensor analysis, the part was cut from the molded part at locations A, B, C, etc., as shown in Figure 16. The sample size used was 10 mm × 10 mm × 2 mm, and the resolution used for the CT scanning was 1 μm.

Table 3. Compression molding parameters used.

| Process Conditions | Actual Values |
|---|---|
| Melt Temperature | 270 °C |
| Mold Temperature | 70 °C |
| Compression time | 60 s |
| Compression pressure | 2000 KN |
| Charge weight | 150 gm |

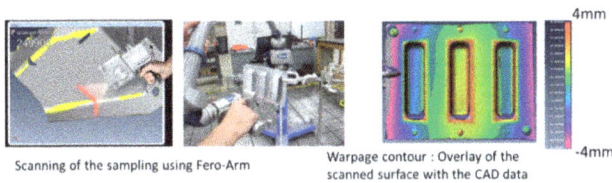

Figure 15. Warpage measurement for the final part.

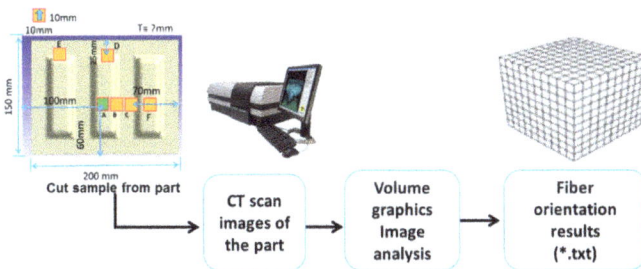

Figure 16. Fiber orientation measurement using CT scan and volume graphic method.

## 5. Results

### 5.1. Warpage Comparison

Figure 17 shows the warpage measurement location from both the measurement and CAE predication. The location of the point was the same for both cases except that the notation used for one was B and the other one was A. After the FARO-ARM laser scanning, the scanned data were superimposed with the original CAD design and the post-processing analysis was conducted using Polywork viewer (2016, Innovmetric Software Inc., Québec, QC, Canada) [36]. At location B, the green line was the warped condition, while the gray line was the initial geometry. Similarly, at location A, the gray line was the original geometry, and the colored contour was the warped shape. The warpage comparison between the CAE prediction at location A and measurement at location B is shown in Figure 17a. The effect of mold temperature to the warpage is shown in Figure 17b. It can be seen that with an increase in the mold temperature, the warpage increased in proportion. The overall error between the simulation and measurement was noted to be around 8%.

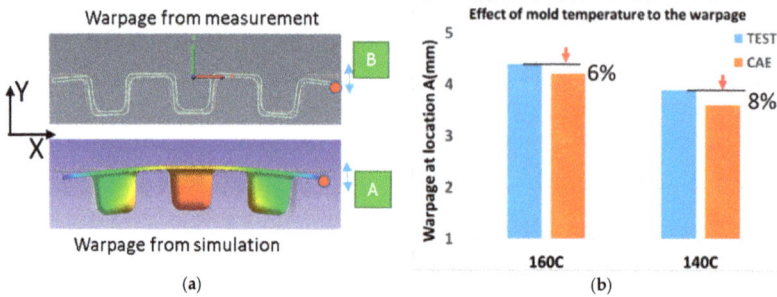

**Figure 17.** Effect of mold temperature to warpage. (**a**) Warpage measurement locations; (**b**) Warpage comparisons.

### 5.2. Fiber Orientation Comparison

The key to successful simulation is the ability to simulate the physics of the event. For the fiber-reinforced analysis, the fiber orientation plays a key role in structural properties of the part and hence the warpage.

Figure 18 shows the fiber orientation measurement results for location A as an overview. It is clearly shown that all the tensors added up to 1.

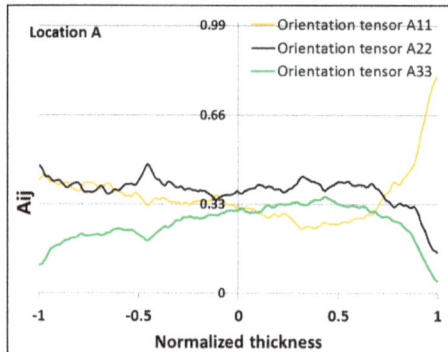

**Figure 18.** Fiber orientation tensors in all three principal directions measured at location A.

We also compared the fiber orientation at selected locations such as A, B, and C (see Figure 19). A11 is usually the direction of the fiber along the X direction, as shown in Figure 19. The comparison of the fiber orientation tensor A11 between the prediction and measurement at these locations is also shown.

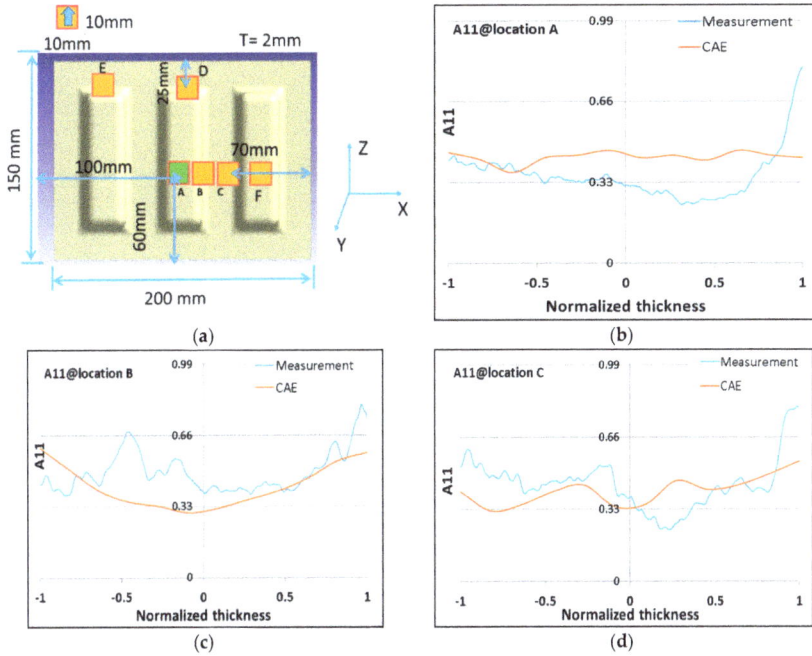

**Figure 19.** Comparison of fiber orientation in locations A, B, and C for A11. (**a**) Sample size and part information; (**b**) A11 at location A; (**c**) A11 at location B; (**d**) A11 at location C.

According to the Department of Energy (DOE) findings, the discrepancy between the simulation and measurement was within the 15% accuracy criterion [37]. Comparison of the fiber orientation results at other locations such as D, E, and F also showed trends similar to A, B, and C. The comparison results are not presented here for brevity. As a reference, Table 4 shows the analytic computed Young's modulus E11 based on fiber orientation from the predicted vs. measured at locations A, B, and C. Such an analysis was based on the Mori–Tanaka Mean Field homogenization scheme [38]. It is clear that the E11 value showed less than a 10% difference between these two.

**Table 4.** Comparison of the analytic modulus based on two different fiber orientation tensors.

| Location/Modulus | E11 (MPa) | | |
| --- | --- | --- | --- |
| | Using Predicted Fiber Orientation | Using Measured Fiber Orientation | Agreement |
| A | 23,855 | 21,972 | 8.6% |
| B | 23,252 | 25,524 | 8.9% |
| C | 22,791 | 24,353 | 6.4% |

*5.3. Fiber Length Comparison*

Figure 20 shows the fiber length distribution comparison. The fiber lengths of the parts were measured using the FASEP method [39]. In the compression molding process simulation, the fiber breakage calculation is used to obtain the fiber length distribution. The fiber breakage algorithm was developed from the Coretech company (Hsinchu County, Taiwan) and implemented in Moldex3D software, so the algorithm is available anywhere [40]. The comparison shows that the fiber length of the final part is within the range of 3–4 mm.

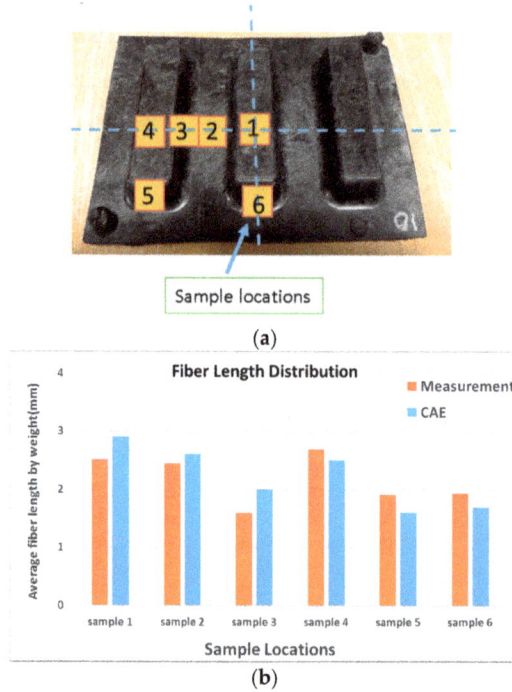

Sample locations

(**a**)

(**b**)

**Figure 20.** Fiber length distribution comparison. (**a**) sample locations; (**b**) fiber length distribution at 6 locations.

*5.4. Stroke Distance Effect*

There is one issue/concern in the compression molding of sheet materials using the proposed approach, i.e., when should the draping stop and how does a partially draped part affect the warpage? In production, the compression molding process occurs continually at the established cycle times, typically 2–3 min depending on the part size. To address this issue, the LS-DYNA analysis was conducted considering three different boundary conditions. As shown in Figure 21, the draping distance can be 100% of the gap between the top punch and the bottom die, or a fraction, i.e., 60%, 80%, etc. The draped part is then transferred to moldex3D for compression molding simulation.

It is shown that the draping distance has an effect on the prepreg shape; therefore, the final warpage of the part is different for each case. From the comparison of warpage measurement results, it was seen that 80% of the stroke distance was best matched with the actual measurement results. When the mold temperature was 140 °C, similar CAE simulations were performed. It is shown that, for 80% of the draping distance, the warpage had matching results between CAE and measurement.

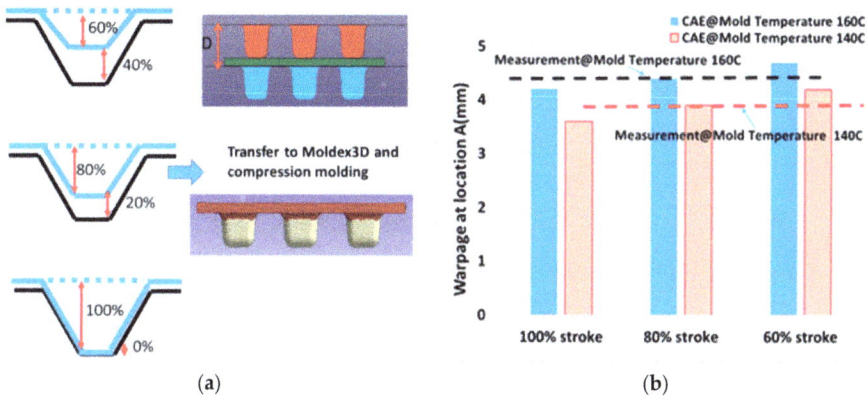

**Figure 21.** Effect of the stroke to the warpage. (**a**) details about stroke; (**b**) warpage comparison.

## 6. Conclusions

A novel two-step CAE method to simulate the compression molding of chopped carbon fiber-reinforced thermoplastic sheet material was developed. First, the draping simulation was carried out to model the draping of a heated carbon thermoplastic sheet on the mold cavity with very little pressure. Next, the draped part was transferred as a prepreg for a further compression molding process. The two-step approach helps with simulating the complex process whereby a solid elastic–plastic material is thermoformed and then compressed to flow like a highly viscous fluid. The effect of the draping distance to the final warpage was discussed and the ideal draping stroke for the given component was demonstrated. The predicted warpage, fiber length, and fiber orientation for the parts made from sheet material using the proposed approach were compared with the measurements from the physical parts. The comparison showed an acceptable match between the simulation and the actual measurements. In this study, it was assumed that the fibers in the sheet form were of in-plane random orientation, and, after draping, they remained as in-plane randomly oriented fibers. The research findings based on these models will benefit lightweight composite parts/components development for the automotive industry.

**Author Contributions:** U.G. conceived the idea, designed the project and helped with the manuscript. Y.S. conducted the experiment and simulation and prepared the data and manuscript. U.K.V. supported the test facility for the experiment. Tim Osswald gave academic suggestions. J.H. and A.Y. gave advice on the simulation and helped improving the software. T.S. helped with the experiment and provided insights into actual parts concerns.

**Conflicts of Interest:** The authors declare no competing interests.

## References

1. Isayev, A. *Injection and Compression Molding Fundamentals*; CRC Press: Boca Raton, FL, USA, 1987.
2. Beardmore, P.; Harwood, J.J.; Kinsman, K.R.; Robertson, R.E. Fiber-reinforced composites engineered structural materials. *Science* **1980**, *208*, 832–840.
3. Biron, M. *Thermoplastics and Thermoplastic Composites*; William Andrew: Norwich, NY, USA, 2012.
4. Long, A.C. (Ed.) *Composites Forming Technologies*; Woodhead Publishing: Cambridge, UK, 2014.
5. Baird, D.G.; Collias, D.I. *Polymer Processing: Principles and Design*; John Wiley & Sons: Hoboken, NJ, USA, 2014.
6. Dumont, P.; Orgéas, L.; Favier, D.; Pizette, P.; Venet, C. Compression moulding of SMC: In situ experiments, modelling and simulation. *Compos. Part A Appl. Sci. Manuf.* **2007**, *38*, 353–368. [CrossRef]
7. Miura, M.; Hayashi, K.; Yoshimoto, K.; Katahira, N. Development of Thermoplastic CFRP for Stack Frame. In *SAE Technical Paper*; SAE International: Detroit, MI, USA, 2016.

8.   Bruce, A.D.; Oswald, T.A. *Compression Molding*; Hanser Publications: Cincinnati, OH, USA, 2003.
9.   Osswald, T.A.; Tucker, C.L. A boundary element simulation of compression mold filling. *Polym. Eng. Sci.* **1988**, *28*, 413–420. [CrossRef]
10.  Oswald, T.A. Numerical Models for Compression Mold Filling Simulation. Ph.D. Thesis, University of Illinois, Champaign, IL, USA, 1987.
11.  Advani, S.G.; Tucker, C.L. A numerical simulation of short fiber orientation in compression molding. *Polym. Compos.* **1990**, *11*, 164–173. [CrossRef]
12.  Sozer, E.M.; Advani, S.G. *Process Modeling in Composites Manufacturing*; CRC Press: Boca Raton, FL, USA, 2010.
13.  Rios, A.; Davis, B.; Gramann, P. Computer Aided Engineering in Compression Molding. In Proceedings of the Composites Fabricators Association, Tampa, FL, USA, 3–6 October 2001.
14.  Barone, M.R.; Caulk, D.A. A model for the flow of a chopped fiber reinforced polymer compound in compression molding. *J. Appl. Mech.* **1986**, *53*, 361–371. [CrossRef]
15.  Wang, P.; Hamila, N.; Boisse, P. Thermoforming simulation of multilayer composites with continuous fibres and thermoplastic matrix. *Compos. Part B Eng.* **2013**, *52*, 127–136. [CrossRef]
16.  Hsiao, S.W.; Kikuchi, N. Numerical analysis and optimal design of composite thermoforming process. *Comput. Methods Appl. Mech. Eng.* **1999**, *177*, 1–34. [CrossRef]
17.  Okine, R.K. Analysis of forming parts from advanced thermoplastic composite sheet materials. *J. Thermoplast. Compos. Mater.* **1989**, *2*, 50–76. [CrossRef]
18.  Song, Y.; Gandhi, U.; Pérez, C.; Osswald, T.; Vallury, S.; Yang, A. Method to account for the fiber orientation of the initial charge on the fiber orientation of finished part in compression molding simulation. *Compos. Part A Appl. Sci. Manuf.* **2017**, *100*, 244–254. [CrossRef]
19.  *Manual of Moldex3D Software*; CoreTech. Inc.: Taipei, Taiwan, 2014.
20.  *Manual of Moldflow Software*; Autodesk. Inc.: New York, NY, USA, 2014.
21.  ASTM D638-02a, Standard Test Method for Tensile Properties of Plastics, West Conshohocken, PA, USA. 2002. Available online: https://www.astm.org/DATABASE.CART/HISTORICAL/D638-02A.htm (accessed on 1 June 2018).
22.  Eberle, A.P.; Baird, D.G.; Wapperom, P. Rheology of non-Newtonian fluids containing glass fibers: A review of experimental literature. *Ind. Eng. Chem. Res.* **2008**, *47*, 3470–3488. [CrossRef]
23.  International Organization for Standardization. *ISO 17 744:2004, Plastics—Determination of Specific Volume as a Function of Temperature and Pressure*; International Organization for Standardization: Geneva, Switzerland, 2004.
24.  ASTM D4440-15, Standard Test Method for Plastics: Dynamic Mechanical Properties Melt Rheology, West Conshohocken, PA, USA. 2015. Available online: https://www.astm.org/Standards/D4440.htm (accessed on 1 June 2018).
25.  ASTM D5930-16, Standard Test Method for Thermal Conductivity of Plastics by Means of a Transient Line-Source Technique, West Conshohocken, PA, USA. 2016. Available online: https://www.astm.org/DATABASE.CART/HISTORICAL/D5930-16.htm (accessed on 1 June 2018).
26.  ASTM E1269-11, Standard Test Method for Determining Specific Heat Capacity Differential Scanning Calorimetry, West Conshohocken, PA, USA. 2011. Available online: https://www.astm.org/Standards/E1269.htm (accessed on 1 June 2018).
27.  Plewa, T.; Linde, T.; Weirs, V.G. Adaptive mesh refinement-theory and applications. *Lect. Notes Comput. Sci. Eng.* **2005**, *41*, 3–5.
28.  Dong, Y.; Lin, R.J.T.; Bhattacharyya, D. Finite element simulation on thermoforming acrylic sheets using dynamic explicit method. *Polym. Polym. Compos.* **2006**, *14*, 307–328.
29.  Lin, G.; Lai, D.L.; Huang, C.T.; Wang, C.C. Numerical Simulation for the Viscoelastic Effects on the Birefringence Variation for an Injected Optical Lens. *SPE ANTEC Indianap.* **2016**, *2016*, 1231–1235.
30.  Huang, J.; Baird, D.G.; McGrath, J.E. Development of fuel cell bipolar plates from graphite filled wet-lay thermoplastic composite materials. *J. Power Sources* **2005**, *150*, 110–119. [CrossRef]
31.  Loos, A.C. *Processing of Composites*; Hanser Publishers: Munich, Germany, 2000; p. 320.
32.  Loos, A.C. Low-cost fabrication of advanced polymeric composites by resin infusion processes. *Adv. Compos. Mater.* **2001**, *10*, 99–106. [CrossRef]
33.  Faro Arm Measuring System. Available online: http://www.faro.com/en-us/products/metrology/measuring-arm-faroarm/overview (accessed on 1 June 2018).

34. Verma, K.; Columbus, D.; Han, B. Development of real time/variable sensitivity warpage measurement technique and its application to plastic ball grid array package. *IEEE Trans. Electron. Packag. Manuf.* **1999**, *22*, 63–70. [CrossRef]

35. *Manual of Volume Graphics Software*; Volume Graphics GmbH: Heidelberg, Germany, 2014.

36. Ebewele, R.O. *Polymer Science and Technology*; CRC Press: Boca Raton, FL, USA, 2000.

37. Nguyen, B.N.; Fifield, L.S.; Gandhi, U.N.; Mori, S.; Wollan, E.J. *Predictive Engineering Tools for Injection-Molded Long-Carbon-Thermoplastic Composites: Weight and Cost Analysis (No. PNNL-25646)*; Pacific Northwest National Lab. (PNNL): Richland, WA, USA, 2016.

38. Papathanasiou, T.D.; Guell, D.C. *Flow-Induced Alignment in Composite Materials*; Woodhead: Cambridge, UK, 1997.

39. FASEP Fiber Length Measurement System. Available online: http://www.fasep.biz/ (accessed on 1 June 2018).

40. Huang, C.T.; Tseng, H.C.; Vlcek, J.; Chang, R.Y. Fiber breakage phenomena in long fiber reinforced plastic preparation. *Conf. Ser. Mater. Sci. Eng.* **2015**, *87*, 012023. [CrossRef]

*Journal of*
*composites science*

MDPI

*Article*

# Fibre Length Reduction in Natural Fibre-Reinforced Polymers during Compounding and Injection Moulding—Experiments Versus Numerical Prediction of Fibre Breakage

Katharina Albrecht [1], Tim Osswald [2], Erwin Baur [3], Thomas Meier [4], Sandro Wartzack [5] and Jörg Müssig [1,*]

[1]  Hochschule Bremen—City University of Applied Sciences Bremen, Biomimetics—The Biological Materials Group, Neustadtswall 30, 28199 Bremen, Germany; katharina.albrecht@hs-bremen.de
[2]  Polymer Engineering Center (PEC), University of Wisconsin-Madison, 1513 University Avenue, Madison, WI 53706, USA; tosswald@wisc.edu
[3]  M-Base Engineering + Software GmbH, Dennewartstr. 27, 52068 Aachen, Germany
[4]  VKT Video Kommunikation GmbH, Sandwiesenstr. 15, 72793 Pfullingen, Germany; thomasmeier@vkt.de
[5]  Friedrich-Alexander-Universität Erlangen-Nürnberg, Martensstrasse 9, 91058 Erlangen, Germany; wartzack@mfk.fau.de
*  Correspondence: jmuessig@bionik.hs-bremen.de; Tel.: +49-421-5905-2747

Received: 22 February 2018; Accepted: 26 March 2018; Published: 28 March 2018

**Abstract:** To establish injection-moulded, natural fibre-reinforced polymers in the automotive industry, numerical simulations are important. To include the breakage behaviour of natural fibres in simulations, a profound understanding is necessary. In this study, the length and width reduction of flax and sisal fibre bundles were analysed experimentally during compounding and injection moulding. Further an optical analysis of the fibre breakage behaviour was performed via scanning electron microscopy and during fibre tensile testing with an ultra-high-speed camera. The fibre breakage of flax and sisal during injection moulding was modelled using a micromechanical model. The experimental and simulative results consistently show that during injection moulding the fibre length is not reduced further; the fibre length was already significantly reduced during compounding. For the mechanical properties of a fibre-reinforced composite it is important to overachieve the critical fibre length in the injection moulded component. The micromechanical model could be used to predict the necessary fibre length in the granules.

**Keywords:** natural fibres; injection moulding; fibre morphology; fibre-reinforced polymers

---

## 1. Introduction

Increasing environmental consciousness leads to new developments in material selection and use. Natural fibre-reinforced compounds (NFC) are a good solution to combine renewable resources with light-weight constructions. Much research has been done to determine the mechanical properties of NFCs, demonstrating interesting potential for different applications [1–7]. The injection-moulding process is a conventional technique in the automotive industry for short fibre-reinforced polymers in large-scale productions, mostly using glass fibres. A new focus is now to also use natural fibre-reinforced injection-moulded polymers. Among others, there are two important factors influencing the mechanical properties of NFCs: the fibre orientation and the fibre morphology (fibre length and aspect ratio) [8–11]. Therefore, it is very important to determine the fibre orientation and to understand the fibre breakage behaviour during processing. Further numerical simulations are necessary to help establish injection-moulded NFCs in the automotive industry. First, successful research steps were taken to perform injection-moulding simulations with NFCs [12–17]. To include the fibre breakage behaviour of natural fibres in commercial injection moulding simulations further research is necessary. Fibre damage takes place during the two process steps, compounding and injection moulding, and are a result of the following:

- fibre-fibre interaction;
- fibre-matrix interaction (Figure 1); and
- fibre-wall interaction [11,18].

The fibre-fibre interaction resulting from collision, spatial hindrance, and friction leads to fibre bending, which may result in fibre breakage. The fibre-fibre interaction is strongly correlated with the fibre mass content in the polymer [19]. The fibre-matrix interaction occurs mostly due to the fact that the outer part of granules is melting faster than the inner part. The fibres often stick out of the solid polymer and high viscous polymer melt flows around; so the fibres are bent, buckled, and sheared off, which can result in fibre breakage (Figure 1) [10,11]. A mathematical approach dealing with fibre instability phenomena, like fibre kinking and fibre splitting, was provided by Merodio and Ogden [20,21].

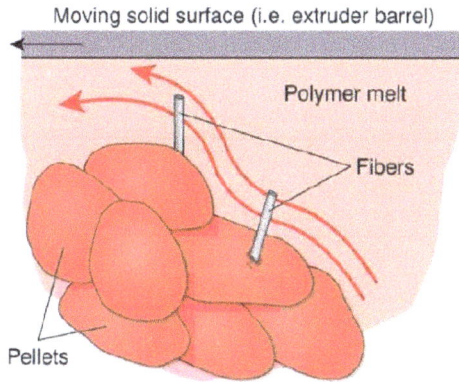

**Figure 1.** Fibre-matrix interaction, fibre damage mechanism at the interface between solid pellets and polymer melt [10].

Since the fibre morphology influences both the flow behaviour in the polymer melt and mechanical properties of the component, it is important to be able to predict fibre breakage. A mechanistic model can be used to simulate single fibres in a polymer melt and to predict fibre breakage. But till now, a mechanistic model has not been used to predict lignocellulosic fibre breakage accurately [22].

For glass fibres extensive research effort was done to analyse fibre breakage during compounding and injection moulding [23]. Several techniques to measure fibre length reduction after processing have been developed [24–31]. The first step for all techniques is the separation of the fibre from the matrix. For glass fibres the most common way is to burn off the polymer [24–28]. Another way is to extract the fibre with a suitable solvent [30–32]. For NFCs it is not possible to burn off the matrix without burning the natural fibres as well. Therefore, an appropriate solvent is necessary to dissolve the matrix without destroying the fibres.

During compounding of glass fibres with a polymer lower screw speed, higher barrel temperature and lower mixing times lead to higher average fibre lengths [23]. For injection moulding lower back pressure, lower injection speed, and more generous gate and runner dimensions lead to higher average fibre lengths [23]. Ramani et al. (1995) showed that most of the fibre length reduction occurs in the mixing section of an extruder [30]. The increase of the fibre mass content from 30% to 40% glass fibres led to higher damage due to higher fibre-fibre interactions [30]. Another important influence on final fibre length is the viscosity of the matrix; a higher viscosity leads to lower average fibre lengths [18].

In contrast to glass fibres, natural fibres (when fibre bundles are used) do not only break (length reduction) but also split (thickness reduction) during processing. Natural fibres are not single fibres, such as glass or carbon fibres, but they occur mostly as fibre bundles. For natural fibres it is of importance to distinguish between the term fibre (single cell) and fibre bundle (several fibres sticking together by pectin substances) [33–35].

Furthermore, the different natural fibre types have different damage behaviour. Recent research activities can be found in literature determining the different breakage mechanisms of flax, sisal, hemp, and miscanthus [22,36,37].

Castellani et al. (2016) found three main mechanisms to differ the fibre breakage behaviour of lignocellulosic fibres:

1. "Fatigue-driven breakage: elements bend many times and then break"
2. "Fragile behaviour: rigid elements bend and break"
3. "Peeling: chunks are removed from fibre bundles" [36].

In Figure 2 the different breakage mechanisms are shown for retted and unretted hemp, sisal, flax, and miscanthus. Further studies assume that the initiation of flax and hemp fibre breakage is related to dislocations or also called "kink bands" [38–40]. Dislocations are defects which are produced during the extraction of fibres out of the stem [38,41,42].

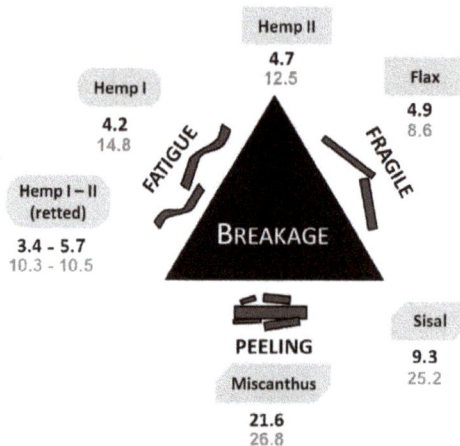

**Figure 2.** Different breakage mechanisms of natural fibres during compounding in a thermoplastic polymer melt. The numbers in black indicate Klason lignin content and, in grey, hemicellulose mass% content [36].

Steuernagel et al. (2013) found an interesting aspect for natural fibres as reinforcement, even with respect to recycling, due to the fact that the fibre bundles not only break, but also split [43]. Hence, one can assume that the aspect ratio can increase during compounding, which can result in higher mechanical properties [43], while glass fibres maintain their thickness and only reduce in length during recycling processes. Due to the splicing of the natural fibres, the surface area increases and a higher interaction with the matrix can be achieved [43].

In the present study the fibre length and thickness reduction of flax and sisal fibre bundles shall be analysed during compounding and injection moulding. Furthermore, a micromechanical model, which was developed for glass fibres, was adapted to predict fibre breakage during injection moulding of flax and sisal fibre bundles in a polypropylene matrix. Therefore, the following research hypotheses were tested:

- During the two process steps, compounding and injection moulding, the fibre length and fibre width of sisal and flax is significantly reduced.
- The mechanistic model can predict the real fibre length reduction and can be used as a product development tool to determine the necessary fibre length in the compound to overachieve the critical fibre length in the injection-moulded component.

## 2. Materials and Methods

### 2.1. Fibres and Matrix

Chopped sisal (*Agave sisalana* P.) fibre bundles were purchased for this study from Cayetano Garcia Del Moral S.L. (Cabra del Santo Cristo, Spain; harvest year: 2012). To guarantee a homogenous fibre dosing the fibres were pelletized by BaVe-Badische Faserveredelung GmbH (Malsch, Germany). Flax (*Linum usitassimum* L.). Fibre bundles were harvested in 2010 in the Netherlands and also pelletized by BaVe-Badische Faserveredelung GmbH before compounding. Subsequently, the used terms "sisal" and "flax" refer to fibre bundles of sisal and flax, extracted from the leaves of *Agave sisalana* P. and the stems of *Linum usitassimum* L., respectively. With both fibre types, compounds with 30 mass% fibres and a polypropylene (PP) (Moplen EP 500 V, LyondellBasell, Frankfurt, Germany) were produced with a twin screw extruder (ZE 34 Basic, KraussMaffei Berstorff GmbH, Hanover, Germany) at the IfBB (Hochschule Hanover, Germany). For a better adhesion between the natural fibres and the matrix, a coupling agent was used with a dosage of 3 wt% (SCONA TPPP 8112 FA, BYK-Chemie GmbH, Wesel, Germany). Furthermore, plates (30 × 16 × 3 mm$^3$) were produced with a multi-tool injection mould on a KM 160-750 EX machine (KraussMaffei Technologies GmbH, Munich, Germany) at IfBB. The compounding and injection moulding process parameters are described in more detail in [44].

### 2.2. Fibre Extraction

To observe the morphological change of sisal and flax during the two process steps, compounding and injection moulding, the fibre bundles were extracted out of the PP matrix. The natural fibres were extracted with an organic solvent out of the granules and the injection-moulded plates. This procedure is described in more detail in [45].

### 2.3. Fibre Morphology Analysis (SEM and Fibre Shape)

The fibre morphology of sisal and flax was analysed before the compounding, after the compounding (extracted from the granules) and after injection moulding (extracted from the plates).

First, the fibre bundles were investigated optically using a JSM-6510 scanning electron microscope (SEM) (JEOL GmbH, Eching, Germany) at an acceleration voltage of 2 kV. Before the investigation, sisal and flax were coated with a thin conductive gold layer for 60 s under a current of 56 mA using a Bal-Tec sputter coater type SCD 005 (Bal-Tec AG, Balzers, Liechtenstein).

Second, the length and width of sisal and flax were analysed by the image analysis software FibreShape 5.1.1 (IST AG, Vilters, Switzerland). Therefore, the original flax and sisal fibre bundles were scanned with an Epson Perfection V700 photo scanner (Seiko Epson Corperation, Suwa, Japan) with a resolution of 1200 dpi. The smaller, extracted fibre bundles were prepared on slide frames (40 × 40 mm$^2$, glass width of 2 mm; Gepe, Zug, Switzerland) and scanned with a Canonscan CS 4000scanner (Canon, New York, NY, USA) with a resolution of 4000 dpi. Before scanning, all samples were conditioned for at least 24 h at 20 °C and 65% relative humidity according to DIN EN ISO 139.

## 2.4. Statistics

The statistical analysis of the Fibre Shape results was done by the programme "R" Version 3.4.0. Kolmogorov-Smirnov-Lilliefors tests (sample size > 700) were performed to evaluate the data regarding a normal distribution with a level of significance of $\alpha = 0.05$. To prove if the fibre bundles were significantly damaged during the processes, Wilcoxon-Mann-Whitney tests were performed with a level of significance of $\alpha = 0.01$. This non-parametric test was chosen due to the fact that the data are not normally distributed and the samples are considered to be unpaired. For large sample sizes even small differences in the data distributions might lead to significant differences. Therefore, the effect size was determined, which is a quantitative measure of the strength of a phenomenon, according to [46,47]. In this study it was measured if the two process steps have none, a small, an intermediate, or a large effect on the change of the fibre morphology.

Even if the distributions of the length and width values are not normally distributed, the mean values were used for the simulation with the mechanistic model. The arithmetic mean is a location parameter for metrically scaled variables, whereas the median is a location parameter for ordinal scaled variables. The ordinal scale refers to measurements that can be ordered in terms of "greater", "less", or "equal"; therefore it is also called rank scale, whereas the metric scale allows to determine the magnitude of differences between the measured values. [48].

## 2.5. Ultra-High-Speed Camera

An ultra-high-speed camera (FASTCAM SA-Z type 2100K-M-32GB-FD, Photron, Tokyo, Japan) was used to optically analyse the differences in the breaking behaviour of flax and sisal. During a tensile test of the fibres bundles, pictures were taken with 200,000 frames per second. The tensile tests were performed with a Fafegraph M testing machine (Textechno, Mönchengladbach, Germany) with a testing speed of 20 mm/s at a gauge length of 20 mm working with a pneumatic clamping system (PVC/PMMA clamps) until the fibres totally failed.

## 2.6. Micromechanical Model

The micromechanical or mechanistic model, which was used in this study, is a particle level simulation and represents each fibre as a chain of segments interconnected by spherical joints. For each time step the force and torque balance are performed on each element of the object (Figure 3, Equations (1) and (2)):

$$\sum_{k=1}^{N_v} F_{ik}^v + F_i^c - F_{i+1}^c + \sum_{j=1}^{N_b} F_{ij}^h = 0 \tag{1}$$

$$M_i^b - M_{i+1}^b - r_i \times F_{i+k}^c + \sum_{k=1}^{N_v} r_{ik} \times F_{ik}^v + \sum_{j=1}^{N_b} [T_{ij}^h + r_{ij} \times F_{ij}^h] = 0 \tag{2}$$

where $F_{ik}^v$ is a force due to the contact between the rods $i$ and $k$, $M_i^b$ and $M_{i+1}^b$ are moments due to bending, $N_v$ is the number of rods which are in contact with rod $i$, and $N_b$ is the number of beads in rod $i$.

The segments experience hydrodynamic effects, fibre-fibre interaction and fibre flexibility, but the model excludes volume effects due to fibre-fibre and fibre-wall contacts (Figure 4). The micromechanical model is used to determine the fibre interaction coefficients for fibre-reinforced polymers. Further it can also predict fibre breakage behaviour in polymer melts, if the mechanical properties of the fibres are known. In a shear cell the forces and interactions are calculated to predict the fibre interaction coefficient $C_I$ and the fibre breakage. In this study, a cluster of fibres in a shear cell is modelled while the fibres flow through the gate of a mould. The commercial computational fluid dynamics software COMSOL Multiphysics® (COMSOL AB, Stockholm, Sweden) was used. The CFD simulation used a non-Newtonian viscosity represented with a Bird-Carreau model [49]. Before the shear flow modelling, the fibre cluster needs to be precompressed until the desired fibre density, representing the real fibre volume, is reached. The mechanistic model and the exact procedure to determine the $C_I$ for flax/PP and sisal/PP are described in more detail in [50–52].

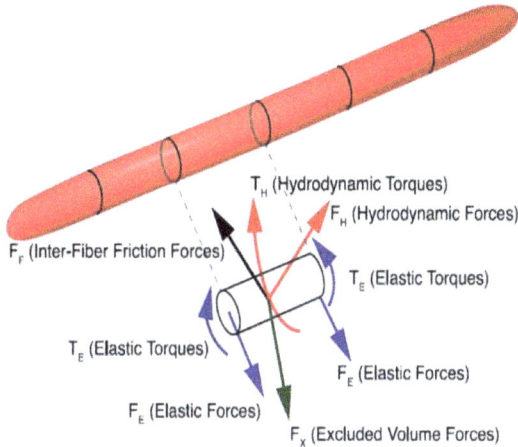

**Figure 3.** Discretization of a fibre as a chain of elements, corresponding forces, and momentum balance for an individual element [50].

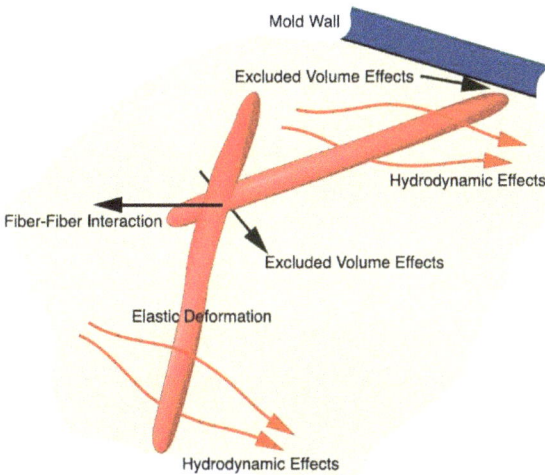

**Figure 4.** Fibre-fibre interaction and further effects which are included in the mechanistic model [50].

## 3. Results and Discussion

### 3.1. Optical Observations via SEM and Ultra-High-Speed Camera

The scanning electron microscopy (SEM) analysis showed a reduction of the length and width of sisal and flax (Figures 5 and 6) during processing. In Figure 5 fibre breakage (A) and fibre splitting (B) was observed for sisal after compounding. In Figure 6 shorter and smaller fibre bundles can be observed after compounding for flax (right) compared to the original flax (left).

Flax fibre bundles show a fragile breaking behaviour, but elementary fibres break by fatigue [36]. The SEM analysis shows that, sisal has more cohesive bundles compared to flax. Castellani et al. [36]

showed that, for sisal, the separation of bundles is difficult, before the separation they break into single fibres and chunks form. The elementary fibres often remain attached to the bundle surface of sisal [36]. Oksman et al. [53] also showed that flax is several times thinner compared to sisal after processing. This can also be seen on the SEM pictures in the present study (Figures 5 and 6). Le Moigne et al. [54] and Oksman et al. [53] showed for flax fibre bundles during compounding a separation into single fibres, whereas sisal fibre bundles tend to separate in thinner bundles and, to a smaller amount, into single fibres.

**Figure 5.** SEM pictures of sisal fibre bundles before (**left**) und after compounding (**right**). A: Fibre bundle breakage. B: Start of fibre bundle splitting. C: Peeling behaviour of sisal.

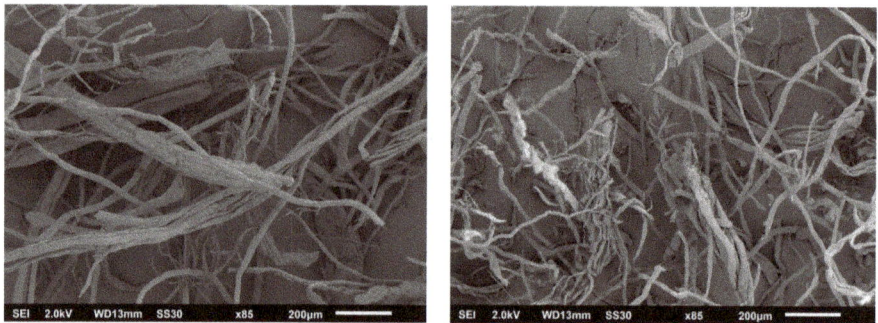

**Figure 6.** SEM pictures of flax fibre bundles before (**left**) und after compounding (**right**).

The ultra-high-speed camera observation showed that the sisal fibre bundle failure starts with a peeling behaviour (Figure 7; 0.085 ms). This peeled part remains attached at one end to the sisal bundles and behaves like a whip. No further change can be observed for the next 10.095 ms. Then, the sisal fibre bundles starts to break and splits into smaller bundles at the breaking point until it is completely broken 0.070 ms later (Figure 7). In contrast, several very small bundles and single fibres start to split apart from the flax fibre bundle (Figure 8; 0.085 ms). Compared to sisal, in the next 10.095 ms more single fibres split apart. For the next 1.525 ms more single fibres split apart, until the flax fibre bundle is completely broken after 11.705 ms (Figure 8).

The ultra-high-speed camera analysis of the tensile tests encourages the breaking mechanisms of flax and sisal found by Castellani et al. [36]. From their compounding experiments they assume a peeling behaviour for sisal, while flax breaks brittle and splits into single fibres [36].

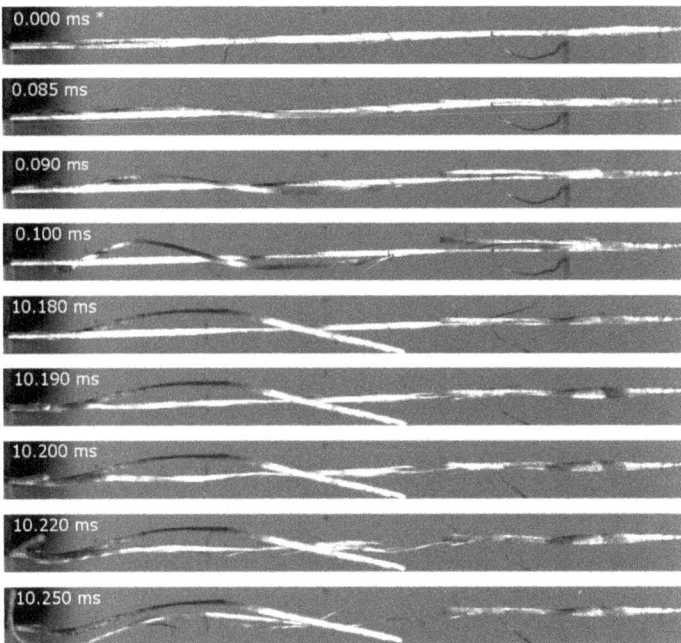

**Figure 7.** Ultra-high-speed camera observation during tensile tests of sisal fibre bundles. *: 0 ms presents the last picture without any visible fibre failure.

**Figure 8.** Ultra-high-speed camera observation during tensile tests of flax fibre bundles. *: 0 ms presents the last picture without any visible fibre failure.

## 3.2. Morphological Analysis via FibreShape

Fibre length and aspect ratio influence the fibre rotation during injection moulding and, therefore, affect the fibre orientation and mechanical properties of the component. Consequently, it is important to analyse the fibre morphology experimentally, which may serve as input parameters for further simulation studies.

The distribution of the measured lengths and widths of sisal and flax is shown in Figures 9 and 10, respectively. There is a significant change of the length and width of sisal and flax during compounding. The test of effect size also shows that there is a large effect of the compounding regarding the fibre morphology of sisal and flax. For sisal, a significant difference was found for fibre length before and after injection moulding (Figure 9 Left) and no significant difference regarding width (Figure 9 Right). However, the test of the size effect demonstrates that there is no effect on the length as a result of the injection moulding process. For the flax length, no significant influence of the injection moulding process could be found (Figure 10 Left), but a significant difference regarding the width (Figure 10 Right). For flax, the median of the width is even a bit higher after injection moulding compared to the objects in the granules. This does not mean that the fibre bundles were re-joined during the process, more likely being an artefact that different fibres can be measured before and after injection moulding as a result of having a new set of flax fibres with apparently thicker fibre bundles after injection moulding. Berzin et al. (2017) also found that each fibre bundle can split and/or break differently during the compounding process, according to its own dimensions [22].

To sum up, compounding has a large effect on fibre bundle breakage and fibre bundle splitting for sisal and flax. During injection moulding in our experiments, sisal and flax are not damaged further.

The aspect ratio is the ratio between the length of an object to the diameter (in our case the object width): L/D. If the fibres just break (reducing the length) during processing and do not split (maintaining the width), the aspect ratio decreases. This is valid for glass and carbon fibres. If the fibre bundles of natural fibres break and split, the aspect ratio does not necessarily need to decrease during processing. For sisal and flax a significantly change of the aspect ratio could be found during compounding and injection moulding. Regarding the effect size, compounding shows an intermediate effect and the injection moulding no effect on the aspect ratio of sisal. Compounding shows a large effect on the length and the width of sisal. It seems that sisal breaks more than it splits during compounding, therefore, the aspect ratio decreases. For flax, the compounding just shows a small effect on the aspect ratio, even though there is a significant difference in the results (Figure 11). Flax fibre bundles break and also split into smaller fibre bundles. The aspect ratio is almost the same before and after compounding for flax, with mean values of 26 and 28, respectively (Table 1).

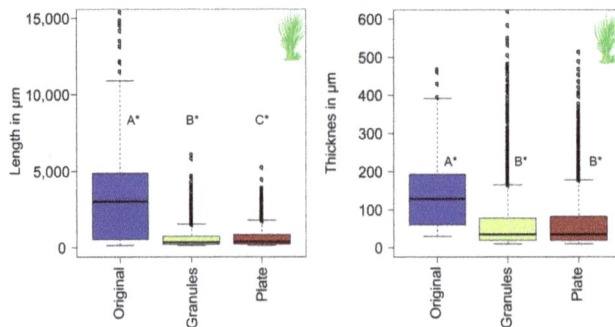

**Figure 9.** Distribution of the length (**left**) and width (**right**) for sisal before processing (original), after compounding (granules), and after injection moulding (plate). The results are shown as box-and-whisker plots (whiskers with a maximum of 1.5 × IQR, outliers shown as circles). Boxplots with * represent not normally distributed samples and different capitals represent significant differences.

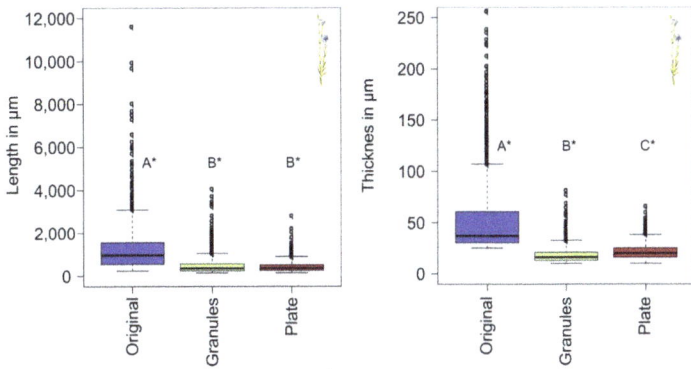

**Figure 10.** Distribution of the length (**left**) and width (**right**) for flax before processing (original), after compounding (granules), and after injection moulding (plate). The results are shown as box-and-whisker plots (whiskers with a maximum of 1.5 × IQR, outliers shown as circles). Boxplots with * represent not normally distributed samples and different capitals represent significant differences.

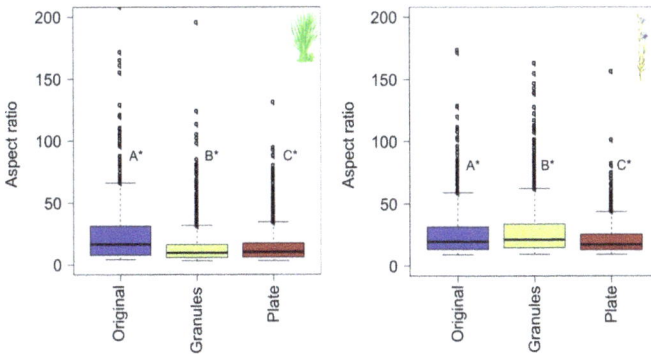

**Figure 11.** Change of the aspect ratio for sisal (**left**) and flax (**right**): before processing (original), after compounding (granules), and after injection moulding (plate). The results are shown as box-and-whisker plots (whiskers with a maximum of 1.5 × IQR, outliers shown as circles). Boxplots with * represent not normally distributed samples and different capitals represent significant differences.

**Table 1.** Mean, standard deviation (SD), and median of the measured fibre length thickness and aspect ratio for flax and sisal. The analysis was done with FibreShape; n is the number of counted objects.

| Flax | Original | | Granules | | Plate | |
|---|---|---|---|---|---|---|
| **n** | 1110 | | 2179 | | 1275 | |
| | Mean ± SD | Median | Mean ± SD | Median | Mean ± SD | Median |
| **Length in μm** | 1272 ± 1137 | 970 | 481 ± 376 | 362 | 438 ± 257 | 369 |
| **Thickness in μm** | 54 ± 41 | 37 | 18 ± 8 | 16 | 21 ± 8 | 20 |
| **Aspect ratio** | 26 ± 18 | 20 | 28 ± 20 | 21 | 21 ± 13 | 18 |
| **Sisal** | Original | | Granules | | Plate | |
| **n** | 709 | | 2436 | | 2468 | |
| | Mean ± SD | Median | Mean ± SD | Median | Mean ± SD | Median |
| **Length in μm** | 3590 ± 4026 | 3034 | 647 ± 702 | 353 | 666 ± 660 | 388 |
| **Thickness in μm** | 139 ± 88 | 129 | 69 ± 82 | 35 | 68 ± 75 | 35 |
| **Aspect ratio** | 25 ± 28 | 17 | 14 ± 12 | 10 | 14 ± 12 | 11 |

For flax fibre bundles a separation into elementary fibres was found during processing, and sisal separates into thinner fibre bundles and just a few single fibres [53,54]. Literature values for the width for single flax fibres ranges from 1.7 to 76 µm and, for flax fibre bundles, from 40 to 620 µm (see [4] for a literature overview). In this study, the original flax has a mean thickness of 54 µm (Table 1) and is reduced to 18 µm during compounding. Already thin flax bundles were split into elementary fibres. Literature values for the width of single sisal fibres ranges from 4 to 47 µm and for sisal fibre bundles from 9 to 460 µm (see [4] for a literature overview). In this study, the original sisal has a mean thickness of 139 µm and is reduced during compounding to 68 µm (Table 1), which is still in the range of the fibre bundles.

Oksman et al. (2009) and Le Moigne et al. (2011) showed a strong reduction on the flax length, but a much higher aspect ratio for flax compared to sisal after compounding [53,54]. In the present study the aspect ratio of flax and sisal is almost the same before processing, at 26 ± 18 and 25 ± 28, respectively. After compounding, the aspect ratio for flax (28 ± 20) is twice as high for sisal (14 ± 12) (Table 1). In further studies regarding the length reduction of natural fibres, the fibres were analysed before compounding and after injection moulding, but not after compounding [55,56]. Thus, the length reduction of the two different processing steps cannot be distinguished. For the validation of the mechanistic model it is important to analyse the fibre length reduction after both process steps, as in our study.

### 3.3. Mechanistic Model

In Figure 12 the fibre breakage results of the mechanistic model are shown for flax and sisal. The fibre length reduction in percent is shown during injection moulding as a function of the length in the compound, before the injection moulding process. These values range from 0%, which means the fibre length has not changed during processing and is still the initial length, to theoretically 100% for fully broken fibres. The results show that sisal with an average length of 2 mm before injection moulding is reduced by 12.5% while entering the gate to an average length of 1.75 mm, whereas flax, with an average length of 2 mm, breaks to an average length of 0.8 mm, which is a reduction of almost 60% (Figure 12). In this study, the length of the fibres/fibre bundles were analysed in the compounds. Sisal has an average value of 647 µm and flax of 481 µm (Table 1). No further length reduction effect could be found during injection moulding for both fibre types. The mechanistic model showed the same results, the fibres do not break during injection moulding, if they are already reduced to 0.6 mm (sisal) and 0.5 mm (flax) during compounding.

The different fibre types vary in their breaking behaviour. Di Giuseppe et al. (2017) performed a study to analysis quantitatively the fibre breakage during compounding by an inner mixer for sisal, hemp, flax, and miscanthus [37]. They determined a "breakage index" for the length the diameter and the aspect ratio for all four fibre types. Sisal has the smallest breakage index, which means sisal breaks less under shear stress compared to the other fibre types. In contrast, flax shows the highest breakage index of all fibre types due to its fragile breaking behaviour [37]. Our simulation shows the same results, with the same original length, and flax has a much higher breakage index compared to sisal under shear stress.

The breakage behaviour is not only dependent on the fibre type, but also influenced by the original dimensions of the fibres; as expected the longer the fibre, higher breakage occurs [37]. This experimental observation can be confirmed with our results using the micromechanical model (Figure 12). The experimental analysis of the fibre length reduction during compounding shows a strong influence of the original fibre length. Sisal, with an original length of 3590 µm (mean), was reduced by 82% during compounding. Flax, with an original length of 1272 µm (mean,) was only reduced by 62% (Table 1), even though flax has a more fragile breaking behaviour. The breakage index is, therefore, not only dependent on the fibre type, but also on the original length.

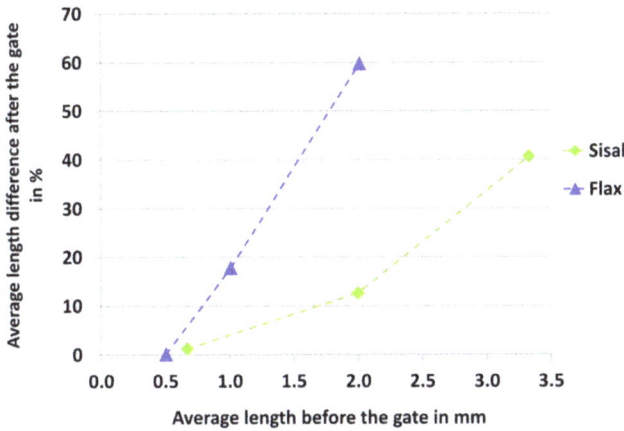

**Figure 12.** Fibre breakage results of the mechanistic model for flax and sisal. The x-axis shows the average fibre length before the injection moulding process, before entering the gate. The y-axis shows the percentage of the length reduction as the fibres enter the mould after the gate during injection moulding.

The different breakage behaviour of the natural fibre does not only effect the length reduction but also the aspect ratio. In contrast to glass and carbon fibres, not only the length, but also the thickness, of natural fibre bundles can be changed while processing. As not only the length of sisal is reduced in the present study by 82%, but also the thickness decreases due to shear forces during compounding, the aspect ratio decreases by 44%. In contrast, the aspect ratio of flax even decreases slightly by 8%. Flax does not only break, but also splits into elementary fibres [36]. The additional splitting of the fibre bundles may have a positive effect on the mechanical properties of the compounds due to higher aspect ratio values, because not only the length, but also the aspect ratio, is affecting the reinforcement effect of natural fibres in the polymer [8,9].

In a further experimental study with a focus on achieving long fibres in the compound would be important to overachieve the critical fibre length in the injection moulding component. Therefore, for each natural fibre type the compounding process needs to be adapted. The micromechanical model can predict the necessary fibre length in the compounds to reach the necessary fibre length in the injection-moulded component. The micromechanical model can be used in this case as a product development tool to realize good mechanical properties in NFC.

Additionally, further studies with the mechanistic model, including the simulation of fibre bundle splitting, would be useful. Therefore, ambitious experimental analyses are necessary to determine the adhesion forces between the single fibres at the middle lamella. For first steps in this direction for flax were conducted by Charlet and Béakou (2011) [57]. The detachment forces at the middle lamella were analysed for two single flax fibres and a numerical modelling of the flax middle lamella was performed [57]. Including fibre splitting into the micromechanical model would be an important step to optimize the model as a product development tool for adequate prediction of mechanical properties in NFC due to exact determined fibre length and aspect ratio in the injection moulding component.

In addition to the prediction of the length reduction during injection moulding, the micromechanical model is primarily used to determine fibre interaction coefficients [47–49,55]. Fibre interaction coefficients are used in commercial simulation software programs (e.g., Moldex3D®, Cadmould®, MoldFlow®) to predict the fibre orientation during injection moulding. The fibre interaction coefficient is unique for each fibre-matrix combination. More information about the prediction of fibre interaction coefficients via micromechanical model is found in [50–52,58]. In the present study, the fibre interaction coefficients $C_I$ were analysed for 30 mass % fibres/PP for flax and sisal (Table 2). Further details can be found in [13,15,50]. $C_I$ is a phenomenological parameter for the fibre-fibre interaction, the higher the value, the higher the interaction between the fibres [10]. The $C_I$ value for sisal/PP is higher compared to flax/PP (Table 2). Sisal interacts more than the flax in the polymer melt; this is most probably because of the fibre/fibre bundle morphology. Sisal is longer and thicker compared to flax in the granules and in the injection moulded plates (Table 1).

**Table 2.** Fibre interaction coefficients $C_I$ for flax and sisal PP.

| PP-Compound (30 Mass % of Fibres) | $C_I$ |
|---|---|
| Flax | 0.0037 |
| Sisal | 0.0059 |

The fibre interaction coefficient $C_I$ for sisal/PP was successfully used to determine the fibre orientation in an injection-moulding simulation using the software Cadmould® (Simcon Kunststofftechnische Software GmbH, Würselen, Germany). Details can be found in [13].

Further experimental studies with longer natural fibres are necessary to validate the fibre breakage prediction with the micromechanical model. Therefore, compounds with long fibres need to be produced, and different compounding process parameters need to be adapted, because high screw speeds and low feed rates increase the fibre fragmentation [59]. In addition to the breakage behaviour, it is important to analyse the possible fibre-matrix separation with long fibres in complex and thin-walled moulds. In the present project, a thin-walled mould of a glove box cover with fine ribs could be filled without separation between the short fibres and the matrix [15].

## 4. Conclusions

- The length and width were significantly reduced during the compounding process for sisal and flax fibres. Statistically, a large effect of the compounding was found for length and width.
- The injection moulding process showed neither any effect on a further length reduction nor further splitting for sisal and flax.
- It is important to generate long fibres/fibre bundles already in the compound to overachieve the critical fibre length in the injection moulded component.
- The micromechanical model can be used as a product development tool to predict the necessary object length in the granules to achieve the necessary object length in the injection moulded component to realize good mechanical properties.
- If sisal and flax fall below a certain length during the compounding process, no further reduction of the fibre length can be observed during injection moulding. In the present study this phenomenon could be shown by experiments and simulations for a length of 500 μm (mean value) for flax and 600 μm (mean value) for sisal.
- Splitting during processing is a very important fact for natural fibre bundles. The reduction of the fibre bundle width has a large influence on the aspect ratio (L/D). This phenomenon is not yet implemented in simulation. Before it can be implemented in the simulation, ambitious experiments are necessary to understand and measure the splitting of fibre bundles at the middle lamella.

**Acknowledgments:** The work was funded within the project "Material and Flow Models for Natural Fibre Reinforced Injection Moulding Materials for Practical Use in the Automotive Industry" by the German Federal Ministry of Food, Agriculture and Consumer Protection (BMEL) through the Fachagentur Nachwachsende Rohstoffe e.V. (FNR, Gülzow, Germany). Project partners were: Ford Forschungszentrum Aachen GmbH, Aachen, Germany; IAC (International Automotive Components), Ebersberg, Germany; LyondellBasell, Frankfurt, Germany; Kunststoffwerk Voerde Hueck and Schade GmbH and Co. KG, Ennepetal, Germany; Simcon Kunststofftechnische Software GmbH, Würselen, Germany; M-Base Engineering + Software GmbH, Aachen, Germany; University of Wisconsin-Madison, Madison, WI, USA; University of Applied Sciences and Arts Hannover, IfBB (Institute for Bioplastics and Biocomposites), Hannover, DE; HSB - City University of Applied Sciences Bremen, Bremen, Germany; Clausthal University of Technology, Institute of Polymer Materials and

Plastic Engineering, Clausthal, Germany, and Fraunhofer LBF, Darmstadt, Germany. The authors would like to acknowledge Katharina Haag (The Biological Materials Group, City University of Applied Sciences Bremen, Bremen, Germany) for the help during high-speed recordings.

**Author Contributions:** Katharina Albrecht, Jörg Müssig, Erwin Baur, and Tim Osswald conceived and designed the experiments; Katharina Albrecht, Jörg Müssig, Thomas Meier, and Erwin Baur performed the experiments; Katharina Albrecht and Jörg Müssig analysed the data; Thomas Meier contributed the ultra-high-speed camera; the micromechanical model was developed for glass fibres and adapted to natural fibres by Tim Osswald and Erwin Baur; Katharina Albrecht wrote the first draft of the paper; Jörg Müssig and Tim Osswald revised the paper; Sandro Wartzack provided scientific support; and all authors contributed the final manuscript.

**Conflicts of Interest:** The authors declare no conflict of interest.

## References

1. Alkbir, M.F.M.; Sapuan, S.M.; Nuraini, A.A.; Ishak, M.R. Fibre properties and crashworthiness parameters of natural fibre-reinforced composite structure: A literature review. *Compos. Struct.* **2016**, *148*, 59–73. [CrossRef]

2. Li, Y.; Mai, Y.-W.; Ye, L. Sisal fibre and its composites: A review of recent developments. *Compos. Sci. Technol.* **2000**, *60*, 2037–2055. [CrossRef]

3. Mertens, O.; Gurr, J.; Krause, A. The utilization of thermomechanical pulp fibers in WPC: A review. *J. Appl. Polym. Sci.* **2017**, *134*, 45161. [CrossRef]

4. Müssig, J.; Fischer, H.; Graupner, N.; Drieling, A. Testing Methods for Measuring Physical and Mechanical Fibre Properties (Plant and Animal Fibres). In *Industrial Application Natural Fibres: Structures, Properties and Technical Applications*; Müssig, J., Ed.; John Wiley & Sons, Ltd.: Hoboken, NJ, USA, 2010; pp. 267–309. [CrossRef]

5. Pickering, K.L.; Efendy, M.G.A.; Le, T.M. A review of recent developments in natural fibre composites and their mechanical performance. *Compos. Part A Appl. Sci. Manuf.* **2016**, *83*, 98–112. [CrossRef]

6. Shah, D.U. Natural fibre composites: Comprehensive ashby-type materials selection charts. *Mater. Des.* **2014**, *62*, 21–31. [CrossRef]

7. Yan, L.; Chouw, N.; Jayaraman, K. Flax fibre and its composites—A review. *Compos. Part B Eng.* **2014**, *56*, 296–317. [CrossRef]

8. Beaugrand, J.; Berzin, F. (2013): Lignocellulosic Fiber Reinforced Composites: Influence of Compounding Conditions on Defibrization and Mechanical Properties. *J. Appl. Polym. Sci.* **2013**, *128*, 1227–1238. [CrossRef]

9. Graupner, N.; Albrecht, K.; Ziegmann, G.; Enzler, H.; Müssig, J. Influence of reprocessing on fibre length distribution, tensile strength and impact strength of injection moulded cellulose fibre-reinforced polylactide (PLA) composites. *Express Polym. Lett.* **2016**, *10*, 647–663. [CrossRef]

10. Osswald, T.A.; Menges, G. Anisotropy Development during Processing. In *Material Science Polymers Engineers*, 3rd ed.; Osswald, T.A., Menges, G., Eds.; Hanser: Munich, Germany, 2012; pp. 263–294. [CrossRef]

11. Rohde-Tibitanzl, M. State of the Art. In *Direct Processing Long Fiber Reinforced Thermoplastic Composites Their Mechanical Behavior Static Dynamic Load*; Carl Hanser Verlag GmbH & Co.: Munich, Germany, 2015; pp. 3–68. [CrossRef]

12. Abdennadher, A. Injection Moulding of Natural Fibre Reinforced Polypropylene: Process, Microstructure and Properties. Ph.D. Thesis, Ecole Nationale Supérieure des Mines de Paris, Frankreich, Paris, France, 2015.

13. Albrecht, K.; Baur, E.; Endres, H.-J.; Gente, R.; Graupner, N.; Koch, M.; Neudecker, M.; Osswald, T.; Schmidtke, P.; Wartzack, S.; et al. Measuring fibre orientation in sisal fibre-reinforced, injection moulded polypropylene-Pros and cons of the experimental methods to validate injection moulding simulation. *Compos. Part A Appl. Sci. Manuf.* **2017**, *95*, 54–64. [CrossRef]

14. Azaman, M.D.; Sapuan, S.M.; Sulaiman, S.; Zainudin, E.S.; Abdan, K. An investigation of the processability of natural fibre reinforced polymer composites on shallow and flat thin-walled parts by injection moulding process. *Mater. Des.* **2013**, *50*, 451–456. [CrossRef]

15. Ford Forschungszentrum Aachen GmbH, IAC Group GmbH, LyondellBasell, Kunststoffwerk Voerde, Simcon Kunststofftechnische Software GmbH, M-Base Engineering und Software GmbH, Hochschule Hannover, Hochschule Bremen, Technische Universität Clausthal, Fraunhofer LBF & University of Wisconsin-Madison. Werkstoff-und Fließmodelle für naturfaserverstärkte Spritzgießmaterialien für den praktischen Einsatz in der Automobilindustrie. Final Report of the Project "NFC-Simulation". 2014. Available online: http://www.fnr-server.de/ftp/pdf/berichte/22005511.pdf (accessed on 23 March 2018). (In German)

16. Nikklä, M.; Filz, P. Injection Moulding and Simulation of Consumer Products with Aqvacomp Composites. In Proceedings of the Biocomposites Conference Cologne (BCC)—7th Conference on Wood and Natural Fibre Composites, Maternushaus, Germany, 6–7 December 2017; nova-Institut GmbH: Maternushaus, Germany, 2017.
17. Wan Abdul Rahman, W.A.; Sin, L.T.; Rahmat, A.R. Injection moulding simulation analysis of natural fiber composite window frame. *J. Mater. Process. Technol.* **2008**, *197*, 22–30. [CrossRef]
18. Fu, S.-Y.; Lauke, B.; Mai, Y.-W. *Science and Engineering of Short Fibre Reinforced Polymer Composites*; Woodhead Publishing in Materials; Elsevier: Amsterdam, The Netherlands, 2009; ISBN 978-1-84569-269-8.
19. Folgar, F.; Tucker, C.L. Orientation Behavior of Fibers in Concentrated Suspensions. *J. Reinf. Plast. Compos.* **1984**, *3*, 98–119. [CrossRef]
20. Merodio, J.; Ogden, R.W. Instabilities and loss of ellipticity in fiber-reinforced compressible nonlinearly elastic solids under plane deformation. *Int. J. Solids Struct.* **2003**, *40*, 4707–4727. [CrossRef]
21. Merodio, J.; Ogden, R.W. Material Instabilities for Fiber-Reinforced Nonlinearly Elastic solids under plane deformation. *Arch. Mech.* **2002**, *54*, 525–552.
22. Berzin, F.; Beaugrand, J.; Dobosz, S.; Budtova, T.; Vergnes, B. Lignocellulosic fiber breakage in a molten polymer. Part 3. Modeling of the dimensional change of the fibers during compounding by twin screw extrusion. *Compos. Part A Appl. Sci. Manuf.* **2017**, *101*, 422–431. [CrossRef]
23. Fu, S.-Y.; Hu, X.; Yue, C.-Y. Effects of fiber length and orientation distributions on the mechanical properties of short-fiber-reinforced polymers—A review. *J. Soc. Mater. Sci.* **1999**, *48*, 74–83. [CrossRef]
24. Bajracharya, R.M.; Manalo, A.C.; Karunasena, W.; Lau, K.-T. Experimental and theoretical studies on the properties of injection moulded glass fibre reinforced mixed plastics composites. *Compos. Part A Appl. Sci. Manuf.* **2016**, *84*, 393–405. [CrossRef]
25. Bijsterbosch, H.; Gaymans, R.J. Polyamide 6—Long glass fiber injection moldings. *Polym. Compos.* **1995**, *16*, 363–369. [CrossRef]
26. Denault, J.; Vu-Khanh, T.; Foster, B. Tensile properties of injection molded long fiber thermoplastic composites. *Polym. Compos.* **1989**, *10*, 313–321. [CrossRef]
27. Fu, S.-Y.; Lauke, B. Characterization of tensile behaviour of hybrid short glass fibre/calcite particle/ABS composites. *Compos. Part A Appl. Sci. Manuf.* **1998**, *29*, 575–583. [CrossRef]
28. Fu, S.-Y.; Lauke, B. Fracture resistance of unfilled and calcite-particle-filled ABS composites reinforced by short glass fibers (SGF) under impact load. *Compos. Part A Appl. Sci. Manuf.* **1998**, *29*, 631–641. [CrossRef]
29. Jaszkiewicz, A.; Meljon, A.; Bledzki, A.K.; Radwanski, M. Gaining knowledge on the processability of PLA-based short-fibre compounds—A comprehensive comparison with their PP counterparts. *Compos. Part A Appl. Sci. Manuf.* **2016**, *83*, 140–151. [CrossRef]
30. Ramani, K.; Bank, D.; Kraemer, N. Effect of screw design on fiber damage in extrusion compounding and composite properties. *Polym. Compos.* **1995**, *16*, 258–266. [CrossRef]
31. von Turkovich, R.; Erwin, L. Fiber fracture in reinforced thermoplastic processing. *Polym. Eng. Sci.* **1983**, *23*, 743–749. [CrossRef]
32. Curtis, P.T.; Bader, M.G.; Bailey, J.E. The stiffness and strength of a polyamide thermoplastic reinforced with glass and carbon fibres. *J. Mater. Sci.* **1978**, *13*, 377–390. [CrossRef]
33. Müssig, J.; Martens, R. Quality Aspects in Hemp Fibre Production—Influence of Cultivation, Harvesting and Retting. *J. Ind. Hemp* **2003**, *8*, 11–32. [CrossRef]
34. Schnegelsberg, G. *Handbuch der Faser–Theorie und Systematik der Faser*; Deutscher Fachverlag: Frankfurt am Main, Germany, 1999; ISBN 978-3871506246. (In German)
35. Vincent, J. A unified nomenclature for plant fibres for industrial use. *Appl. Compos. Mater.* **2000**, *7*, 269–271. [CrossRef]
36. Castellani, R.; Di Giuseppe, E.; Beaugrand, J.; Dobosz, S.; Berzin, F.; Vergnes, B.; Budtova, T. Lignocellulosic fiber breakage in a molten polymer. Part 1. Qualitative analysis using rheo-optical observations. *Compos. Part A Appl. Sci. Manuf.* **2016**, *91*, 229–237. [CrossRef]
37. Di Giuseppe, E.; Castellani, R.; Budtova, T.; Vergnes, B. Lignocellulosic fiber breakage in a molten polymer. Part 2. Quantitative analysis of the breakage mechanisms during compounding. *Compos. Part A Appl. Sci. Manuf.* **2017**, *95*, 31–39. [CrossRef]
38. Baley, C. Influence of kink bands on the tensile strength of flax fibers. *J. Mater. Sci.* **2004**, *39*, 331–334. [CrossRef]

39. Duc, A.L.; Vergnes, B.; Budtova, T. Polypropylene/natural fibres composites: Analysis of fibre dimensions after compounding and observations of fibre rupture by rheo-optics. *Compos. Part A Appl. Sci. Manuf.* **2011**, *42*, 1727–1737. [CrossRef]

40. Hughes, M. Defects in natural fibres: Their origin, characteristics and implications for natural fibre-reinforced composites. *J. Mater. Sci.* **2012**, *47*, 599–609. [CrossRef]

41. Bos, H.; Van Den Oever, M.; Peters, O. Tensile and compressive properties of flax fibres for natural fibre reinforced composites. *J. Mater. Sci.* **2002**, *37*, 1683–1692. [CrossRef]

42. Hernandez-Estrada, A.; Gusovius, H.-J.; Müssig, J.; Highes, M. Assessing the Susceptibility of Hemp Fibre to the Formation of Dislocations during Processing. *Ind. Crops Prod.* **2016**, *85*, 382–388. [CrossRef]

43. Steuernagel, L.; Ziegmann, G.; Meiners, D. Recycling of fiber reinforced thermoplastics—Natural fibers vs. glass fibers. In Proceedings of the CompositesWeek@Leuven and TexComp-11 Conference, Leuven, Belgium, 16–20 September 2013.

44. Neudecker, M.; Endres, H.-J. Processing and Manufacturing of Natural Fiber Reinforced Plastics to Specimens for Generating Simulation Data (NFC-Simulation). In Proceedings of the ANTEC®2014—Technical Conference & Exhibition, Las Vegas, NV, USA, 28–30 April 2014; Society of Plastics Engineers: Bethel, CT, USA, 2014; pp. 707–711.

45. Albrecht, K.; Osswald, T.; Wartzack, S.; Müssig, J. Natural fibre-reinforced, injection moulded polymers for light weight constructions—Simulation of sustainable materials for the automotive industry. In Proceedings of the 20th International Conference Engineering Design (ICED15), Politecnico di Milano, Italy, 27–30 July 2015; Volume 4, pp. 313–322.

46. Cohan, J. *Statistical Power Analysis for the Behavioral Sciences*, 2nd ed.; Printed in the USA; Lawrence Erlbaum Associates: Lawrence, NJ, USA, 1988; ISBN 978-0-12-179060-8.

47. Lenhard, W.; Lenhard, A. Calculation of Effect Sizes. Bibergau (Germany): Psychometrica. Available online: https://www.psychometrica.de/effect_size.html (accessed on 4 July 2017). [CrossRef]

48. Bamberg, G.; Baur, F.; Krapp, M. *Statistik. 14. Überarbeitete Auflage. Oldenbourgs Lehr-und Handbücher der Wirtschafts-u. Sozialwissenschaften*; Oldenbourg Verlag: München, Germany, 2008; ISBN 978-3486272185.

49. López, L.; Ramírez, D.; Osswald, T.A. Fiber Attrition and Orientation Productions of a Fiber Filled Polymer through a Gate—A Mechanistic Approach. In Proceedings of the ANTEC®2013—Technical Conference & Exhibition, Cincinnati, OH, USA, 22–24 April 2013; Society of Plastics Engineers: Bethel, CT, USA, 2013; pp. 2163–2167.

50. Baur, E.; Goris, S.; Ramírez, D.; Schmidtke, P.; Osswald, T. Mechanistic model to determine fiber orientation simulation material parameters. In Proceedings of the ANTEC®2014—Technical Conference & Exhibition, Las Vegas, NV, USA, 28–30 April 2014; Society of Plastics Engineers: Bethel, CT, USA, 2014; pp. 1605–1610.

51. Ramírez, D. Study of Fiber Motion in Moding Processes by Means of a Mechanistic Model. Ph.D. Thesis, University of Wisconsin-Madison, Madison, WI, USA, 2014.

52. Walter, I.; Goris, S.; Teuwsen, J.; Tapia, A.; Perez, C.; Osswald, T.A. A direct particle level simulation coupled with the Folgar-Tucker RSC Model to predict fiber orientation in injection molding of long glass fiber reinforced thermoplastics. In Proceedings of the ANTEC®2017—Technical Conference & Exhibition, Anaheim, CA, USA, 8–10 May 2017; Society of Plastics Engineers: Bethel, CT, USA, 2017.

53. Oksman, K.; Mathew, A.P.; Langström, R.; Nyström, B.; Joseph, K. The influence of fibre microstructure on fibre breakage and mechanical properties of natural fibre reinforced polypropylene. *Compos. Sci. Technol.* **2009**, *69*, 1847–1853. [CrossRef]

54. Le Moigne, N.L.; van den Oever, M.; Budtova, T. A statistical analysis of fibre size and shape distribution after compounding in composites reinforced by natural fibres. *Compos. Part A Appl. Sci. Manuf.* **2011**, *42*, 1542–1550. [CrossRef]

55. Hamma, A.; Kaci, M.; Ishak, Z.A.M.; Pegoretti, A. Starch-grafted-polypropylene/kenaf fibres composites. Part 1: Mechanical performances and viscoelastic behaviour. *Compos. Part A Appl. Sci. Manuf.* **2014**, *56*, 328–335. [CrossRef]

56. Muthuraj, R.; Misra, M.; Defersha, F.; Mohanty, A.K. Influence of processing parameters on the impact strength of biocomposites: A statistical approach. *Compos. Part A Appl. Sci. Manuf.* **2016**, *83*, 120–129. [CrossRef]

57. Charlet, K.; Béakou, A. Mechanical properties of interfaces within a flax bundle—Part I: Experimental analysis. *Int. J. Adhes. Adhes.* **2011**, *31*, 875–881. [CrossRef]

58. Pérez, C.; Ramírez, D.; Osswald, T.A. Mechanistic model simulation of a compression molding process: Fiber orientation and fiber-matrix separation. In Proceedings of the ANTEC®2015—Technical Conference & Exhibition, Orlando, FL, USA, 23–25 March 2015; Society of Plastics Engineers: Bethel, CT, USA, 2015.
59. Berzin, F.; Vergnes, B.; Beaugrand, J. Evolution of lignocellulosic fibre lengths along the screw profile during twin screw compounding with polycaprolactone. *Compos. Part A Appl. Sci. Manuf.* **2014**, *59*, 30–36. [CrossRef]

*Journal of*
*composites science*

MDPI

*Article*

# Mechanical Properties and Wear Behavior of a Novel Composite of Acrylonitrile–Butadiene–Styrene Strengthened by Short Basalt Fiber

**Mohammed Y. Abdellah [1,2,\*], Hesham I. Fathi [3,4], Ayman M. M. Abdelhaleem [5] and Montasser Dewidar [6]**

1   Mechanical Engineering Department, Faculty of Engineering, South Valley University, Qena 83523, Egypt
2   Mechanical Engineering Department, College of Engineering and Islamic Architecture, Umm Al-Qura University, Makkah 21421, KSA
3   Faculty of Engineering, Aswan University, Aswan 81542, Egypt; hesham.elqady@aswu.edu.eg
4   Chemicals and Petrochemicals Engineering Department, Egypt-Japan University of Science and Technology, New Borg El-Arab City, Alexandria 21934, Egypt
5   Mechanical Design and Production Engineering Department, Faculty of Engineering, Zagazig University, Zagazig 44519, Egypt; aymanmns@hotmail.com
6   Department of Mechanical Engineering, Faculty of Engineering, Kafr El-sheikh University, Kafr El-sheikh 33516, Egypt; Dewidar5@hotmail.com
\*   Correspondence: mohammed_yahya42@yahoo.com; Tel.: +966-546-240-463

Received: 12 May 2018; Accepted: 4 June 2018; Published: 7 June 2018

**Abstract:** Polymer matrix composites (PMC) have a competitive and dominant role in a lot of industries, like aerospace and automobiles. Short basalt fiber (SBF) is used to strengthen acrylonitrile–butadiene–styrene (ABS) polymers as a composite. The composite material is fabricated using injection molding with a new technique to obtain a uniform distribution for the ABS matrix at an elevated temperature range from 140 °C to 240 °C. Four types of specimen were produced according to the mechanically mixed amounts of SBF, which were (5, 10, 15, 20) wt %. The produced material was tested for tension, hardness and impact to measure the enhancement of the mechanical properties of the ABS only and the ABS reinforced by SBF composite. Wear tests were carried out using a pin on disc at a velocity of 57.5 m/s at three normal loads of 5, 10 and 15 kN. Tensile strength increased with up to 5 wt % of SBF, then decreased with an increasing amount of SBF reinforcement, while surface hardness increased with increasing SBF. The impact strength was found to degrade with the whole increment of SBF. Wear resistance increased with the increasing SBF reinforcement amount at all applied normal loads.

**Keywords:** acrylonitrile–butadiene–styrene (ABS); short basalt fiber (SBF); wear resistance; hardness; impact energy

## 1. Introduction

Composite materials nowadays play an increasing role in a lot of applications, such as the aerospace and automobile industry, and finally in biomedical applications. Acrylonitrile–butadiene–styrene (ABS) polymer plays an important role in many engineering plastics and composite materials, and has a rubber polybutadiene structure [1]. Natural fibers, like basalt fiber, nowadays plays a competitive role and receives special consideration compared to conventional glass and carbon fiber, because of their low cost and close mechanical and chemical properties [2–5].

Jiang et al. [6] investigated the effect of strengthening ABS with calcium carbonate particles of micro and nano size. They concluded that the microcrystalline cellulose/acrylonitrile–butadiene–styrene (MCC/ABS) composite had a higher relative elastic modulus and lower relative tensile and impact strength when compared with ABS alone, while in Sept et al. [7], it was proved that particle size and heat treatment have little effect on the mechanical, thermal and physical properties of ABS.

Abdel-Haleem et al. [8] strengthen ABS using basalt fiber (BF). They used an injection molding machine to manufacture the BF/ABS composite with a varying wt % of BF. They investigated the effect of the basalt fiber content on the tensile and impact strength, and on hardness. They concluded that both tensile strength and Rockwell hardness improved with the increased wt % of BF, while the impact strength decreased.

Lopresto et al. [9] compared the mechanical performance of basalt fiber and glass fiber, and found that basalt fiber produces a higher Young's modulus, compressive strength, and impact strength. They reported that low-cost basalt fiber is more competitive in glass fiber applications.

Difallah et al. [10] investigated the mechanical and tribological properties of ABS filled with graphite powder. It was observed that increasing the graphite content in the ABS matrix led to a drastic decrease in tensile strength, while it produced a good enhancement of the tribological properties and friction coefficient; this was possibly due to the fact that graphite powder is a solid lubricants.

Colombo et al. [11] used basalt fiber to strengthen the polymer matrix. They needed to study the effect of basalt fiber on the mechanical properties. The results showed that high mechanical properties were achieved for both tensile and compressive behavers, moreover, the failure modes were found to be more tight and concise.

Jeevanantham et al. [12] experimentally investigated the tensile and flexural strength of polyurethane reinforced by basalt fiber composite laminate. The composite laminate was fabricated using a match plate mold under 1500 $kg_f/cm^2$ for 4 h at room temperature. They studied the effect of varying weight fractions of basalt fiber. It resulted in good tensile and flexural strength up to a certain weight fraction.

Chen et al. [13] carried out quasi-static and dynamic tests on unidirectional basalt-fiber-reinforced polymer (BFRP) to study the tensile properties of such a material. They observed that the tensile properties of BFRP were strain rate sensitive, they also extracted a good empirical formulation for predicting tensile properties. Khosravi and Farsani [14] attempted to enhance the mechanical properties of a unidirectional basalt fiber/epoxy composite using a modified nanoclay. Scalici et al. [15] carried out work on a basalt-fiber-reinforced composite using different vacuum-assisted impregnation techniques.

Wu et al. [16] investigated the durability of a basalt-fiber-reinforced epoxy composite in a different corrosive medium. It was found that the basalt fiber showed relatively strong resistance to water and salt corrosion, reasonable resistance to acid corrosion, and bad resistance in alkaline liquids.

Many studies have been carried out on continuous and short basalt fiber as a reinforcement phase [17–31].

Basalt fiber is natural fiber that is produced by basalt rock, this makes their chemical, physical and mechanical properties sensitive to the type of raw material, technology, and method of obtaining the final product [9,32]. Therefore, reinforcement plastic with this type of material needs a lot of investigation and studies.

Sudeepan et al. [33] studied the effect of varying micro-size zinc oxide (Zno) contents in a matrix of ABS polymer on the mechanical and tribological properties. It was concluded that tensile and flexural modulus and hardness generally increase with increasing filler content, whereas tensile and flexural strength increased up to 15 wt %, before decreasing again. They also reported the enhancement of the wear and friction coefficient with the filler content.

Zhang et al. [34] fabricated a short-basalt-fiber-reinforced polyamide product and filled it with $MOS_2$ and graphite contents using a hot press molding technique. The tribological properties of the composite product were investigated with a model ring-on block test rig, whereas the wear mechanism was studied based on scanning electron microscope examination. It was concluded that the tribological properties were improved with the filler contents.

Ha et al. [35] experimentally investigated the corrosion and tribological properties of basalt-fiber-reinforced composites. The corrosion effect was based on time and $H_2SO_4$ concentration. It was summarized that there was a general increase in specimen weight after corrosion in $H_2SO_4$ due to basalt fiber precipitate. The friction coefficient and wear behavior were irregular, while afterwards corrosion friction coefficient was two or three times greater.

The present study had three major main goals, which can be summarized as follows:

(1)  Fabrication of a novel composite material using ABS reinforced for the first time with short basalt fiber (SBF).
(2)  Measure the mechanical characteristic properties and durability of the new composite.
(3)  Measure the wear behaviors and wear mechanism of the novel composite.

The authors in [8] carried out a similar study, but using long basalt fiber, and without studying the tribological properties.

The paper structured as follows: in the first section the materials are described, in the second section the manufacturing technique for the composite sample is summarized, in the third section the mechanical and wear tests are explained, and finally the results and discussion are followed by the conclusion, where the best results are considered.

## 2. Materials and Methods

The work was performed using ABS that was supplied by LG Chem, Ltd., Hwachi-Dong, Yeosu-City, Jeonaranam-Do, 550-280, Korea. It is of RX710 grade, has a melting flow index of 21 g/10 min and a density of 1.04 g/cm$^3$ [36]. The reinforcement of SBF was of 3:50 µm length and less than 6.5 µm in diameter, and was supplied by ROCKAL Co. (Cairo, Egypt) [37]. Figure 1 shows the raw materials of the composite product; their specifications are listed in Table 1. The basalt fiber as an inorganic material reinforcement has no homogeneity in melt [38]. It has many forms: fibrous, short or it may yarn [39].

**Figure 1.** Composite raw material. (**A**) acrylonitrile-butadiene–styrene (ABS); (**B**) Short basalt fiber (SBF).

**Table 1.** Specification of the raw material of the composite.

| Raw Material | Density (g·cm$^{-3}$) | Melt Flow Index (g/10 min) | Particle Size (µm) | Diameter (µm) |
|---|---|---|---|---|
| Acrylonitrile–butadiene–styrene (ABS) | 1.04 | 21 | - | - |
| Short basalt fiber (SBF) | - | - | 3:50 | Less than 6.5 |

*Sample Preparation*

The specimens were manufactured using reciprocating screw injection molding (HAITIAN PL 1200, Haitian International Holdings Limited, Ningbo, China). The working range of temperatures of the barrel along the injection molding machine was 140 °C, 170 °C, 220 °C and 240 °C, as listed in Table 2. The ABS was dried at 12 °C for nearly 5 h to avoid possible moisture degradation. The SBF was mechanically mixed with ABS by (5, 10, 15, and 20) wt %. The direction of the polymer plastic flow was controlled to be unidirectional for all produced specimens, this is to avoid the weld line, which can lead to weakness and cracks. The mold of the injection molding machine is explained in [8]. The standard specimens were all 4 mm thickness for the tension test, 5 mm for the impact test and 6 mm for the wear test.

**Table 2.** Conditions of the injection molding machine.

| Parameter Pressure in (Bar) | First Stage | Second Stage | Third Stage | Fourth Stage | Fifth Stage |
|---|---|---|---|---|---|
| Injection | 140 | 140 | 140 | 130 | 130 |
| Closing | 80 | 80 | 80 | 30 | 100 |
| Holding | 100 | 100 | 100 | - | - |
| Opening | 60 | 60 | 60 | 60 | 60 |
| Charging | 100 | 100 | 100 | - | - |

## 3. Mechanical and Wear Testing

All the produced specimens of SBF-reinforced ABS composite were tested in tension according to the ASTM D3039/[40] at room temperature and 50% relative humidity. The tensile test was carried out using the one-ton capacity universal testing machine at 5 mm/min crosshead speed. The dog bone tensile test specimen is shown in Figure 2 with dimensions. The Charpy impact test was performed on all specimens accord to ASTM D6110-17 [41] at room temperature using a 5 kg hummer (see standard sample Figure 2b). The Rockwell hardness test, which is based on the Rockwell C-method, was carried out using a Digital Rockwell Hardness Tester, HRS-150 (Beijing United Test Co., Ltd., Beijing, China), the test was performed at room temperature according to the ASTM D785-08 standard [42]. The wear test was carried out on all produced specimens according to the ASTM G132-96 standard [43] at room temperature and 57.5 m/s using a pin on disc device, the counter face of the abrasive disc was of surface roughness (p1000). The weight loss was measured for each specimen at various normal load (5, 10 and 15) kN before and after the test. A microstructure examination of the fractured SBF-reinforced ABS composite with the tension and damage surface in the wear test were investigated using JEOL JSM.5500LV (JEOL Ltd., Tokyo, Japan) scanning electron microscopy at an acceleration voltage of 25 kV after immersion in nitrogen gas medium. The fractured and damaged surface was coated with a very thin layered of gold. Modes of failure were observed optically using a high-resolution digital camera for tension test samples.

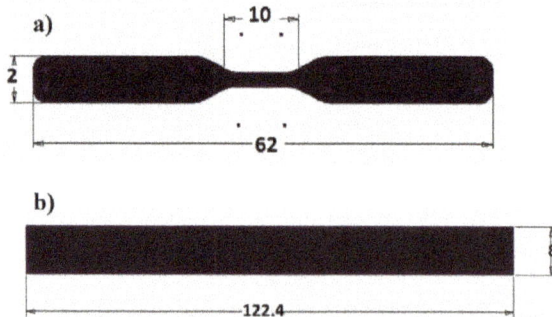

**Figure 2.** Standard specimen (**a**) Tension test; (**b**) Impact test. (Unit: mm).

## 4. Results and Discussion

*4.1. Tension Test*

Figures 3 and 4a show a stress and strain diagram of SBF-reinforced ABS; it is clear that the composite tensile strength increases with the increasing wt % of the SBF up to 5 wt %, then it decreases. This trend might be due to the large area size and very weak interface [1]. The ductility of the newly-produced material drastically reduced with the increasing amount of SBF, as is shown in the percentage length elongation curve in Figure 4b; this can be attributed to the small area of SBF, which causes weak interfaces with the ABS [6]. With the lower content of reinforcement, there was a larger interfacial area, which led to the specimen being able to withstand a larger load capacity. The Young's modulus increased up to 5 wt % then decreased, this may be due to the low strain when the strain was evaluated (see Table 3). The 3D plot in Figure 5 relates the elongation and tensile strength of the wt % of SBF. The failure modes for all tension test specimens were net tension with some surface microcracks, which is observed near the fracture process zone (see Figure 6). Scanning electron microscope (SEM) examinations show a uniformly-distributed matrix of ABS and smooth surface failure without microcracks (see Figure 7a). However, the surface becomes rough with the increasing wt % of ABS (Figure 7b–e); this can be attributed to the breaking of the bonds between the SBF and ABS, as these bonds cause fiber bridging and pull the matrix to a location of joining. Fiber breaking (marked with rectangular lines) at the interface between the ABS was observed, while fiber pull-out (marked with circular lines) clearly appeared. For the 20 wt % SBF, the fracture surface shows a random distribution of SBF, which means good mixing with the matrix. Each test was repeated five times and the stem diameter variations (SDV) was calculated and is listed in Table 3.

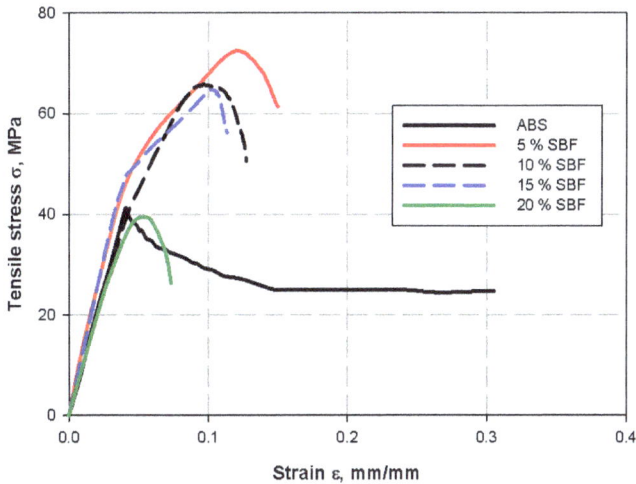

**Figure 3.** Stress–strain diagram for short basalt fiber (SBF)-reinforced acrylonitrile–butadiene–styrene (ABS).

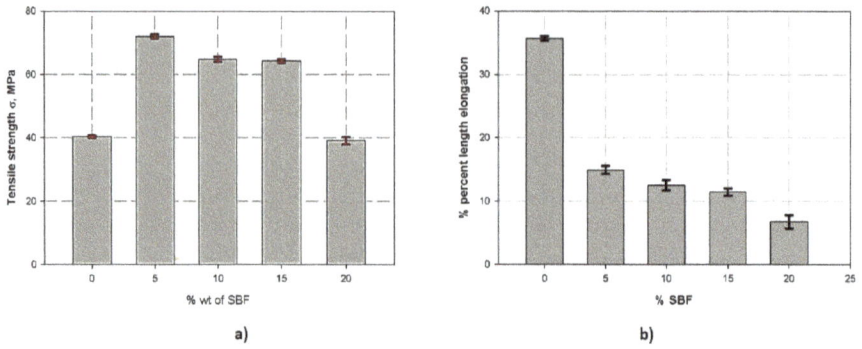

**Figure 4.** Variation with wt % (**a**) Tensile strength; (**b**) Percent elongation.

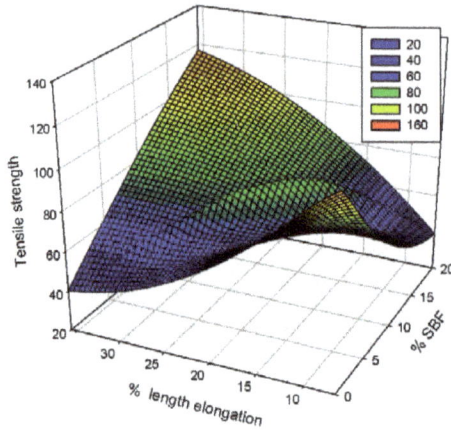

**Figure 5.** A 3D curve relating the tensile test results with wt % of SBF reinforcement.

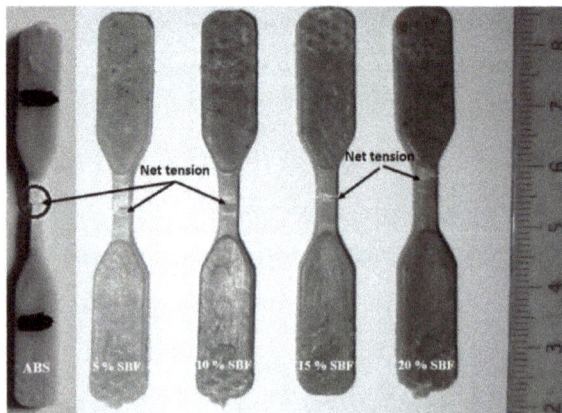

**Figure 6.** Failure modes in the tensile test specimens.

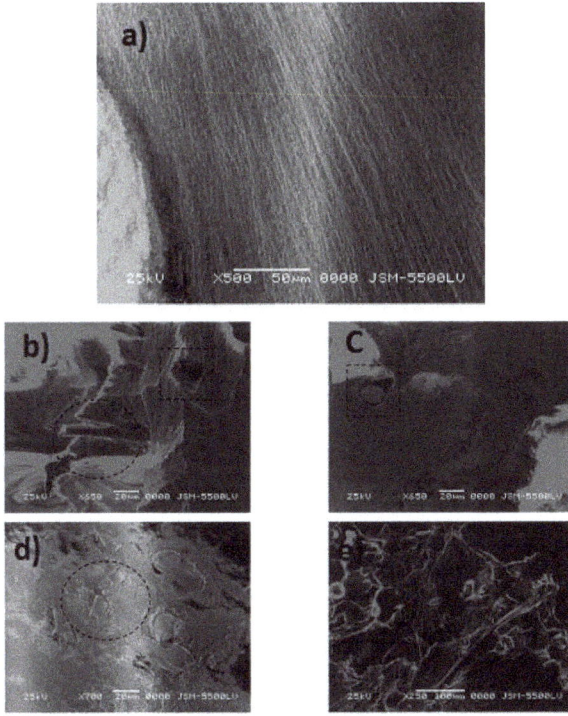

**Figure 7.** Scanning electron microscope (SEM) microstructure examination of fracture plane (**a**) ABS; (**b**) 5 wt % SBF; (**c**) 10 wt %; (**d**) 15 wt %; (**e**) 20 wt %.

**Table 3.** Tensile test results.

| Samples | Tensile Strength (MPa) | SDV | Elongation at Break (%) | SDV | Young's Modulus (MPa) |
|---|---|---|---|---|---|
| ABS | 40.32 | 0.36 | 35.7 | 0.36 | 1450 |
| ABS + 5%SBF | 72.69 | 0.63 | 14.94 | 0.62 | 1481 |
| ABS + 10%SBF | 64.82 | 0.82 | 12.53 | 0.81 | 1099.7 |
| ABS + 15%SBF | 64.29 | 0.58 | 11.46 | 0.57 | 1117.2 |
| ABS + 20%SBF | 39.07 | 1.16 | 6.79 | 1.062 | 871.3 |

*4.2. Hardness and Impact Tests*

Figure 8 shows results of the Rockwell hardness test (RH). It is clearly illustrated that with the increasing wt % of SBF in the ABS matrix, the (RH) increased. This is due to the increasing amount of brittle SBF, which resists penetration through the material's composite surface. The impact strength represented by the released energy stored in the material was reduced with the increasing wt % of SBF reinforcement (see Figure 9), due to decreasing ductility. There are two mechanisms for impact strength enhancement in a composite material, one increases the interfacial bond between the polymer and reinforcement, and the other adds a flexible and elastic phase that increases the impact [44]. There are two reasons for impact strength degradation, one is the weakness of interfacial bonding between brittle and high stiffness SBF and ABS, and the other is the cavitation or inner distance between the SBF increase, with an increasing amount of reinforcement phase [45–47]. The fracture surfaces under the impact load were analyzed using SEM photos, which are shown in Figure 10. It is observed that the interaction between the SBF and ABS matrix was poor, with an increasing fiber contents, therefore, a rough surface could be observed. The relation between the Rockwell hardness and impact strength with the variation of the wt % of SBF is shown in Figure 11.

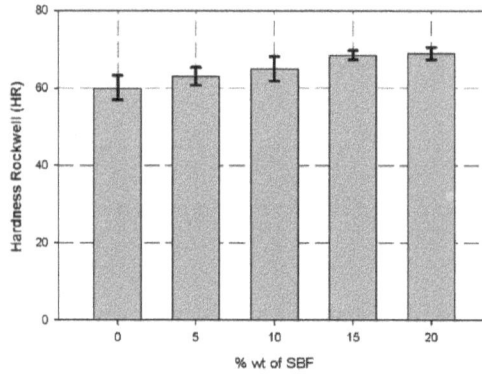

**Figure 8.** Vickers hardness variation with wt % of ABS.

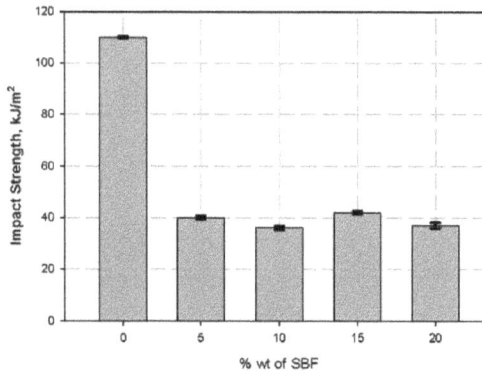

**Figure 9.** Impact fracture toughness variation with wt % SBF.

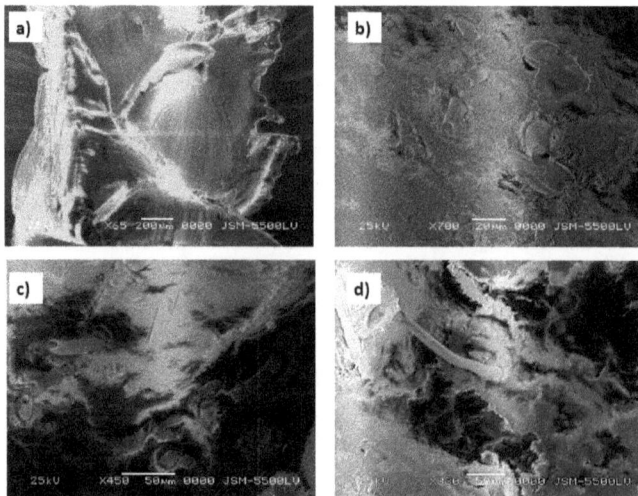

**Figure 10.** SEM microstructure examination of fracture plane (**a**) 5 wt %; (**b**) 10 wt %; (**c**) 15 wt %; (**d**) 20 wt % content SBF.

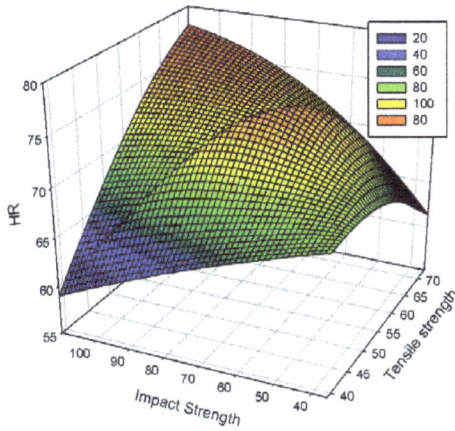

**Figure 11.** Relation between the impact fracture and hardness with wt % SBF variation.

### 4.3. Wear Test

The reinforcement phase plays a great role in the tribological behavior and wear of the polymer material, as reported in [48]. The ABS plays an important role in a lot of applications, such as bearing and slide material. Figure 12 shows the weight loss variation with the increasing wt % of SBF content reinforcement. The SBF is a ceramic material that can be an abrasive agent. This abrasive agent enhances and increases the wear resistance of the composite material. Figure 13 shows a change of weight loss with the variation of normal applied loads and the variation of the wt % of the SBF. It can be clearly observed that with the increasing normal load, the weight loss increased, while with the increasing wt % of SBF, the weight loss decreased, which means that the wear resistance was enhanced. A total of 20 wt % of SBF results in a reduction in weight loss of nearly 78%, 82% and 84% for 5 kN, 10 kN, and 15 kN normal loads, respectively. The SEM micrograph of the damaged surface due to wear action is depicted in Figure 14 for ABS and 20 wt % SBF. The shear action is shown directly over the ABS (see Figure 14a), while for 20 wt % SBF the shear action interacted directly with the brittle fibers; therefore, an increasing wt % of SBF reduces the weight loss or wear rate (see Figure 13b). It was also found that the SBF is randomly distributed in the ABS matrix. There is some surface damage and breakage in the SBF fiber; this may be due to weak interfacial bonding between the SBF and ABS matrix.

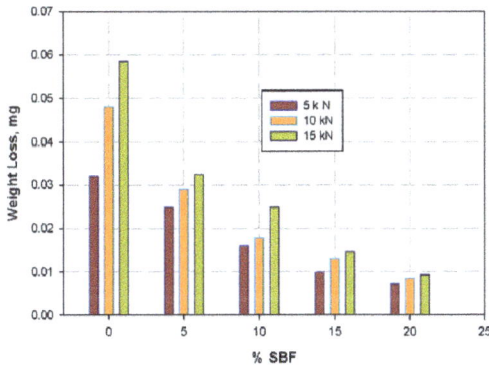

**Figure 12.** Wear resistance variation with wt % of SBF.

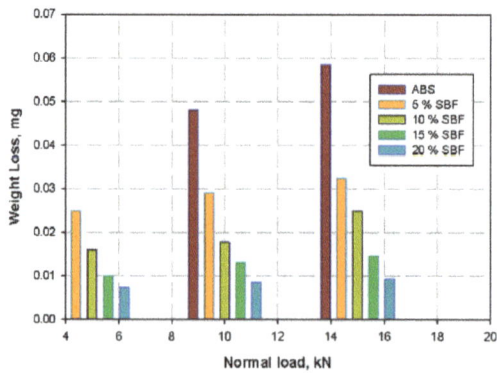

**Figure 13.** Wear resistance variation with normal load at a variation of wt % of SBF.

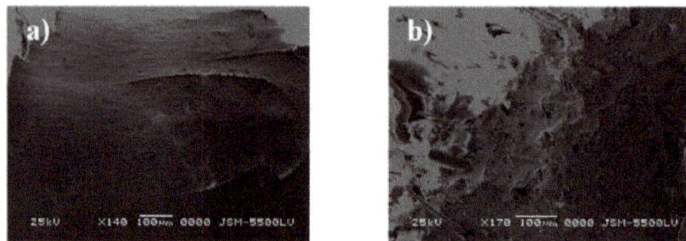

**Figure 14.** Shear mode in wear test. (a) ABS; (b) SBF.

## 5. Conclusions

Inorganic fillers have been used widely to enhance the mechanical and tribological properties of polymer products at lower cost. When SBF is used as a reinforcement in an ABS matrix, it is an inorganic material found to have weak interfacial bonding with the ABS. Increasing the wt % of the SBF content in the ABS matrix enhanced both the surface hardness and wear strength, while the weakness of the interfacial bonding between the fiber and the ABS led to decreased impact strength. Moreover, the tensile strength as well as the percent elongation in general decreased with the increasing wt % of SBF. It can be clearly concluded that the SBF is randomly distributed in the ABS matrix. The SBF failure mechanism may be fiber pull-out, fiber breakage and an absence of matrix carking. To enhance the tensile strength of a composite with an ABS base matrix, the SBF content should not exceed 5 wt %.

**Author Contributions:** Conceptualization, Validation, Formal Analysis, writting and Methodology, M.Y.A. and H.I.F., Software and Writing-Review & Editing, A.M.M.A. and M.D.

**Funding:** This research received no external funding.

**Acknowledgments:** Faculty of engineering energy at university of Aswan due to their continuous help during preforming these works.

**Conflicts of Interest:** The authors declare no conflict of interest.

## References

1.  Li, Y.; Shimizu, H. Improvement in toughness of poly(L-lactide) (PLLA) through reactive blending with acrylonitrile–butadiene–styrene copolymer (ABS): Morphology and properties. *Eur. Polym. J.* **2009**, *45*, 738–746. [CrossRef]
2.  Sarasini, F.; Tirillò, J.; Valente, M.; Valente, T.; Cioffi, S.; Iannace, S.; Sorrentino, L. Effect of basalt fiber hybridization on the impact behavior under low impact velocity of glass/basalt woven fabric/epoxy resin composites. *Compos. Part A Appl. Sci. Manuf.* **2013**, *47*, 109–123. [CrossRef]

3. Czigány, T. Trends in fiber reinforcements-the future belongs to basalt fiber. *Express Polym. Lett.* **2007**. [CrossRef]

4. Deák, T.; Czigány, T. Chemical composition and mechanical properties of basalt and glass fibers: A comparison. *Text. Res. J.* **2009**, *79*, 645–651. [CrossRef]

5. Czigány, T.; Vad, J.; Pölöskei, K. Basalt fiber as a reinforcement of polymer composites. *Period. Polytech. Eng. Mech. Eng.* **2005**, *49*, 3.

6. Jiang, L.; Lam, Y.; Tam, K.; Chua, T.; Sim, G.; Ang, L. Strengthening acrylonitrile-butadiene-styrene (ABS) with nano-sized and micron-sized calcium carbonate. *Polymer* **2005**, *46*, 243–252. [CrossRef]

7. Sepet, H.; Tarakcioglu, N.; Misra, R. Determination of the mechanical, thermal and physical properties of nano-CaCO3 filled high-density polyethylene nanocomposites produced in an industrial scale. *J. Compos. Mater.* **2016**, *50*, 3445–3456. [CrossRef]

8. Abdelhaleem, A.M.; Abdellah, M.Y.; Fathi, H.I.; Dewidar, M. Mechanical properties of ABS embedded with basalt fiber fillers. *J. Manuf. Sci. Prod.* **2016**, *16*, 69–74. [CrossRef]

9. Lopresto, V.; Leone, C.; de Iorio, I. Mechanical characterisation of basalt fibre reinforced plastic. *Compos. Part B Eng.* **2011**, *42*, 717–723. [CrossRef]

10. Difallah, B.B.; Kharrat, M.; Dammak, M.; Monteil, G. Mechanical and tribological response of ABS polymer matrix filled with graphite powder. *Mater. Des.* **2012**, *34*, 782–787. [CrossRef]

11. Colombo, C.; Vergani, L.; Burman, M. Static and fatigue characterisation of new basalt fibre reinforced composites. *Compos. Struct.* **2012**, *94*, 1165–1174. [CrossRef]

12. Jeevanantham, R.; Venketaramanamurthy, V.; Rajeswari, D. Mechanical and wear characterization of basalt fiber reinforced polyurethane composites. *Int. J. Adv. Eng. Technol.* **2016**. [CrossRef]

13. Chen, W.; Hao, H.; Jong, M.; Cui, J.; Shi, Y.; Chen, L.; Pham, T.M. Quasi-static and dynamic tensile properties of basalt fibre reinforced polymer. *Compos. Part B Eng.* **2017**, *125*, 123–133. [CrossRef]

14. Khosravi, H.; Eslami-Farsani, R. Enhanced mechanical properties of unidirectional basalt fiber/epoxy composites using silane-modified $Na^+$-montmorillonite nanoclay. *Polym. Test.* **2016**, *55*, 135–142. [CrossRef]

15. Scalici, T.; Pitarresi, G.; Badagliacco, D.; Fiore, V.; Valenza, A. Mechanical properties of basalt fiber reinforced composites manufactured with different vacuum assisted impregnation techniques. *Compos. Part B Eng.* **2016**, *104*, 35–43. [CrossRef]

16. Wu, G.; Wang, X.; Wu, Z.; Dong, Z.; Zhang, G. Durability of basalt fibers and composites in corrosive environments. *J. Compos. Mater.* **2015**, *49*, 873–887. [CrossRef]

17. Matkó, S.; Anna, P.; Marosi, G.; Szep, A.; Keszei, S.; Czigany, T.; Pölöskei, K. Use of reactive surfactants in basalt fiber reinforced polypropylene composites. In *Macromolecular Symposia*; WILEY-VCH Verlag: Weinheim, Germany, 2003; pp. 255–268.

18. Czigány, T. Basalt fiber reinforced hybrid polymer composites. In *Materials Science Forum*; Trans Tech Publications: Zürich, Switzerland, 2005; pp. 59–66.

19. Bashtannik, P.; Ovcharenko, V.; Boot, Y.A. Effect of combined extrusion parameters on mechanical properties of basalt fiber-reinforced plastics based on polypropylene. *Mech. Compos. Mater.* **1997**, *33*, 600–603. [CrossRef]

20. Liu, Q.; Shaw, M.T.; Parnas, R.S.; McDonnell, A.M. Investigation of basalt fiber composite aging behavior for applications in transportation. *Polym. Compos.* **2006**, *27*, 475–483. [CrossRef]

21. Jancar, J. Effect of interfacial shear strength on the mechanical response of polycarbonate and PP reinforced with basalt fibers. *Compos. Interfaces* **2006**, *13*, 853–864. [CrossRef]

22. Artemenko, S. Polymer composite materials made from carbon, basalt, and glass fibres. Structure and properties. *Fibre Chem.* **2003**, *35*, 226–229. [CrossRef]

23. Wei, B.; Cao, H.; Song, S. Tensile behavior contrast of basalt and glass fibers after chemical treatment. *Mater. Des.* **2010**, *31*, 4244–4250. [CrossRef]

24. Manikandan, V.; Jappes, J.W.; Kumar, S.S.; Amuthakkannan, P. Investigation of the effect of surface modifications on the mechanical properties of basalt fibre reinforced polymer composites. *Compos. Part B Eng.* **2012**, *43*, 812–818. [CrossRef]

25. Park, J.-M.; Shin, W.-G.; Yoon, D.-J. A study of interfacial aspects of epoxy-based composites reinforced with dual basalt and SiC fibres by means of the fragmentation and acoustic emission techniques. *Compos. Sci. Technol.* **1999**, *59*, 355–370. [CrossRef]

26. Mangat, P.S. Strength and deformation characteristics of an acrylic polymer-cement composite. *Matér. Constr.* **1978**, *11*, 435–443. [CrossRef]

27. Girgin, Z.C.; Yıldırım, M.T. Usability of basalt fibres in fibre reinforced cement composites. *Mater. Struct.* **2016**, *49*, 3309–3319. [CrossRef]
28. Alaimo, G.; Valenza, A.; Enea, D.; Fiore, V. The durability of basalt fibres reinforced polymer (BFRP) panels for cladding. *Mater. Struct.* **2016**, *49*, 2053–2064. [CrossRef]
29. Abdellah, M.Y.; Gelany, A.; Mohamed, A.F.; Khoshaim, A.B. Protection of limestone coated with different polymeric materials. *Am. J. Mech. Eng.* **2017**, *5*, 51–57. [CrossRef]
30. Abdellah, M.Y. Delamination modeling of double cantilever beam of unidirectional composite laminates. *Fail. Anal. Prev.* **2017**, *17*, 1011–1018. [CrossRef]
31. Abdellah, M.Y.; Hassan, M.K.; El-Ainin, H.A. Plasticity and formability controlling of cast iron using thermo-mechanical treatment. *Am. J. Mater. Eng. Technol.* **2014**, *2*, 38–42.
32. Fiore, V.; Scalici, T.; di Bella, G.; Valenza, A. A review on basalt fibre and its composites. *Compos. Part B Eng.* **2015**, *74*, 74–94. [CrossRef]
33. Sudeepan, J.; Kumar, K.; Barman, T.K.; Sahoo, P. Study of mechanical and tribological properties of ABS/ZnO polymer composites. *Adv. Mater. Manuf. Charact.* **2015**, *5*, 1–11. [CrossRef]
34. Zhang, X.; Pei, X.; Wang, Q. Friction and wear properties of basalt fiber reinforced/solid lubricants filled polyimide composites under different sliding conditions. *J. Appl. Polym. Sci.* **2009**, *114*, 1746–1752. [CrossRef]
35. Mohamed, K.H.; Mohammed, Y.A.; Azabi, S.K.; Marzouk, W.W. Investigation of the mechanical behavior of novel fiber metal laminates. *Int. J. Mech. Mechatron. Eng.* **2015**, *15*, 112–118.
36. LG Chem. Available online: http://www.lgchem.com.tr/index.php (accessed on 25 October 2015).
37. Ahmed, F.M.; Mohammed, Y.A.; Mohammed, K.H. Relaxation and compressive characteristic in composite glass fiber reinforced pipes. *Int. J. Sci. Eng. Res.* **2015**, *6*.
38. Militký, J.; Kovačič, V.; Bajzík, V. Mechanical properties of basalt filaments. *Fibres Text. Eastern Eur.* **2007**, *15*, 64–65.
39. Militký, J.; Kovačič, V.R.; Rubnerová, J. Influence of thermal treatment on tensile failure of basalt fibers. *Eng. Fract. Mech.* **2002**, *69*, 1025–1033. [CrossRef]
40. ASTM D3039/D 3039M. *Standard Test Method for Tensile Properties of Polymer Matrix Composite Materials*; ASTM International: West Conshohocken, PA, USA, 1995.
41. ASTM D6110-17. *Standard Test Method for Determining the Charpy Impact Resistance of Notched Specimens of Plastics*; ASTM International: West Conshohocken, PA, USA, 2017.
42. ASTM D785-08. *Standard Test Method for Rockwell Hardness of Plastics and Electrical Insulating Materials*; ASTM International: West Conshohocken, PA, USA, 2005.
43. ASTM G132-96. *Standard Test Method for Pin Abrasion Testing*; ASTM international: West Conshohocken, PA, USA, 2013.
44. Chotirat, L.; Chaochanchaikul, K.; Sombatsompop, N. On adhesion mechanisms and interfacial strength in acrylonitrile–butadiene–styrene/wood sawdust composites. *Int. J. Adhes. Adhes.* **2007**, *27*, 669–678. [CrossRef]
45. Wu, S. Phase structure and adhesion in polymer blends: A criterion for rubber toughening. *Polymer* **1985**, *26*, 1855–1863. [CrossRef]
46. Margolina, A.; Wu, S. Percolation model for brittle-tough transition in nylon/rubber blends. *Polymer* **1988**, *29*, 2170–2173. [CrossRef]
47. Nakamura, Y.; Yamaguchi, M.; Okubo, M.; Matsumoto, T. Effect of particle size on the fracture toughness of epoxy resin filled with spherical silica. *Polymer* **1992**, *33*, 3415–3426. [CrossRef]
48. Bahadur, S.; Gong, D. The action of fillers in the modification of the tribological behavior of polymers. *Wear* **1992**, *158*, 41–59. [CrossRef]

*Journal of*
*composites science*

MDPI

*Article*

# Multi-Objective Patch Optimization with Integrated Kinematic Draping Simulation for Continuous–Discontinuous Fiber-Reinforced Composite Structures

**Benedikt Fengler [1], Luise Kärger [1,\*], Frank Henning [1] and Andrew Hrymak [2]**

[1]   Karlsruhe Institute of Technology (KIT), 76131 Karlsruhe, Germany; benedikt.fengler@kit.edu (B.F.);
      frank.henning@ict.fraunhofer.de (F.H.)
[2]   Western University, London, ON N6A 5B9, Canada; ahrymak@uwo.ca
\*    Correspondence: luise.kaerger@kit.edu; Tel.: +49-721-608-45386

Received: 26 February 2018; Accepted: 21 March 2018; Published: 30 March 2018

**Abstract:** Discontinuous fiber-reinforced polymers (DiCoFRP) in combination with local continuous fiber reinforced polymers (CoFRP) provide both a high design freedom and high weight-specific mechanical properties. For the optimization of CoFRP patches on complexly shaped DiCoFRP structures, an optimization strategy is needed which considers manufacturing constraints during the optimization procedure. Therefore, a genetic algorithm is combined with a kinematic draping simulation. To determine the optimal patch position with regard to structural performance and overall material consumption, a multi-objective optimization strategy is used. The resulting Pareto front and a corresponding heat-map of the patch position are useful tools for the design engineer to choose the right amount of reinforcement. The proposed patch optimization procedure is applied to two example structures and the effect of different optimization setups is demonstrated.

**Keywords:** multi-objective optimization; draping simulation; local reinforcement; evolutionary algorithm; patch optimization

---

## 1. Introduction

Discontinuous fiber reinforced polymers (DiCoFRP) offer a greater degree of design freedom than continuous fiber reinforced polymers (CoFRP). However, CoFRP offer higher properties in terms of stiffness and strength. Hence, the combination of both material classes provides the potential to utilize both the mechanical properties of CoFRP as well as the high degree of design freedom of DiCoFRP (cf. Figure 1) [1].

During the product development process, optimization techniques are applied to create design proposals within a given design space. The main objective of the optimization is to utilize the material performance, where different approaches have been developed for CoFRP and DiCoFRP. The most common approach for laminate optimization of CoFRP structures is a three-step process presented by Altair [2] and implemented in their commercial optimization tool OptiStruct. The approach consists of a free size, a size and a stacking optimization. A more general approach is the combination of stacking sequence with shape optimization within one single optimization run [3]. Due to the higher degree of design freedom, DiCoFRP require different optimization strategies. Therefore, topology optimization is used to create design proposals [4]. Topology optimization can also be used for nonlinear structural behavior, which can be caused by geometrical or material nonlinearity [5]. Traditional topology optimization was designed for isotropic materials. To regard the anisotropy of DiCoFRP, which is both process and topology dependent, some adjustments in the optimization procedure are necessary. In [6,7] an approach to optimize DiCoFRP parts is presented, where local fiber volume content and local thickness are used as optimization parameters. Another field of DiCoFRP optimization is the shape optimization, which is based on given parameters [8]. In contrast to topology optimization, shape optimization does not create a new total design but creates parameter-based changes of the component's form. Shape optimization methods can be split into CAD-based (Computer-Aided Design) methods and mesh-based methods [3,8].

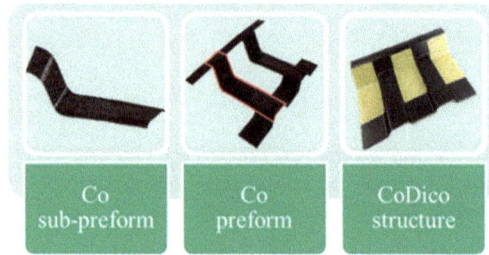

**Figure 1.** Example of a local continuous fiber reinforced composites (CoFRP, black) reinforced discontinuous fiber reinforced composites (DiCoFRP, white) structure (obtained from [1]).

For a common component design process, it is often not purposeful nor sufficient to consider only one objective for the optimization. The most common approach to consider more than one objective is the use of the weighted sum method. However, this method has the drawback that weighting factors include user preferences in the optimization procedure [8,9]. Consequently, the result of the optimization is not necessarily the overall optimal solution. Thus, the optimization should be able to handle more than one objective to create a Pareto front as the basis of decision-making.

The manufacturability of a virtually optimized composite structure is a crucial precondition for the usability of optimization results. In multi-objective optimization, the consideration of manufacturing constraints ensures that all compared solutions are producible. This reduces the possibility of errors in the Pareto front. Therefore, manufacturing constraints such as large areas of identical thickness, smooth changes in thickness and maximum number of contiguous layers with identical material orientation are often included in structural optimization [3,10–12]. Zuo et al. [10] used a symmetry constraint to improve the quality of the topology optimization results. Zhou et al. [13] introduced a method to incorporate a tool opening direction constraint as well as an extrusion direction constraint. However, these approaches do not consider the interaction between process and structural design. For that purpose, an information transfer between the different simulation steps needs to be utilized, as realized in continuous composite CAE chains [14,15] (Computer-Aided Engineering). Due to the enormous computational effort, optimization based on process and structural simulation has scarcely been addressed so far. Kaufmann et al. [16] introduce a draping knowledge database to include manufacturing costs in the design process of slightly curved laminated CoFRP structures. To consider optimized manufacturable fiber orientations in structural simulation, Kärger et al. [17,18] introduce a continuous CAE workflow with integrated draping optimization, but do not consider structural optimization. Bulla et al. [19] propose a ply-wise structural optimization, where initial draping simulations are performed for the non-optimized plies to assume conceivable fiber orientations. This approach, however, does not consider the backward influence of the optimized ply structure on the draping results.

The optimization of combined DiCoFRP and CoFRP has hardly been studied. An approach for the size and position optimization of a single reinforcement patch on slightly curved geometries is shown by Zehnder and Ermanni [20]. Thereby, the patch contour is created by a parametrized CAD model and used as basis input for mesh generation. Contrary to our approach, the method does not incorporate a draping simulation. Hence, geometry-dependent patch alignment and fiber orientations are not considered directly within the finite-element-based optimization framework, but intermediate mesh generation via CAD is necessary. With the structural optimization of a patch layup for a sailing boat, Zehnder et al. [21] present another use case, where they did not focus on manufacturability or draping, but on finding the optimum ply layup. Therefore, the structure is divided into several patches and the local stacking sequence is used as optimization parameter. Mathias et al. [22] used a genetic algorithm for the optimization of composite repair patches for 2D aluminum components. Mejlej et al. [23] present the optimization of local reinforcements for the automated fiber placement process, based on two different genetic algorithms. Another strategy to combine CoFRP with DiCoFRP is the use of local tow reinforcements. Jansson et al. [24] present a genetic algorithm coupled with a surrogate model to find the optimum layup for a given structure supported by a local unidirectional tow path. None of these optimization approaches includes draping simulation methods to consider the geometry-dependent position and orientation of the local CoFRP reinforcements.

Evolutionary algorithms have been successfully applied to several different engineering use cases [4]. Skordos et al. [25] used the combination of an evolutionary algorithm with a kinematic draping simulation to improve the draping quality of a given layup. In their approach, the draping starting point and direction are used as design parameters, while the shear angle is used as optimization objective. A more advanced draping optimization approach is proposed by Chen et al. [26]. They use an FE-based draping simulation and optimize the arrangement of lateral clamping by applying evolutionary algorithms to minimize the shear angle. Those process optimization approaches, however, do not consider the structural performance. Those use cases show that engineering problems can be solved efficiently with evolutionary algorithms. In the present work, an optimization strategy for unidirectional CoFRP patch reinforcements on complexly curved DiCoFRP structures is presented. Manufacturing constraints in terms of fiber orientations and manufacturable patch positions and sizes are taken into account during the optimization procedure. To model fiber orientations with dependence on varying patch positions, a kinematic draping simulation is integrated into the optimization algorithm. To automatically consider multiple objectives regarding patch usage and structural performance within one single optimization run, a genetic algorithm is used and combined with the kinematic draping simulation. The defined optimization problem combines shape and topology optimization, where the change of patch size is a kind of shape optimization, while the positioning of the patch can be assigned to the field of topology optimization.

## 2. Draping Simulation

### 2.1. Suitability of Existing Draping Simulation Methods for Optimization

Since the fiber structure determines the mechanical properties of the final part, it is important to predict the fiber architecture by draping simulation. In principle, the available draping simulation approaches can be separated in two fields, kinematic (also known as mapping approaches) and mechanical methods [27]. Both approaches have been developed with a different focus. Mechanical approaches are based on constitutive models and offer higher prediction accuracy of the actual material behavior [28–31]. Therefore, a good knowledge of the material behavior is essential. Compared to this, kinematic approaches offer only an approximate prediction of the material behavior and, thus, do not require material characterization. Moreover, kinematic approaches are usually very efficient. Due to the lack of exact modeling of the material behavior, it is not possible to predict wrinkling by kinematic draping simulation. The drawbacks and advantages of kinematic methods have been extensively studied by multiple authors [32–36]. A main disadvantage of kinematic draping simulation is the dependence of the simulation result on the choice of the starting point and of the initial path of the kinematic algorithm [37,38].

Due to the high number of simulations performed during optimization, a very efficient draping algorithm is needed. Since kinematic draping simulation requires small computation times, it is, from a computational point of view, more suitable for optimization workflows than constitutive-based draping simulation. Furthermore, the optimization shall work as a suggestion for the designer and not as a final verification. Hence, a mechanical draping simulation must be performed afterwards to ensure manufacturability of the optimized patch setup without forming defects like wrinkling, fiber fracture or gapping. This is similar to topology optimization, where a verification simulation is needed to ensure the load-bearing capacity of the optimized topology [8]. The focus of the draping simulation within the presented optimization workflow is the calculation of the final patch geometry, while the prediction of forming defects is not yet relevant. Therefore, kinematic draping simulation is chosen to be sufficient for the aspired purpose. The drawbacks regarding the dependence on the initial starting point are not highly important, since all possible starting points are taken into account by the proposed optimization workflow, see Section 3.2.

*2.2. Kinematic Draping Algorithm to Be Embedded in the Optimization Workflow*

The general assumption of kinematic draping simulation is the inextensibility of the material in fiber direction. Depending on the type of fiber reinforcement, the deformation of the material is assumed as pure shear or simple shear along the fiber direction. While woven fabrics deform in pure shear, unidirectional reinforcements are assumed to deform in simple shear, as shown in Figure 2. In simple shear, the distance between adjacent fibers is assumed to be constant [27].

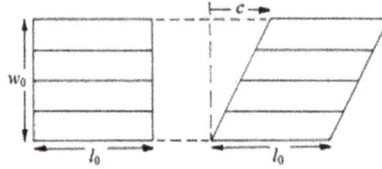

**Figure 2.** Assumptions of the simple shear deformation behavior of unidirectional fiber reinforcements: constant fiber length $l_0$ and constant fiber distance (from [27]).

Starting point, patch orientation, patch length and width are the necessary input parameters for the kinematic draping simulation. Another required input is the geometry of the initial component, on which the patch is going to be draped. The first step is to create the initial paths A and B on the part surface, starting from the given starting point (cf. Figure 3a). Thereby, A is the initial path in fiber direction, while B is perpendicular to the fiber direction.

For the calculation of the next node $a^{<i+1>}$ from the current node $a^{<i>}$, a number of constraints need to be fulfilled. Firstly, the distance between both nodes is set to be equal to the defined mesh size $m$

$$\left| a^{\langle i+1 \rangle} - a^{\langle i \rangle} \right| = m. \tag{1}$$

Furthermore, the node $a^{<i+1>}$ has to be on the initial component surface

$$a_3^{\langle i+1 \rangle} = F(a_1^{\langle i+1 \rangle}, a_2^{\langle i+1 \rangle}), \tag{2}$$

where $a_1^{\langle i+1 \rangle}$, $a_2^{\langle i+1 \rangle}$ and $a_3^{\langle i+1 \rangle}$ are the $x$, $y$, $z$ components of the new node and $F(x, y)$ represents the surface of the initial component. Additionally, it has to be ensured that the given global patch direction $\alpha$ is maintained. With these assumptions all nodes along the initial paths A and B can be calculated.

*J. Compos. Sci.* **2018**, *2*, 22

(a)

(b)

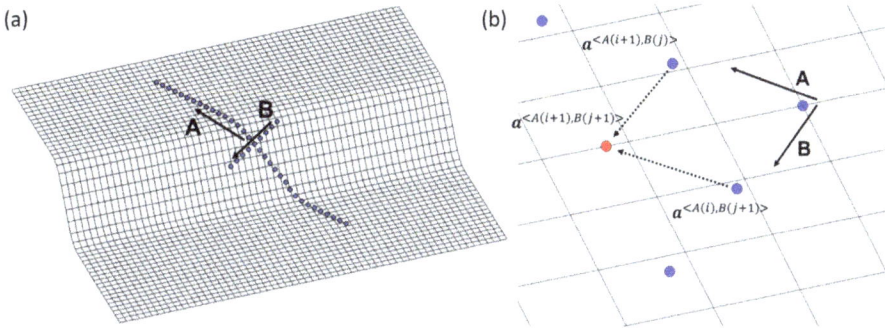

**Figure 3.** Node calculation, starting from the initial paths A and B (**a**), and for all further nodes (**b**).

Based on these initial paths, all other nodes can be calculated. From the assumption of simple shear (Figure 2) the following equations can be derived for the parallel paths

$$\left| a^{\langle A(i+1),\ B(j+1)\rangle} - a^{\langle A(i+1),B(j)\rangle} \right| = m,$$ (3)

$$\left| a^{\langle A(i+1),B(j+1)\rangle} - a^{\langle A(i),B(j+1)\rangle} \right| = m, \text{ with } i = 1, 2,\dots \frac{l_0}{m},\ j = 1, 2,\dots \frac{w_0}{m}.$$ (4)

Equations (3) and (4) represent that the distance between the fibers as well as the fiber length is constant during the draping process. Hereby $a^{\langle A(i+1),B(j)\rangle}$ is the next node on path A, while $a^{\langle A(i),B(j+1)\rangle}$ is the closest node on path B (cf. Figure 3b, 10) the initial patch length and $w_0$ the initial patch width. Again, the node has to be on the part surface, which is represented by Equation (2).

Rearrangement of Equations (2)–(4) leads to the following system of equations

$$0 = \left( a_1^{\langle A(i+1),\ B(j+1)\rangle} - a_1^{\langle A(i+1),B(j)\rangle} \right)^2 + \left( a_2^{\langle A(i+1),\ B(j+1)\rangle} - a_2^{\langle A(i+1),B(j)\rangle} \right)^2$$
$$+ \left( a_3^{\langle A(i+1),\ B(j+1)\rangle} - a_3^{\langle A(i+1),B(j)\rangle} \right)^2 - m^2$$

$$0 = \left( a_1^{\langle A(i+1),\ B(j+1)\rangle} - a_1^{\langle A(i),B(j+1)\rangle} \right)^2 + \left( a_2^{\langle A(i+1),\ B(j+1)\rangle} - a_2^{\langle A(i),B(j+1)\rangle} \right)^2$$
$$+ \left( a_3^{\langle A(i+1),\ B(j+1)\rangle} - a_3^{\langle A(i),B(j+1)\rangle} \right)^2 - m^2$$ (5)

$$0 = a_3^{\langle A(i+1),\ B(j+1)\rangle} - F(a_1^{\langle A(i+1),\ B(j+1)\rangle}, a_2^{\langle A(i+1),\ B(j+1)\rangle}),$$

which has to be solved for all nodes. The presented kinematic draping approach is implemented in MatLab and embedded in the optimization workflow.

## 3. Multi-Objective Patch Optimization Workflow

### 3.1. Genetic Algorithms for Multi-Objective Optimization

The field of evolutionary algorithms consist of several different strategies, like genetic algorithms, evolutionary strategies, genetic programming, and evolutionary programming [39]. The most popular type of evolutionary algorithm is the genetic algorithm, which is why both are often used synonymously. Contrary to other evolutionary algorithm (e.g., evolutionary strategies), for genetic algorithm mutation is not the primary search operator and therefore a recombination step is included, to propose solutions based on prior solutions. Since a genetic algorithm is applied here, this section will only focus on this method.

A genetic algorithm consists of the three main steps recombination, mutation, and replacement (also known as selection) [40]. The genetic algorithm workflow, adapted for the purpose of this work, is shown in Figure 7. A set of possible solutions is called population, where one parameter set is called individual. The initial population is created with a random distribution within the search space. After a fitness calculation of each individual, a replacement process is applied to select the individuals for the next generation. In our work, we will demonstrate the influence of the used replacement mechanism. Therefore, an elitist method is compared with a below-limit method. For the elitist method, only the best solutions are passed on from one generation to the next. Referred to Figure 4, this means, only front 1 with solution $q_1$ will be preserved, since fitness 1 and 2 for solution $q_1$ are better than for solution $q_2$. This method is likely to lower the diversity in a population, but converges faster to a final front. With the below-limit method also a number of dominated individuals is passed on from one generation to the next. In the given example in Figure 4, front 2 would also be kept in the population. This helps to maintain diversity in the set of solutions, but convergence is usually slower than for the elitist method.

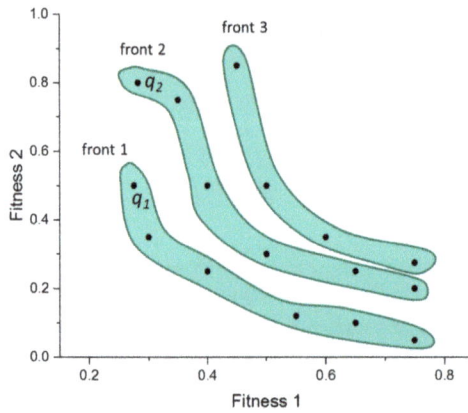

**Figure 4.** Classification of intermediated results in different fronts, here front 1 is equal to the Pareto front.

In the recombination phase, a crossover operator is applied to create new individuals (offspring) based on existing individuals. The optimization algorithm is used with two different crossover operators. Both operators use a binary representation for the design variables and a bit-switching crossover. Bit-switching means that a number of bits is exchanged between the parents to create new offspring. The number of used parents is two for both crossover types. Each crossover type is characterized by the crossover parameter $\mu$, which represents the number of bit switching operations. The first used crossover operator is the multi-point parametrized binary (cf. Figure 5). Here each design variable is treated individually, hence the total number of switched bits is related to the number of design variables. If the crossover parameter $\mu$ is set to one, for each parameter one bit is switched.

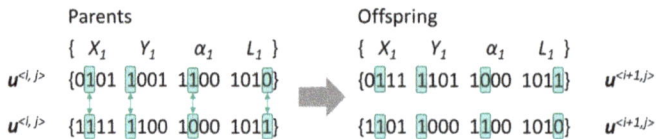

**Figure 5.** Workflow for the multi-point parameterized binary crossover for $\mu = 1$.

The second crossover parameter used in this work is the multi-point binary crossover, where the vector of design variables $u$ is represented as one bit string (cf. Figure 6). This means that the number of crossover operations is independent of the number of design variables. Since the crossover is applied to the whole vector of design variables at once, it is possible that some variables are changed at more than one bit, while other variables remain unchanged.

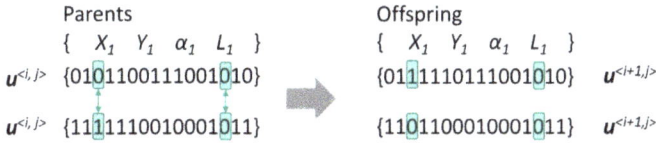

Parents                                              Offspring
$\{ \quad X_1 \quad Y_1 \quad \alpha_1 \quad L_1 \quad \}$                  $\{ \quad X_1 \quad Y_1 \quad \alpha_1 \quad L_1 \quad \}$
$u^{<i,j>} \quad \{0101100111001010\}$                $\{0111110111001010\} \quad u^{<i+1,j>}$

$u^{<i,j>} \quad \{1111110010001011\}$                $\{1101100010001011\} \quad u^{<i+1,j>}$

**Figure 6.** Workflow of the multi-point binary crossover for $\mu = 2$.

The diversity in the population is maintained by the incorporation of a mutation step. The mutation scheme applied here uses also the binary representation of the design variables. Thereby one random bit of a design variable is changed with a given probability $\beta$.

There are two general approaches to create new populations: the generational and the steady-state approach. In the generational approach, the new generation created only with the offspring replaces the parents' generation. In contrast, the steady-state approach applies a replacement mechanism to select the individuals form the parents' generation and the offspring for the next generation. The steady-state strategy is applied here, since it is more likely to converge [41]. Furthermore, a niching pressure is used to ensure a diversity along the Pareto front and a more regular distribution. Therefore, the solutions are pre-sampled before the actual replacement mechanism is encountered. This means that in a given area only one solution remains "selectable" during the replacement process.

### 3.2. Patch Optimization Workflow with Embedded Kinematic Draping Simulation

The multi-objective patch optimization problem is defined by

$$Minimize\ F(u) = [F_1(u),\ F_2(u)]^T,\ u \in U, \tag{6}$$

where the objective function $F_1$ is the global strain energy, the objective function $F_2$ is the patch usage in terms of patch length, and

$$u = [x, y, l, \alpha]^T \tag{7}$$

is the vector of design variables consisting of patch position $(x, y)$, patch length $l$ and angle $\alpha$ of the patch orientation, see Figure 7b. The optimization is performed within the following boundary conditions:

$$x_{min} < x_i < x_{max}$$

$$y_{min} < y_i < y_{max} \tag{8}$$

$$l_{min} < l_i < l_{max},\ with\ stepsize\ l_{step}$$

$$\alpha_{min} < \alpha_i < \alpha_{max},\ with\ stepsize\ \alpha_{step}$$

The minimization of the strain energy can be seen as a maximization of the total part stiffness. For linear–elastic material behavior, the strain energy can be defined by

$$W_\varepsilon^{global} = \int_\Omega \sigma^T * \varepsilon\, dV = \frac{1}{2} \sum_{i=1}^{n} P_i^T * d_i, \tag{9}$$

where $P_i$ and $d_i$ are the nodal forces and nodal displacements of a finite element model with $n$ element nodes.

The objective of the optimization is to find a solution u which minimizes all objective functions. Here, the objective functions $F_1$ and $F_2$ depend on each other, which means that improving one objective will impair another. For such cases, Villfredo Pareto described the idea to find a set of optimal solutions, the so-called Pareto set or Pareto front. Contrary to most multi-objective optimizations, no weighting function will be used in our approach. In this way, all combinations of optimal solutions stay preserved and solutions do not get lost due to enforced weighting of the multiple objective functions.

The general approach of the patch optimization workflow is shown in Figure 7a. Firstly, the initial population of design variable sets $u_i$ is created with uniform random distribution for each parameter within the given design space. Secondly, the initial fitness calculation is performed, where the kinematic draping simulation (see Section 2.2) is integrated into the fitness calculation step of the multi-objective genetic algorithm (Figure 7c). The draping simulation is used to create the necessary input for the structural simulation, in terms of position and fiber orientation of the formed CoFRP patch. Furthermore, the draping simulation is used to return the fitness value $F_2$. Fitness value $F_2$ represents the amount of used patch length. Since the random selection of design variables $u$ may create patch lengths, which overlap the boundaries of the geometry, two different scenarios have to be distinguished:

1. The patch fits completely on the DiCoFRP component: Here the patch usage is equal to the design parameter, representing the patch length.
2. The patch does not fit completely on the DiCoFRP component: Here the patch is cut at the boarder of the component and the resulting length is returned as $F_2$.

Subsequent to draping simulation, a structural simulation computes the fitness value $F_1$, which represents the compliance of the component, quantified by the global strain energy. The structural simulation is performed by using the finite element software Abaqus.

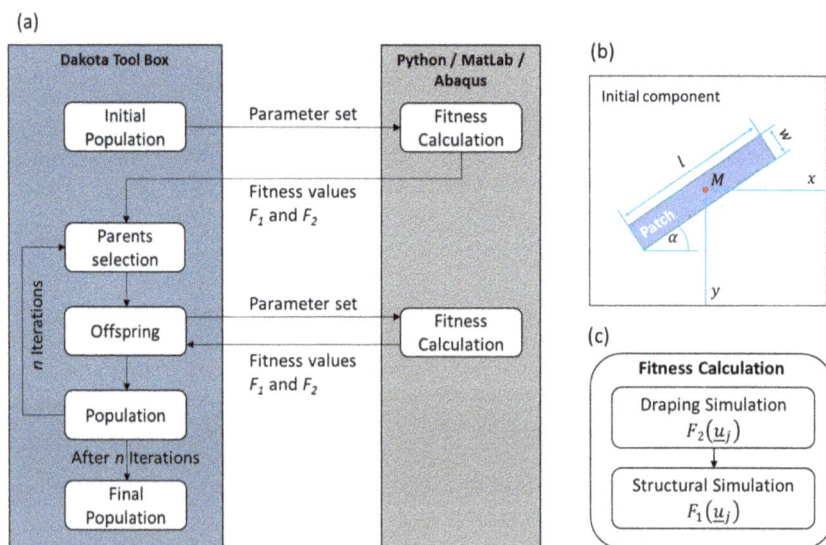

**Figure 7.** Integration of the fitness calculation in the genetic optimization algorithm of the Dakota tool box (**a**), used to solve the patch optimization problem (**b**) by applying the two-step fitness calculation (**c**).

For the optimization algorithm, the open access software toolkit Dakota is used, an implementation of the Sandia National Laboratories [42]. The Dakota toolkit provides diverse operators for genetic algorithms (as the ones described in Section 3.1) as well as an interface between analysis code and optimization methods. Within the optimization loop, two termination criteria can be used: A maximum number of iterations $n$ and a convergence metric.

### 3.3. Convergence Criteria

Since multi-objective optimization has the goal to predict a wide range of the Pareto front, more than one criterion is needed to properly characterize the performance. Hence, three criteria are introduced to evaluate the convergence rate, the quantity of solutions, and the diversity:

1.  Dominated area for the evaluation of the convergence rate. The dominated area is defined as the area enclosed by the current front and the extreme points $k_1$ and $k_2$ and therefore represents the progress of the optimization, cf. Figure 8. The extreme points are defined by

$$k_i = [F_1(u_i), F_2(u_i)]^T, \ i \in 1,2 \tag{10}$$

$$k_1 = \{k|F_1(u_1) > F_1(u) \ \forall \ u \in U\} \tag{11}$$

$$k_2 = \{k|F_2(u_2) > F_2(u) \ \forall \ u \in U\}. \tag{12}$$

2.  Number of solutions along the Pareto front.
3.  Patch length distribution. The patch length is modeled as a discrete design variable. Therefore, the percentage share of the patch length partitions along the Pareto front is used to describe the distribution along the front.

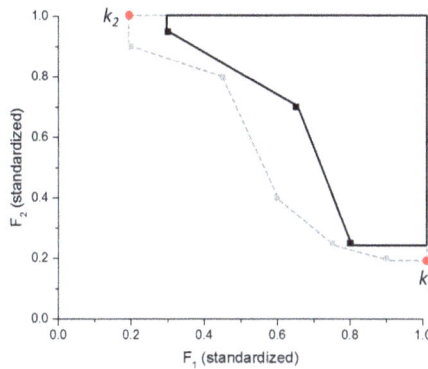

**Figure 8.** Dominated area calculation for generation $i$ (black solid line) and $i + 1$ (grey dotted line), and the maximum points $k_1$ and $k_2$ (red dots).

To evaluate the global strain objective in conjunction with the patch usage objective, the patch position (including patch length) is visualized by means of a heat-map, which weights the resulting fitness values. Therefore, the results along the Pareto front are summarized in a weighted form. The weighting factor $\omega_i$ is calculated by

$$\omega_i = \left( \frac{F_1(u_i)}{F_{1,\max}(u_{\max1})} \cdot \frac{F_2(u_i)}{F_{2,\max}(u_{\max2})} \right)^{-1}, \ i = 1 \dots n, \tag{13}$$

with $n$ as the number of individuals on the Pareto front. $F_{1,\max}$ is the maximum value observed for fitness 1, and $F_{2,\max}$ the maximum value observed for fitness 2. The weighting is normalized and plotted on a uniform mesh, which is only used for visualization purposes (cf. Figure 15). Beside the performance evaluation, the heat map is a useful tool for the designer to find the areas in which the reinforcement is most effective. Thereby, areas with high patch concentration (cf. red regions in Figure 15) should be preferred over those with a lower patch concentration (cf. blue regions in Figure 15).

## 4. Application Examples

### 4.1. Problem Description and Optimization Parameter

The proposed algorithm will be demonstrated by two example structures. The first demonstrator part is a flat plate (Figure 9 left), the second test case is a curved structure (Figure 9 right). For both structures a bending load is applied. The design area for the patch placement is in both cases the top side. All optimization runs are conducted with the fix number of two patches. Both problems are modeled with shell elements and linear elastic material behavior.

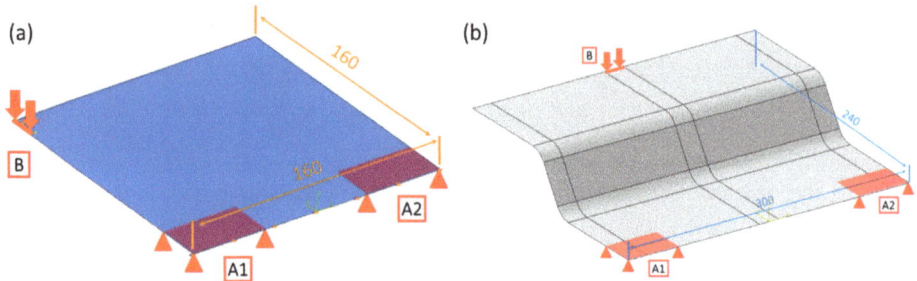

**Figure 9.** Loading conditions and geometrical dimensions of the optimization test cases: Flat plate with bending load (**a**) and curved structure with bending load (**b**), boundary conditions are represented by the red areas (A1 and A2).

The plate structure A is used to demonstrate the general capability of the method to solve the patch optimization problem. Furthermore, it is used to compare the crossover types and select the more suitable one for the detailed studies with the curved structure B. Table 1 gives an overview of the conducted optimization runs, where the optimization operators for mutation, replacement and niching (cf. Section 3.1) are systematically varied for structure B. All configurations of B are compared with a basis setup B-Opt 1.

**Table 1.** Overview of the used optimization parameter (A = Optimization problem "plate", B = optimization problem "curved structure").

| Name | Structure | Crossover | Mutation | Replacement | Niching |
|---|---|---|---|---|---|
| A-Opt 1 | Plate | Parameterized binary | High | Elitist | No |
| A-Opt 2 | Plate | Multi-point binary | High | Elitist | No |
| B-Opt 1 | Curved shell | Multi-point binary | Low | Elitist | No |
| B-Opt 2 | Curved shell | Multi-point binary | High | Elitist | No |
| B-Opt 3 | Curved shell | Multi-point binary | Low | Below limit | No |
| B-Opt 4 | Curved shell | Multi-point binary | Low | Elitist | Yes |

### 4.2. Results and Discussion for the Flat Plate Example

For the crossover comparison, the basis setup A-Opt 1 (Table 1) is used as benchmark. To characterize the distribution within the search space, all solutions are arranged in groups with patch-length steps of 25 mm, from 0 mm to 425 mm maximum patch length for example A. The division into the seventeen groups will also be used for A-Opt-2. Since the resulting patch length for example B is longer, a division into 26 groups of 25 mm each is used. The distribution of the results within the search space for example A is shown in Figure 10a. The presented figure shows that solutions with smaller total patch length are more likely to appear. This is also demonstrated by the percentage share of all A-Opt 1 results in Figure 10c. In the given example, the shortest possible patch length is 25 mm, therefore patch usage group

1 remains empty. Furthermore, four reference points are given in Figure 10a, which correspond to the patch configurations shown in Figure 10b. These points shall help to evaluate the optimization results.

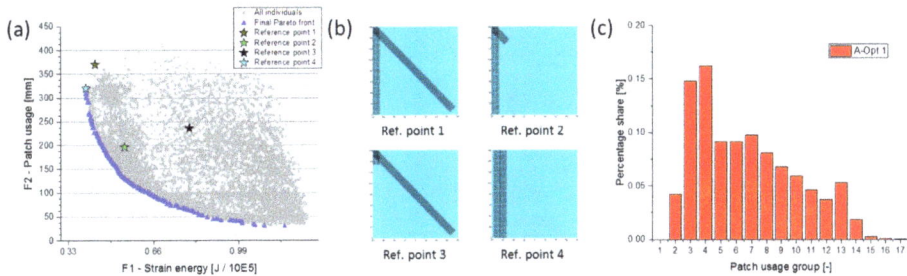

**Figure 10.** (a) Distribution of all calculated individuals in the search space of A-Opt 1 in comparison with four reference solutions. The blue points represent the final Pareto front; (b) Patch configurations of four reference solutions; (c) The red bars illustrate the resulting percentage share of all individuals (blue and grey dots in (a)) for each patch-usage group with a step-length of 25 mm.

The configurations A-Opt 1 and A-Opt 2 are used to show the general ability of the proposed algorithm to solve the patch optimization problem. Furthermore, the results will be used to select the crossover type (parameterized binary or multi-point binary, cf. Section 3.1) used for the curved structure. Comparing the Pareto fronts, both optimizations converge to a similar front (cf. Figure 11a) which allows the assumption that the proposed approach runs stable.

The convergence rate, represented by the increase of the dominated area (Figure 11b), is significantly higher for the multi-point binary crossover (A-Opt 2) than for the parameterized binary crossover (A-Opt 1). This is due to the fact that the multi-point binary crossover can change a design variable on several bits, while other design variables remain unchanged. Consequently, the multi-point binary crossover has a more pronounced global search than the parametrized binary crossover, which only applies one change per variable. Comparing the number of solutions found on the final Pareto front, the multi-point binary crossover also has a better performance. Both heat maps show the similar result that a reinforcement on the left side of the plate will create the best result in terms of the total strain energy (Figure 11e,f). This can also be seen by the proposed reference points (cf. Figure 10b) where configuration 4 is the best solution, in terms of strain energy minimization. Comparing the distribution of the solutions on the Pareto front of A-Opt 1 (cf. Figure 11d) with the distribution of all solutions of A-Opt 1 (Figure 10c), it can be seen that the qualitative distribution is very similar. As for A-Opt 1, the solutions along the Pareto front of A-Opt 2 also show a tendency to smaller total patch lengths. However, the more global variability of A-Opt 2 yields slightly longer patches on average.

Summarizing the crossover effects, the multi-point binary crossover offers a better performance in terms of convergence rate and number of found solutions. Since both final Pareto fronts are quite similar, the multi-point binary crossover is used to perform the optimization runs with the curved structure.

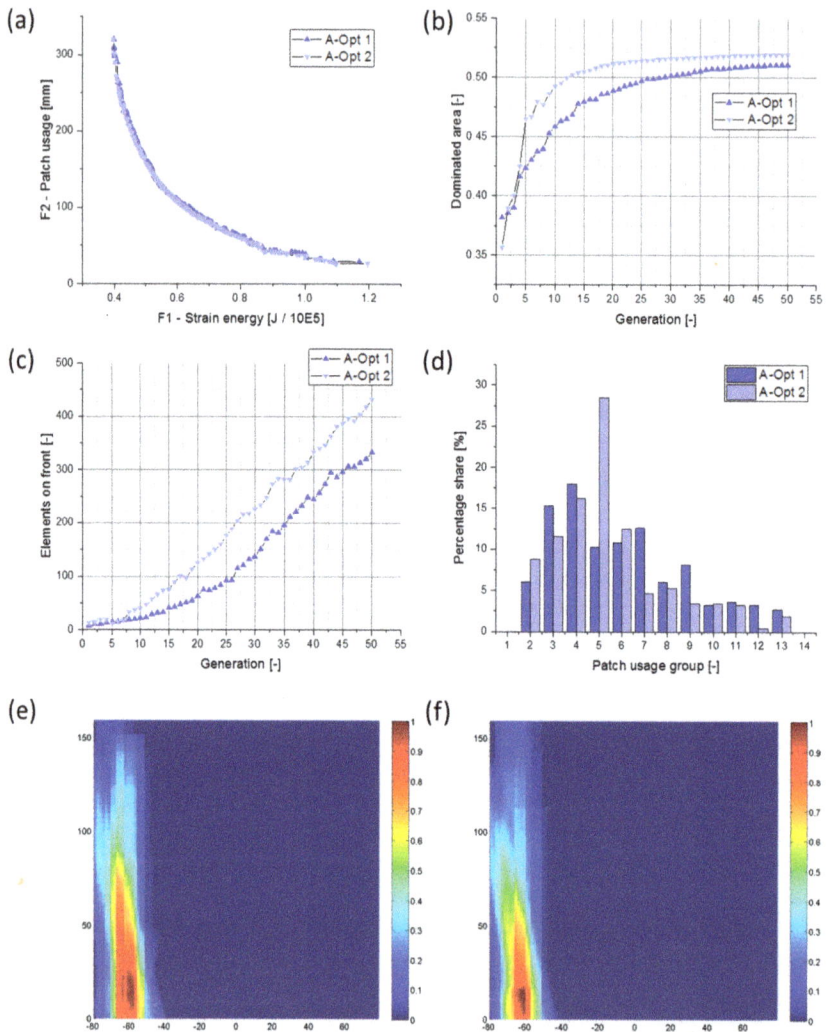

**Figure 11.** Comparison of the results of the two crossover processes A-Opt 1 (parameterized binary crossover) and A-Opt 2 (multi-point binary crossover): (**a**) Comparison of the final Pareto fronts, (**b**) the dominated area for the evaluation of the convergence behavior, (**c**) development of the number of elements on the current Pareto front for each generation, (**d**) percentage share of the solutions along the Pareto front (blue dots in (**a**)) for each length-group with a step-length of 25 mm, (**e**) heat map for A-Opt 1, (**f**) heat map for A-Opt 2.

### 4.3. Results and Discussion for the Curved Structure Example

For the second optimization example B, the multi-point binary crossover is kept constant, while the optimization parameters for mutation, replacement and niching are systematically varied to show their influence on the performance of the optimization. Like the plate structure A, the Pareto fronts obtained for the different setups (cf. Table 1) are compared with predefined reference points to evaluate the results (cf. Figure 12). Thereby, the patch reinforcement of reference point 4 ends at the constraint area, while the reinforcement from point 3 is limited by the part geometry.

Comparing the different mutation settings, the higher mutation (B-Opt 2) creates a slightly better result regarding stiffness maximization for large patch lengths (cf. Figure 13). This is due to the fact that the changes applied by the mutation operator work similar to the crossover process, and therefore increase the search power of the optimization. The dominated area does not converge faster compared to the low mutation (Figure 13b), but starts at a higher value. When looking at the number of found solutions, as a second quality criterion, the number of found solutions at the final front is higher for the higher mutation rate (cf. Figure 13c). Comparing the heat maps, it can be seen that both optimization runs end up in similar results (cf. Figure 16a,b). The better performance in terms of finding the extreme point regarding stiffness maximization is not visible in the heat map (Figure 16b), since only a small proportion of solutions is in the extreme area, and is therefore underrepresented within the heat map. This also shows that the heat map is an additional tool, but should not be considered as the only basis of decision, but always in combination with the Pareto front.

Comparing the results obtained by calculation B-Opt 1, B-Opt 3 (replacement type below-limit, where also dominated individuals are passed to the next generation), and B-Opt 4 (with niching), no significant change regarding the Pareto front and the convergence rate can be seen (cf. Figures 14 and 15 each (a) and (b)). The difference in the results of the different setups becomes more obvious comparing the heat maps (cf. Figure 16). Here can be seen that the distribution of the resulting patch positions is more balanced (reflected in a more pronounced symmetrical picture in the heat map), when more fronts are taken into account (B-Opt 3) or a niching pressure is applied (B-Opt 4). This effect results from a higher diversity of the found solutions along the Pareto front.

For the given example, a reinforcement covering the position of the structural kink has a large effect on the total component's stiffness. This becomes evident when comparing reference points 2 and 3 (cf. Figure 12) as well as in the heat maps, represented by the red areas (cf. Figure 16c,d). The red areas in the heat maps for B-Opt 1 and B-Opt 2 are smaller compared to those of B-Opt 3 and B-Opt 4, resulting from a lower diversity, and therefore an allocation of solutions with smaller patch length.

Summarizing the results obtained with the curved structure, all optimizations converge to a similar final Pareto front, showing that the proposed approach is capable to solve the patch optimization problem. Furthermore, it could be seen that a higher mutation rate is likely to improve the ability of finding extreme points regarding stiffness maximization. However, the diversity of the solutions is still low, which results in a more asymmetrical heat map. Utilizing the heat maps, the best results could be achieved by incorporating a niching pressure and using the below limit replacement method. This was demonstrated by a more symmetrical picture in the heat map. The effect on the convergence rate was found to be small for the proposed configurations, except for the higher mutation rate, which creates a slightly higher convergence rate, represented by the dominated area. Based on this, a higher mutation rate is suggested if the finding of extreme solutions (maximum stiffness) has a higher preference. Additionally, a niching pressure and a below limit replacement are helpful to achieve a more balanced Pareto front and therefore more suitable heat maps.

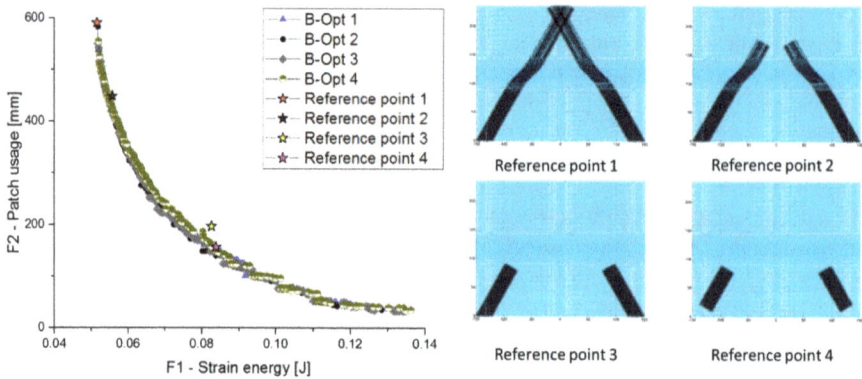

**Figure 12.** Comparison of the Pareto fronts obtained for the curved structure, and predefined reference points to classify the results.

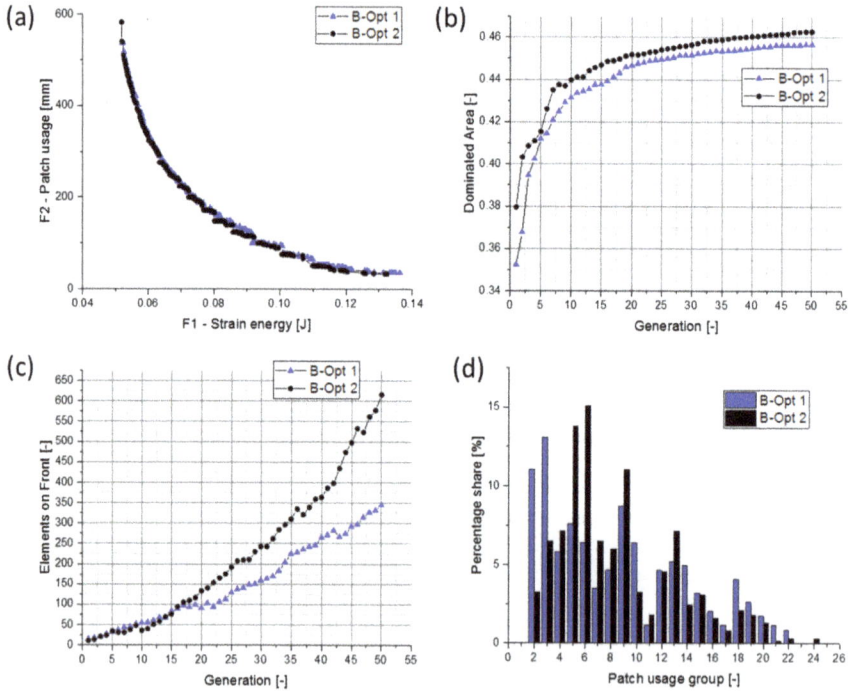

**Figure 13.** Comparison of the results for the variation of the mutation (B-Opt 1: low mutation rate, B-Opt 2: high mutation rate): (**a**) comparison of the final Pareto fronts, (**b**) the dominated area for the evaluation of the convergence behavior, (**c**) development of the elements on the current Pareto front for each generation, (**d**) percentage share of the solutions along the Pareto front for each length-group.

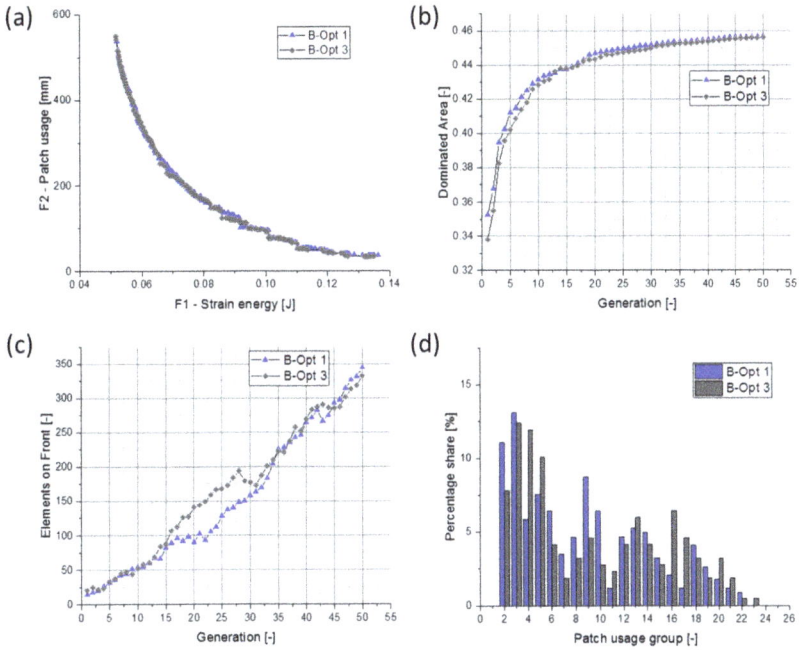

**Figure 14.** Comparison of the results for the variation of the replacement type (B-Opt 1: elitist replacement, B-Opt 3: below-limit replacement): (**a–d**) as in Figure 13.

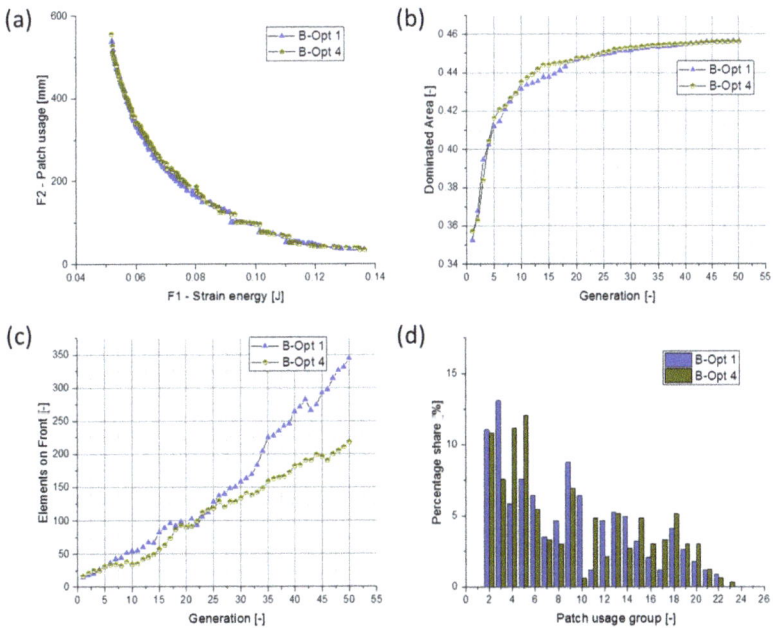

**Figure 15.** Comparison of the results for no-niching and niching (B-Opt 1: no-niching, B-Opt 4: niching): (**a–d**) as in Figure 13.

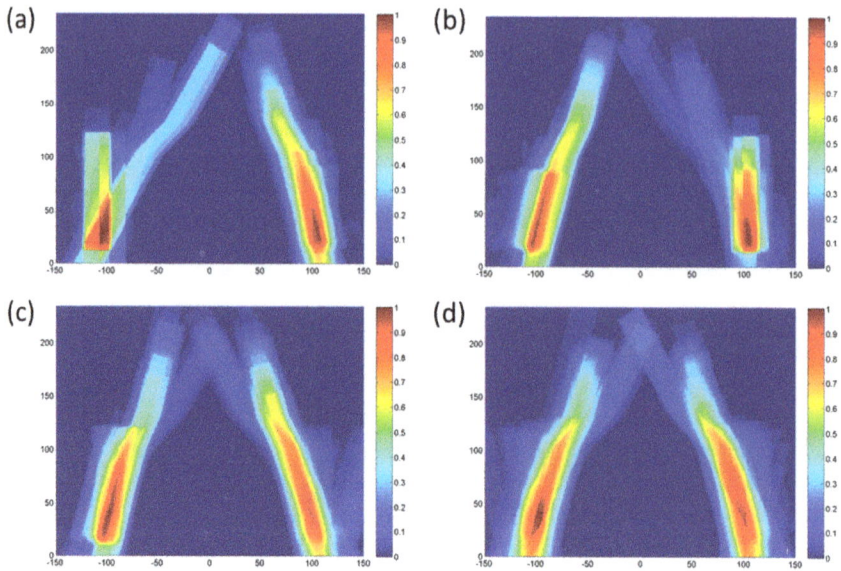

**Figure 16.** Top view of the generated heat-maps for the results obtained along the Pareto front (red: high patch concentration, blue: low patch concentration): (**a**) basic setup (B-Opt 1), (**b**) higher mutation rate (B-Opt 2), (**c**) change of replacement type (B-Opt 3), (**d**) niching pressure (B-Opt 4).

## 5. Conclusions

To optimize local continuous fiber reinforcements, the consideration of manufacturing constraints creates an essential benefit for the quality of the optimization results. By integrating draping simulation in the structural optimization workflow, more feasible optimization results can be achieved. The computed optimal size, position and orientation of the continuous fiber tapes can directly be implemented in the manufacturing process. For that purpose, the presented tape optimization workflow incorporates an efficient kinematic draping simulation method. Furthermore, the tape optimization algorithm is performed with multiple objectives to consider the minimal material consumption and the maximal structural stiffness simultaneously; therefore, a genetic algorithm with which a non-weighting approach could be realized. The resulting Pareto front, created by the multi-objective optimization procedure, is a helpful tool for the design engineer in the decision-making process. A heat-map, resulting from the Pareto front, has been introduced to further support the designer by visualizing areas that are important for patch reinforcement. Furthermore, the heat-maps are a useful tool, when interpreting the performance of the algorithm in terms of providing results with a wide diversity in the search space.

Comparing the influence of different settings for the genetic algorithm, the operators' crossover, mutation, replacement, and niching and their influence on the algorithm's performance have been demonstrated for two different use cases. Both examples show that for the evaluation of multi-objective optimization problems more than one performance criteria is necessary. Thus, convergence performance, number of found solutions, and the distribution within the search space are used to describe and compare the general behavior.

The first example was used to initially verify the general capability of the optimization algorithm and to compare two different crossover approaches. It could be demonstrated that the multi-point binary crossover converges fast and is likely to find a larger number of solutions on the Pareto front.

A curved structure was used a second example. Thereby, the influence of different mutation rate, replacement type, and niching pressure on the algorithm's performance was demonstrated. By comparing the Pareto fronts, it could be demonstrated that the proposed approach is capable of

*J. Compos. Sci.* **2018**, *2*, 22

accurately solving the patch optimization problem, since all optimization runs converge to a similar front. A higher mutation rate has shown a positive effect by finding extreme points regarding stiffness maximization. This is underrepresented within the heat map, since only a small proportion of solutions is in the extreme area. The niching pressure improved the diversity along the front significantly, which could be seen in a more pronounced symmetrical heat map. This underlines the importance of using more than one criteria for the evaluation of the optimization results, like a combination of Pareto front and heat map. Overall, a higher mutation rate is suggested when the finding of extreme solutions has a higher preference. Additionally, a niching pressure and a below-limit replacement are suggested when focusing on a more balanced distribution of the results along the front.

**Acknowledgments:** The research documented in this manuscript has been funded by the German Research Foundation (DFG) within the International Research Training Group "Integrated engineering of continuous-discontinuous long fiber reinforced polymer structures" (GRK 2078). The support by the German Research Foundation (DFG) is gratefully acknowledged. Additionally, the work is part of the research of the Young Investigator Group (YIG) "Tailored Composite Materials for Lightweight Vehicles", funded by the Vector Stiftung.

**Author Contributions:** Benedikt Fengler performed the major part of the work in terms of method development and implementation. He also wrote the first draft of the paper. Luise Kärger initiated the research subject, supervised the method development, supported the discussion of simulation results and thoroughly revised the paper. Andrew Hrymak supervised the method development and result discussion during a research stay at Western University. Frank Henning supervised the work in terms of composite process knowledge and relevance of the addressed subjects.

**Conflicts of Interest:** The authors declare no conflict of interest.

## References

1. Bücheler, D.; Trauth, A.; Damm, A.; Böhlke, T.; Henning, F.; Kärger, L.; Seelig, T.; Weidenmann, K.A. Processing of continuous-discontinuous-fiber-reinforced thermosets. In Proceedings of the SAMPE Europe Conference, Stuttgart, Germany, 14–16 November 2017.
2. Zhou, M.; Kemp, M.; Yancey, R.; Mestres, E.; Mouillet, J. Applications of Advanced Composite Simulation and Design Optimization. *Altair Eng.* **2011**.
3. Allaire, G.; Delgado, G. Stacking sequence and shape optimization of laminated composite plates via a level-set method. *J. Mech. Phys. Solids* **2016**, *97*, 168–196. [CrossRef]
4. Saitou, K.; Izui, K.; Nishiwaki, S.; Papalambros, P. A Survey of Structural Optimization in Mechanical Product Development. *J. Comput. Inf. Sci. Eng.* **2005**, *5*, 214–226. [CrossRef]
5. Huang, X.; Xie, Y.M. Topology optimization of nonlinear structures under displacement loading. *Eng. Struct.* **2008**, *30*, 2057–2068. [CrossRef]
6. Qian, C.C.; Harper, L.T.; Turner, T.A.; Warrior, N.A. Structural optimisation of random discontinuous fibre composites: Part 1—Methodology. *Compos. Part A Appl. Sci. Manuf.* **2015**, *68*, 406–416. [CrossRef]
7. Qian, C.C.; Harper, L.T.; Turner, T.A.; Warrior, N.A. Structural optimisation of random discontinuous fibre composites: Part 2—Case study. *Compos. Part A Appl. Sci. Manuf.* **2015**, *68*, 417–424. [CrossRef]
8. Harzheim, L. *Strukturoptimierung: Grundlagen und Anwendungen*, 2nd ed.; Europa-Lehrmittel: Haan-Gruiten, Germany, 2014.
9. Marler, R.T.; Arora, J.S. The weighted sum method for multi-objective optimization: New insights. *Struct. Multidiscip. Optim.* **2010**, *41*, 853–862. [CrossRef]
10. Zuo, K.; Chen, L.; Zhang, Y.; Yang, J. Manufacturing- and machining-based topology optimization. *Int. J. Adv. Manuf. Tech.* **2006**, *27*, 531–536. [CrossRef]
11. Sørensen, S.N.; Lund, E. Topology and thickness optimization of laminated composites including manufacturing constraints. *Struct. Multidiscip. Optim.* **2013**, *48*, 249–265. [CrossRef]
12. Zein, S.; Madhavan, V.; Dumas, D.; Ravier, L.; Yague, I. From stacking sequences to ply layouts: An algorithm to design manufacturable composite structures. *Compos. Struct.* **2016**, *141*, 32–38. [CrossRef]
13. Zhou, M.; Fleury, R.; Shyy, Y.; Thomas, H.; Brennan, J. Progress in Topology Optimization with Manufacturing Constraints. In Proceedings of the 9th AIAA/ISSMO Symposium on Multidisciplinary Analysis and Optimization, Atlanta, GA, USA, 4–6 September 2002.

14. Hohberg, M.; Kärger, L.; Henning, F.; Hrymak, A. Process Simulation of Sheet Molding Compound (SMC) as key for the integrated Simulation Chain. In Proceedings of the Simulation of Composites—Ready for Industry 4.0? Hamburg, Germany, 24–27 October 2016; pp. 61–67.

15. Kärger, L.; Bernath, A.; Fritz, F.; Galkin, S.; Magagnato, D.; Oeckerath, A.; Schön, A.; Henning, F. Development and validation of a CAE chain for unidirectional fibre reinforced composite components. *Compos. Struct.* **2015**, *132*, 350–358. [CrossRef]

16. Kaufmann, M.; Zenkert, D.; Åkermo, M. Cost/weight optimization of composite prepreg structures for best draping strategy. *Compos. Part A Appl. Sci. Manuf.* **2010**, *41*, 464–472. [CrossRef]

17. Kärger, L.; Galkin, S.; Dörr, D.; Schirmaier, F.; Oeckerath, A.; Wolf, K. Continuous CAE chain for composite design, established on an HPC system and accessible via web-based user-interfaces. In Proceedings of the NAFEMS World Congress, Stockholm, Sweden, 11–14 June 2017.

18. Kärger, L.; Galkin, S.; Zimmerling, C.; Dörr, D.; Linden, J.; Oeckerath, A.; Wolf, K. Forming optimisation embedded in a CAE chain to assess and enhance the structural performance of composite components. *Compos. Struct.* **2018**. [CrossRef]

19. Bulla, M.; Beauchense, E.; Ehrhart, F.; Trickov, V. A holistic simulation driven composite design process. *Proc. NAFEMS Semin. Simul.* **2014**, in press.

20. Zehnder, N.; Ermanni, P. Optimizing the shape and placement of patches of reinforcement fibers. *Compos. Struct.* **2007**, *77*, 1–9. [CrossRef]

21. Zehnder, N.; Ermanni, P. A methodology for the global optimization of laminated composite structures. *Compos. Struct.* **2005**, *72*, 311–320. [CrossRef]

22. Mathias, J.; Balandraud, X.; Grediac, M. Applying a genetic algorithm to the optimization of composite patches. *Comput. Struct.* **2006**, *84*, 823–834. [CrossRef]

23. Mejlej, V.G.; Falkenberg, P.; Türck, E.; Vietor, T. Optimization of Variable Stiffness Composites in Automated Fiber Placement Process using Evolutionary Algorithms. *Procedia CIRP* **2017**, *66*, 79–84. [CrossRef]

24. Jansson, N.; Wakeman, W.D.; Månson, J.E. Optimization of hybrid thermoplastic composite structures using surrogate models and genetic algorithms. *Compos. Struct.* **2007**, *80*, 21–31. [CrossRef]

25. Skordos, A.A.; Sutcliffe, M.P.F.; Klintworth, J.W.; Adolfsson, P. Multi-Objective Optimisation of Woven Composite Draping using Genetic Algorithms. In Proceedings of the 27th International Conference SAMPE Europe, Paris, France, 27–29 March 2006.

26. Chen, S.; Harper, L.T.; Endruweit, A.; Warrior, N.A. Formability optimisation of fabric preforms by controlling material draw-in through in-plane constraints. *Compos. Part A Appl. Sci. Manuf.* **2015**, *76*, 10–19. [CrossRef]

27. Lim, T.; Ramakrishna, S. Modelling of composite sheet forming: A review. *Compos. Part A Appl. Sci. Manuf.* **2002**, *33*, 515–537. [CrossRef]

28. Spencer, A.J.M. Theory of fabric-reinforced viscous fluids. *Compos. Part A Appl. Sci. Manuf.* **2000**, *31*, 1311–1321. [CrossRef]

29. Boisse, P.; Aimène, Y.; Dogui, A.; Dridi, S.; Gatouillat, S.; Hamila, N.; Khan, M.A.; Mabrouki, T.; Morestin, F.; Vidal-Sallé, E. Hypoelastic, hyperelastic, discrete and semi-discrete approaches for textile composite reinforcement forming. *Int. J. Mater. Form.* **2010**, *3*, 1229–1240. [CrossRef]

30. Dörr, D.; Schirmaier, F.J.; Henning, F.; Kärger, L. A viscoelastic approach for modeling bending behavior in finite element forming simulation of continuously fiber reinforced composites. *Compos. Part A Appl. Sci. Manuf.* **2017**, *94*, 113–123. [CrossRef]

31. Schirmaier, F.J.; Dörr, D.; Henning, F.; Kärger, L. A macroscopic approach to simulate the forming behaviour of stitched unidirectional non-crimp fabrics (UD-NCF). *Compos. Part A Appl. Sci. Manuf.* **2017**, *102*, 322–335. [CrossRef]

32. Sharma, S.B.; Sutcliffe, M.P.F. Draping of woven fabrics: Progressive drape model. *Plast. Rubber Compos.* **2013**, *32*, 57–64. [CrossRef]

33. Yang, B.; Jin, T.; Bi, F.; Li, J. A Geometry Information Based Fishnet Algorithm for Woven Fabric Draping in Liquid Composite Molding. *Mater. Sci.* **2014**, *20*, 513–521. [CrossRef]

34. Smiley, A.J.; Pipes, R.B. Analysis of the Diaphragm Forming of Continuous Fiber Reinforced Thermoplastics. *J. Thermoplast. Compos. Mater.* **1988**, *1*, 298–321. [CrossRef]

35. Long, A.C.; Rudd, C.D. A simulation of reinforcement deformation during the production of preforms for liquid moulding processes. *Proc. Inst. Mech. Eng. Part B J. Eng. Manuf.* **1994**, *208*, 269–278. [CrossRef]

36.  Pickett, A.K.; Creech, G.; Luca, P.D. Simplified and advanced simulation methods for prediction of fabric draping. *Rev. Eur. Elem.* **2012**, *14*, 677–691. [CrossRef]
37.  Van West, B.P.; Pipes, R.B.; Keefe, M. A Simulation of the Draping of Bidirectional Fabrics over Arbitrary Surfaces. *J. Text. Inst.* **1990**, *81*, 448–460. [CrossRef]
38.  Vanclooster, K.; Lomov, S.V.; Verpoest, I. Simulating and validating the draping of woven fiber reinforced polymers. *Int. J. Mater. Form.* **2008**, *1*, 961–964. [CrossRef]
39.  Bäck, T. *Evolutionary Algorithms in Theory and Practice: Evolution Strategies, Evolutionary Programming, Genetic Algorithms*; Oxford University Press: Oxford, UK, 1996.
40.  Coello, C.A. A Short Tutorial on Evolutionary Multiobjective Optimization. In *International Conference on Evolutionary Multi-Criterion Optimization*; Springer: Berlin/Heidelberg, Germany, 2001; pp. 21–40.
41.  Giger, M. Representation Concepts in Evolutionary Algorithm-Based Structural Optimization. Ph.D. Thesis, Zürich, Switzerland, 2007.
42.  Adams, B.M.; Bauman, L.E.; Bohnhoff, W.J.; Dalbey, K.R.; Ebeida, M.S.; Eddy, J.P.; Eldred, M.S.; Hough, P.D.; Hu, K.T.; Jakeman, J.D.; et al. Dakota, A Multilevel Parallel Object-Oriented Framework for Design Optimization, Parameter Estimation, Uncertainty Quantification, and Sensitivity Analysis: Version 6.0 User's Manual. July 2014. Updated November 2015 (Version 6.3). Sandia Technical Report SAND2014-4633. Available online: https://dakota.sandia.gov/content/citing-dakota (accessed on 13 March 2017).

Journal of
*Composites Science*

MDPI

*Article*

# Prediction of Young's Modulus for Injection Molded Long Fiber Reinforced Thermoplastics

Hongyu Chen [1] and Donald G. Baird [1,2,*]

[1]  Department of Chemical Engineering, Virginia Tech, VA 24061, USA; hongyu86@vt.edu
[2]  Macromolecules Innovation Institute, Virginia Tech, VA 24061, USA
*  Correspondence: dbaird@vt.edu; Tel.: +1-540-231-5998

Received: 26 June 2018; Accepted: 2 August 2018; Published: 6 August 2018

**Abstract:** In this article, the elastic properties of long-fiber injection-molded thermoplastics (LFTs) are investigated by micro-mechanical approaches including the Halpin-Tsai (HT) model and the Mori-Tanaka model based on Eshelby's equivalent inclusion (EMT). In the modeling, the elastic properties are calculated by the fiber content, fiber length, and fiber orientation. Several closure approximations for the fourth-order fiber orientation tensor are evaluated by comparing the as-calculated elastic stiffness with that from the original experimental fourth-order tensor. An empirical model was developed to correct the fibers' aspect ratio in the computation for the actual as-formed LFTs with fiber bundles under high fiber content. After the correction, the analytical predictions had good agreement with the experimental stiffness values from tensile tests on the LFTs. Our analysis shows that it is essential to incorporate the effect of the presence of fiber bundles to accurately predict the composite properties. This work involved the use of experimental values of fiber orientation and serves as the basis for computing part stiffness as a function of mold filling conditions. The work also explains why the modulus tends to level off with fiber concentration.

**Keywords:** long glass fibers; elastic properties; fiber orientation; fiber length; fiber bundles

## 1. Introduction

During the last decade, an enhanced demand for lightweight materials in automotive applications has resulted in the growth of the use of thermoplastic-discontinuous fiber composites [1]. The increased growth of the use of these thermoplastic matrix composite systems is due to the combination of mechanical properties and melt processability. Long-fiber (lengths > 1 mm) thermoplastic composites (LFTs) possess significant advantages over short fiber (<1 mm) composites in terms of their mechanical properties while retaining their ability to be injection molded [2]. The goal of this research is to improve the stiffness properties predictions for injection molded LFTs. During the plasticating stage of injection molding, significant fiber attrition will occur leading to a broad fiber length distribution (FLD) [3]. Fiber orientation distribution (FOD) is another highly anisotropic feature of the final injection molded parts induced by the mold filling process [4]. Mechanical properties of LFTs are highly dependent on these microstructural variables imparted by the injection molding process [5].

The computation of the elastic stiffness for the aligned and monodispersed short fiber composites was well studied by a large range of people. Tucker et al. [6] reviewed the micromechanical models for this type of composite. By comparing the standard micromechanical models with their finite element method, the authors have shown that the Halpin-Tsai equation gives reasonable estimates for stiffness, but the best predictions come from the Mori-Tanaka model based on the Eshelby's equivalent inclusion method. A similar conclusion was reached by Hine [7]. Ingber and Papathanasiou [8] found the variant of the Halpin-Tsai model is in very good agreement with their boundary element method (numerical simulation) for their entire range of fiber volume contents and aspect ratios, although they had no experimental data for comparison. These models for aligned and monodispersed fiber composites can predict the properties of a representative volume element that can subsequently be averaged to include effects of fiber length and orientation distributions of a real injection molded material. Hine et al. [7] carried out a numerical simulation using a distribution of fiber lengths generated by the Monte-Carlo technique. Garesci et al. [5] applied the fitted probability density function to get the averaged property from each single fiber length. A relatively concise way is to replace the FLD with some sort of mean fiber length [7,9,10]. The most widely used orientation-averaging scheme is to use second and/or fourth order orientation tensors developed by Tucker and Advani [4] to average the property constants. Hine et al. [11] have shown that the results determined by the constant strain orientation averaging method (assuming the units have the same strain and average their stiffness constants) were in good agreement with their numerical simulations.

There exists very little modeling work for predicting stiffness properties on injection molded LFTs [5]. The scenario of the LFTs is different from the works that studied short fiber composite. LFTs injection molding pellets prepared by the pultrusion technique have received much attention. In particular, the fibers are in the form of aligned fiber bundles coated by thermoplastic matrix, which produces pellets with much higher fiber contents than those of more conventional 'short-fiber' compounds. During the compounding process, filamentization of fiber bundles and fracture of the resultant monofilaments into elements of a lower aspect ratio lead to the dispersion of fibers into the polymer matrix [12]. For high content fiber composites, the presence of fiber bundles seems to be unavoidable which could result in a reinforcement with a much lower aspect ratio and effective stiffness than well-dispersed fibers, consequently giving a lower and even decreasing stiffness [13]. However, the existing stiffness models assume the fibers are fully and evenly dispersed in the matrix. The predicted values keep increasing with the fiber content, which is not true within commercial fiber concentration for injection molded LFTs [13,14]. In this paper, we report on the development of an empirical model to correct the fibers' aspect ratio for the actual as-formed LFTs with fiber bundles under high fiber content. After the correction, the analytical predictions show good agreement with the experimental stiffness values from tensile tests on the LFTs for the whole fiber content range investigated. Our analysis shows that it is essential to incorporate the effect of the presence of fiber bundles to accurately predict the composite properties.

## 2. Analytical Modelling Details

### 2.1. Orientation

A single rigid fiber can be represented by a unit vector, $\mathbf{p}$, parallel to the fiber's long axis. The average orientation state of a fiber composite can be descried by even-ordered structural tensors [4]. In this study, the second and fourth order orientation used in the modeling are defined as:

$$\mathbf{A} = \langle \mathbf{pp} \rangle = \int \mathbf{pp} \psi(\mathbf{p}) \, d\mathbf{p} \tag{1}$$

$$\mathbf{A}_4 = \langle \mathbf{pppp} \rangle = \int \mathbf{pppp} \psi(\mathbf{p}) \, d\mathbf{p} \tag{2}$$

where $\psi(\mathbf{p})$ is the probability distribution function for orientation, and the bracket $\langle \cdot \rangle$ denotes the average quantity over a volume domain. The second order orientation tensor, $\mathbf{A}$, can be measured, while the fourth order orientation tensor, $\mathbf{A}_4$, can either be obtained from experiments or estimated in terms of $\mathbf{A}$ using various closure approximation methods [4]. In this study, several approximation closures are implemented to calculate $\mathbf{A}_4$, and the stiffness results evaluated from these approximations are compared with that obtained from the experimentally measured $\mathbf{A}_4$.

## 2.2. Fiber Length

Due to the compounding process, the injection molded LFTs will end up with a very broad fiber length distribution. The actual fiber length information can be described by the experimentally obtained probability of finding a fiber with length $l_i$ given by:

$$pl(l_i) = \frac{N_i}{\sum N_i} \qquad (3)$$

where, $N_i$ is the measured number of fibers with length $l_i$. A more concise approach is to replace the FLD with a single length, normally the number or weight average length defined, respectively, as:

$$L_n = \frac{\sum N_i l_i}{\sum N_i} \qquad (4)$$

$$L_w = \frac{\sum N_i l_i^2}{\sum N_i l_i} \qquad (5)$$

## 2.3. Elastic Properties

As mentioned in the introduction, both the Halpin-Tsai (HT) and Eshelby-Mori-Tanaka (EMT) methods are extensively studied for the unidirectional, or short fiber reinforced composites. The Halpin-Tsai method is the most widely used micromechanical model because of its ease of implementation [15,16]. The corresponding equations are derived from the self-consistent ideas of Hill [17], and the final implementation is semi-empirical in nature. Several authors have concluded that a constant strain assumption works better than the constant stress assumption, that is to say, stiffness averaging surpasses the compliance averaging in the computation [11,18]. So in this article, the HT equations are used to predict the compliance matrix, and then the stiffness matrix is obtained from the inverse of the compliance matrix. Finally, the material property is calculated by averaging the stiffness constants based on FLD and FOD.

In this study, the EMT method is also used and compared with the HT method. The EMT method is a combination of Eshelby's equivalent inclusion method and Mori-Tanaka's back stress analysis, and so, this model is valid even for large volume fraction of fibers. In particular, the equivalent inclusion method of Eshelby is applied in the computation of energy-release rate in terms of the equivalent eigenstrains defined in the fiber and crack [19]. As a result, the EMT method provides the overall stiffness of the composite weakened by fiber-end cracks.

To include the FLD effect on the elastic properties, we use the EMT or HT method to calculate the stiffness of a 'reference' unidirectional fiber composite using either the experimental fiber length probability $pl(l_i)$:

$$C_{ijkl} = \frac{\sum C_{ijkl}^*(l_i/d) * pl(l_i)}{\sum pl(l_i)} \qquad (6)$$

or using an average fiber length $L_{avg}$ ($L_n$ or $L_w$) evaluated from the FLD:

$$C_{ijkl} = C_{ijkl}^* (L_{avg}/d) \qquad (7)$$

where $C_{ijkl}$ is the stiffness matrix having a specific fiber aspect ratio $l_i/d$ or $L_{avg}/d$, and $d$ is the single fiber diameter.

To get the stiffness of the actual injection molded LFTs, a mean tensor averaging procedure is used [20]. Specifically, the stiffness of the calculated unidirectional fiber composite including the FLD effect is averaged over the as-formed fiber orientation state as follow:

$$\overline{C}_{ijkl} = B_1 A_{ijkl} + B_2\left(A_{ij}\delta_{kl} + A_{ij}\delta_{kl}\right) + B_3\left(A_{ik}\delta_{jl} + A_{il}\delta_{jk} + A_{jl}\delta_{ik} + A_{jk}\delta_{il}\right) \tag{8}$$
$$+B_4\delta_{ij}\delta_{kl} + +B_5\left(\delta_{ik}\delta_{jl} + \delta_{il}\delta_{jk}\right)$$

where $\delta_{ij}$ denotes the Kronecker delta and the scalar parameters $B_i$ ($i = 1$–5) are the invariants of the stiffness tensor of the calculated unidirectional fiber composite given as:

$$\begin{aligned} B_1 &= C_{11} + C_{22} - 2 \times C_{12} - 4 \times C_{66} \\ B_2 &= C_{12} - C_{23} \\ B_3 &= C_{66} + \tfrac{1}{2}(C_{23} - C_{22}) \\ B_4 &= C_{23} \\ B_5 &= \tfrac{1}{2}(C_{22} - C_{23}) \end{aligned} \tag{9}$$

The composite specimen can be considered as a stacking sequence of thin layers which might have a different fiber orientation state. The stiffness of each layer is calculated based on the overall FLD and the characterized orientation state using the above-mentioned method. Finally, the classical lamination theory [21] is then applied to calculate the overall effective engineering stiffness of the composite [20].

## 3. Materials and Methods

The composites under investigation were 10 wt%, 30 wt%, 40 wt%, and 50 wt% glass fibers in a polypropylene matrix. The material was received from SABIC Innovative Plastics (Ottawa, IL, USA) as 12.5 mm long pellets created through a pultrusion process in 30 wt% and 50 wt% formulations. Samples with 10 wt% fibers were diluted with neat polypropylene, while 30 wt% and 50 wt% pellets were used to create the 40 wt% composites. The pellets contain a unidirectional bundle of fibers that must be dispersed during the injection molding process, specifically in the plasticating unit. Center-gated disk geometries were formed by injection molding as shown in Figure 1. In this study, the Hele-Shaw region (60% disk radius) and the advancing front region (85% disk radius) were investigated for fiber length, fiber orientation, and mechanical properties. Tensile specimens were cut from the injection-molded samples, and the young's modulus was measured according to ASTM D3039 [22] for polymer matrix composite materials.

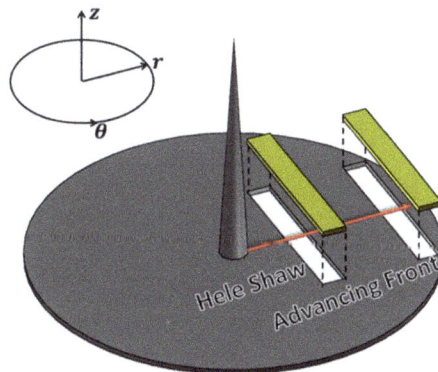

**Figure 1.** The injection-molded glass/ppcenter-gated disk: The Hele-Shaw region (60% disk radius) and advancing front region (85% disk radius) were investigated in this study.

For fiber length measurement, a method based on Reference [23] was modified as follows: Instead of directly injecting epoxy into the samples after the burning off of the polymer matrix, a needle coated with epoxy was inserted into the sample all the way through the thickness direction shown in Figure 2a. As a result, fibers at this location were collected by the epoxy on the needle as shown in Figure 2b. Due to this sampling method, there was biased toward longer fibers, and the measured fiber length was then corrected based on the length of each fiber and the diameter of the needle with epoxy coating. The needle with the fibers was re-burned to get rid of the epoxy. Loose glass fibers were then dispersed on a desktop scanner and imaged at 3200 dpi. At least 3000 fibers per sample were measured using our in-house developed MATLAB® codes. The fiber length distribution followed the typical log-normal distribution commonly observed for fiber composites. Three samples were used to produce the averages in Table 1. The number average ($L_n$) and weight average ($L_w$) fiber lengths were calculated according to Equations (4) and (5), respectively. It can be seen that the average fiber lengths in these samples are reduced with increasing fiber content.

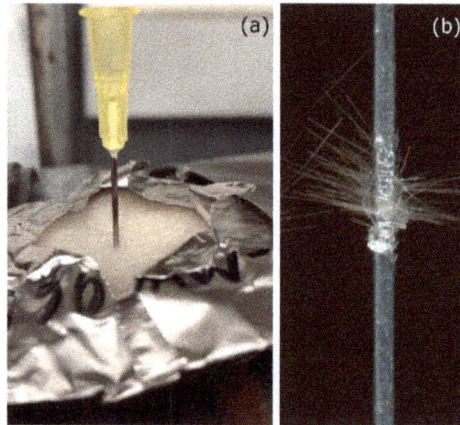

**Figure 2.** The modified fibers sampling method: (**a**) A needle coated with epoxy inserted into the desired location, and (**b**) the pulled out needle with the fibers attached on the surface of the epoxy.

**Table 1.** Fiber lengths information.

| Fiber Content | Hele-Shaw Region | | Advancing-Front Region | |
|---|---|---|---|---|
| | $L_n$ (mm) | $L_w$ (mm) | $L_n$ (mm) | $L_w$ (mm) |
| 10 wt% ($v_f = 0.038$) | $1.51 \pm 0.081$ | $3.59 \pm 0.71$ | $1.76 \pm 0.12$ | $3.21 \pm 0.37$ |
| 30 wt% ($v_f = 0.135$) | $1.14 \pm 0.078$ | $3.41 \pm 0.41$ | $1.32 \pm 0.063$ | $3.52 \pm 0.26$ |
| 40 wt% ($v_f = 0.197$) | $0.98 \pm 0.080$ | $2.67 \pm 0.28$ | $1.03 \pm 0.086$ | $2.81 \pm 0.35$ |
| 50 wt% ($v_f = 0.268$) | $0.87 \pm 0.061$ | $2.42 \pm 0.32$ | $0.882 \pm 0.062$ | $2.54 \pm 0.21$ |

Measurements of fiber orientation were also made to further investigate its relationship to the stiffness performance at the same locations. Orientation measurements were taken along the *r*-*z* plane, such that *r* denotes the flow direction with the velocity gradient in *z*. Samples were polished using modified metallographic techniques and oxygen plasma etched to enhance the contrast of the glass fiber and polypropylene matrix. Details of the sample preparation and orientation measurement procedure can be found in the References [24,25]. Figure 3 shows the measured through thickness fiber orientation for various glass fiber concentrations at the Hele-Shaw region. For the fiber orientation tensor, **A**, the diagonal components are the most important. They describe the alignment of the population with respect to the axis of the coordinate system. A value that approaches one indicates increased alignment in that direction. Only the θ direction component is presented here, because the young's modulus was measured along this transverse direction. At 30 wt%, 40 wt%, and 50 wt%, the through thickness fiber orientation distributions are very similar showing the characteristic shell–core–shell layer structures. Generally, the θ direction component reaches its largest value near the center of the disk, because that the center of the disk is dominated by extensional flow in this tangential direction. However, at 10 wt%, the distribution deviates significantly from the rest. The through thickness fiber orientation distribution is relatively 'flat' compared to those with higher fiber concentrations. This might be due to the concentration effects on the fiber orientation dynamics. At 10 wt%, the degree of fiber-fiber interaction is much less, that is to say, hindrance to fiber alignment is much less, as a result, the tangential direction alignment is quite dominant.

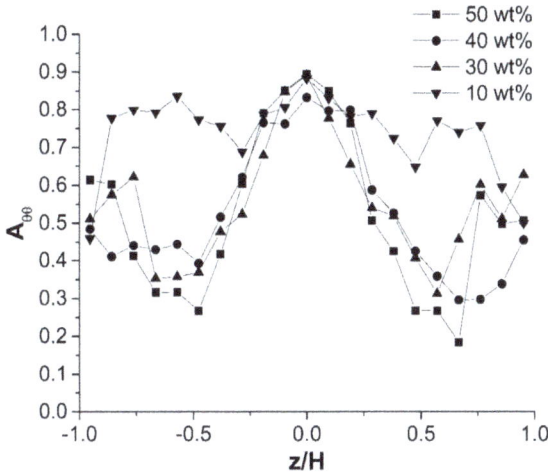

**Figure 3.** Measured θ direction fiber orientation distributions through the thickness direction for 10 wt%, 30 wt%, 40 wt% and 50 wt% glass fiber at the Hele-Shaw region.

## 4. Results and Discussion

In all the calculations in this article, the measured FLD or the corresponding average fiber length ($L_n$ or $L_w$) was input into Equations (6) or (7), while the experimental second-order tensors were used in Equation (8). To analyze the accuracy of various closure approximations for predicting the elastic properties, the fourth order orientation tensors were evaluated by the linear (LIN), quadratic (QUA), hybrid (HYB), Invariant-based optimal fitting (IBOF), and improved orthotropic (ORW3) closure approximations [4,18,26,27]. Then, the as-calculated elastic stiffness using each of these estimated fourth order tensors was compared with that using the original experimental 'true' fourth-order tensor (TRU) obtained from Equation (2). The comparisons of the effective engineering modulus along the tangential direction are presented in Figures 4 and 5 applying the methods of EMT and HT, respectively. The predicted results and the general pattern with the HT and EMT methods in the studied fiber content range are very similar. Both models show a similar linear increase in the transverse modulus with increasing fiber content. However, all the values calculated from the EMT model, no matter what length parameter and closure approximation, are slightly greater than those from the HT model. Moreover, the differences between the two methods become more notable as fiber content increases. In Tucker and Liang's [6] review of the stiffness predictions for unidirectional fiber composites, for composites with an aspect ratio larger than 10, the EMT also has predicted greater values of the dominant modulus than the HT model. To answer the question which closure method or methods are the best for stiffness prediction purposes, the results calculated using the experimental fourth order orientation tensors are used as criteria. It seems that, for the entire fiber content range and all the scenarios using different fiber length parameters (FLD, $L_n$, and $L_w$), the magnitudes of IBOF and

ORW3 predictions are the most comparable to the criteria. Another aspect of this paper is to examine the effects of the length parameters (FLD, $L_n$, and $L_w$) on the stiffness predictions for injection molded LFTs. It is seen that, the predictions of the $L_n$ parameters lie between the largest values generated by the $L_w$ parameters and the smallest predictions from the measured FLD. This result indicates that, for the purpose of replacing the FLD by an average fiber length in the computation, the $L_n$ might outperform the $L_w$ in terms of generating a better match with the FLD.

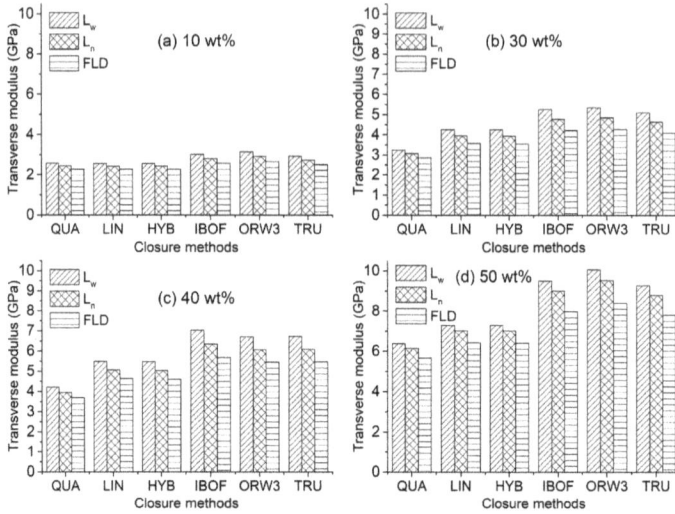

**Figure 4.** Predicted transverse modulus at the Hele-Shaw region using Eshelby-Mori-Tanaka (EMT) model for (**a**) 10 wt%, (**b**) 30 wt%, (**c**) 40 wt%, and (**d**) 50 wt% glass fiber polypropylene composites. Various closure approximations and length parameters were used in the calculations.

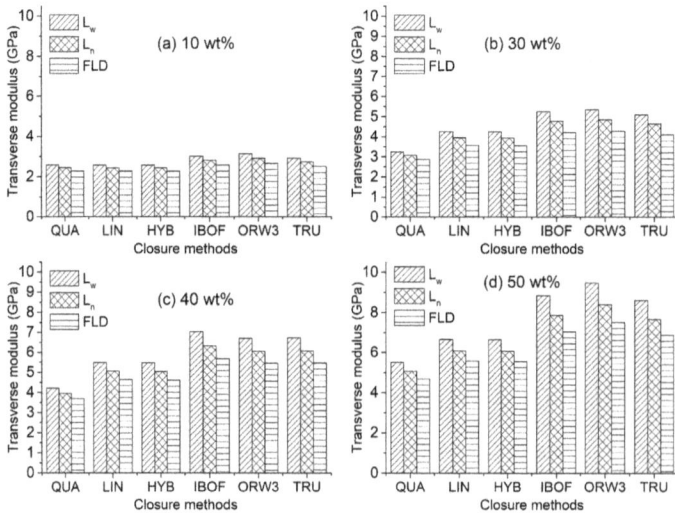

**Figure 5.** Predicted transverse modulus at the Hele-Shaw region using Halpin-Tsai (HT) model for (**a**) 10 wt%, (**b**) 30 wt%, (**c**) 40 wt%, and (**d**) 50 wt% glass fiber polypropylene composites. Various closure approximations and length parameters were used in the calculations.

The predictions are also compared with the transverse (θ direction) tensile test results shown in Figure 6. At most locations through the thickness, the fibers are predominantly oriented toward this direction as shown in Figure 3. Here, both HT and EMT models are applied with the measured FLD and experimental fourth order orientation tensor. Only at the lowest 10 wt% ($v_f = 0.0385$) fiber concentration do the predictions match well with the experimental results. As concentration increases, the deviations of the predictions from the experimental data turn out to be more significant. Several authors [13,28] experimentally observed that any incremental increase in fiber content appears to bring a lower improvement in properties than the previous one. That is to say, the mechanical performance of the injection molded LFTs will reach a plateau or even decrease at very high fiber concentration range. There are several possible reasons for the degradation of the mechanical properties. First, due to the non-homogeneous nature of these materials, problems can arise during their manufacture, which result in void content/porosity in the final parts [29]. Second, there is also a possibility of poor adhesion between the glass fibers and the matrix [30]. Finally, at higher fiber content, fiber bundles are very common in the injection molded LFTs [13]. There are two major effects of the presence of fiber bundles on the mechanical performance of the composites. First, the clumping of fibers will reduce the effective fiber aspect ratios in the reinforcement. Second, fiber bundles have an effect on stress concentration. Specifically, the failed fiber will induce stress concentration in those un-failed neighbor fibers within the bundles. This stress concentration occurs during the nonlinear stage of the tensile test [19,31]. Therefore, it is valid to ignore stress concentration and exclusively consider the effects of reduced aspect ratio on modulus. In this study, the density of the 50 wt% injection molded samples was measured by the pycnometry method described in Reference [32]. The density given by the supplier is 1.33 g/mm³ and the measured value of the injection molded center-gated-disk (CGD) is 1.327 ± 0.0175 g/mm³, which means the void content/porosity in the final parts is negligible. There is no information about the adhesion between the glass fibers and the matrix from the supplier. In this study, we assume the adhesion is perfect to simplify the problem, which most likely not be true. However, the 10 wt%, 30 wt%, 40 wt%, and 50 wt% materials have the same surface treatments for the fibers (they are the same series using the same formulations). Therefore, it is legitimate to only include the effects of the clumping of fibers on the level-off of the elastic properties of the injection molded LFTs as fiber content increases.

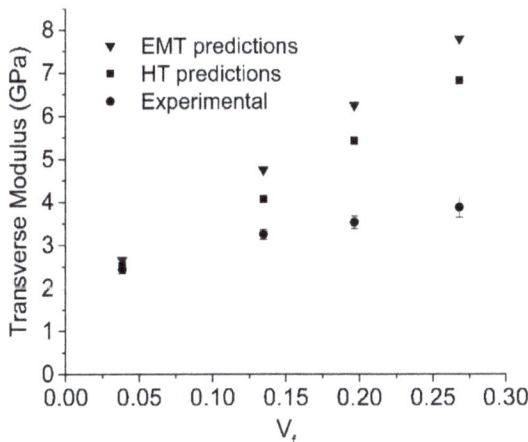

**Figure 6.** A comparison between the predictions using both HT and EMT methods, and the experimental tensile test results at various fiber content.

The cross-sectional microscopic images on the *r-z* plane with glass fiber foot-prints are shown in Figure 7. It is seen that the clumping of fibers turned out to be worse as fiber concentration increased. To include the effects of reduced aspect ratio on the modulus due to the existence of fiber bundles, an empirical model was proposed to modify the effective fiber bundle diameter $d_c$ in the stiffness computation.

$$d_{effective} = d_0 \times d_c (l, v_f)$$

$$d_c = 1 + \frac{\left(\frac{\sqrt{a_r}}{\exp\left(\frac{v_c}{v_f}\right)}\right)^n}{1+\exp\left(-v_f/(1/a_r)\right)}$$

$$a_r = \frac{l}{d_0}$$

where, $d_{effective}$ is the effective bundle diameter, $d_0$ the single fiber diameter, $d_c$ a correction coefficient, $a_r$ the fiber aspect ratio, $l$ the fiber length, and $v_f$ the fiber volume fraction. There are two empirical parameters, which need to be determined: $v_c$ is a critical volume fraction, and $n$ is an exponent index. Both $v_c$ and $n$ can determine the slope and upper boundary of this empirical function. We believe the correcting coefficient $d_c$ is a function of both fiber volume content and fiber length. At very low concentration (dilute concentration), the fibers have a much less chance to contact with each other and form bundles. The value of $d_c$ should approach 1.0 at low fiber volume content. In addition, $d_c$ should also keep increasing with $v_f$ until reach an upper boundary. So, we modified the form of the logistic function or logistic curve ('S' shape) and proposed our empirical model in the form of Equation (10) [33].

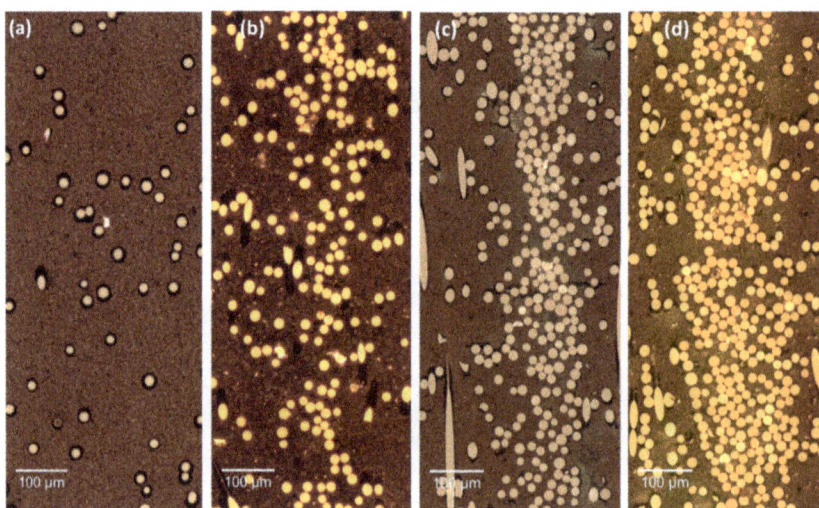

**Figure 7.** Cross-sectional microscopic images at the *r-z* plane for (**a**) 10 wt%, (**b**) 30 wt%, (**c**) 40 wt%, and (**d**) 50 wt% glass fiber polypropylene composites.

This empirical model was used to correct the bundles' size and fit both HT and EMT models to the tensile test results in the Hele-Shaw region with the non-linear least squares fitting method. The measured FLD and experimental fourth order orientation tensor were also used in the calculation. The empirical parameters of the $d_c$ obtained by the fitting of both HT and EMT models are shown in Table 2. The comparisons of the fitted results with experiments at the Hele-Shaw region is shown in Figure 8a. After the application of the empirical model, the predictions turns out to be much more accurate when compared with tensile test results. However, this model might over-predict the bundles' size, because the perfect adhesion between glass fibers and matrix are assumed which might not be true. The empirical parameters obtained from the Hele-Shaw region were applied to calculate the modulus at the advancing-front. The comparisons among the as-calculated predictions, corrected predictions, and the experimental results are shown in Figure 8b. The predictions also show significant improvement after the diameter correction.

**Table 2.** Fitted parameters of the empirical model.

| Model | $v_c$ | $n$ |
|-------|-------|-----|
| EMT | 0.15 | 1.47 |
| HT | 0.21 | 1.90 |

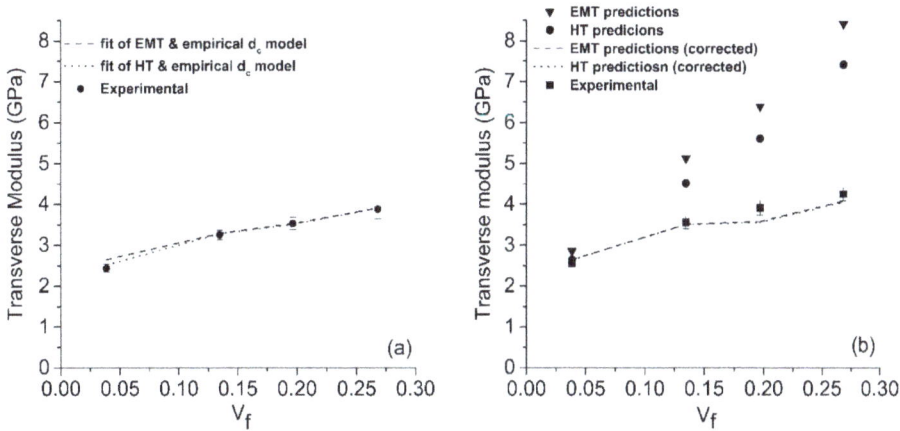

**Figure 8.** A comparison between the fit of EMT and HT with the experimental tensile test results with various fiber content, at the (**a**) Hele-Shaw region. Comparisons among the as-calculated predictions, the corrected predictions using the fitted parameters from Hele-Shaw region and experimental data at (**b**) the advancing front region.

Figure 9 shows the fiber bundles' size as a function of fiber length at various fiber concentration using Equation (10). The two empirical parameters were obtained by fitting the predictions from the EMT method to experimental tensile data. At low concentration (10 wt%), the magnitude of $d_c$ barely increased in the length range from 0.06 mm to 12 mm. The slopes of the lines increase notably with concentration. At higher concentration (30, 40, 50 wt%) the values of $d_c$ show a rapid increase with fiber length. The trends of $d_c$ can be explained qualitatively from the fiber breakage aspect. At higher fiber content, the dominant fiber breakage mechanisms are fiber–fiber and fiber–machine interactions [34]. The residual fiber length exhibited a linear decrease following an increase in the fiber content [14,35]. Several authors also suggested that the fiber breakage rate was proportional to the fiber length or fiber aspect ratio [12,36]. The calculated bundle size, $d_c$, for those longer fibers is very large, especially for the 50 wt% fiber content. Qualitatively, under high fiber content, those long fibers have a great chance to contact the neighboring fibers and the machine (screw and wall of barrel). Fiber bundles reduce the effective fiber length or aspect ratio, significantly, allowing the preservation of longer fibers for higher content fiber composites during the intensive injection molding process.

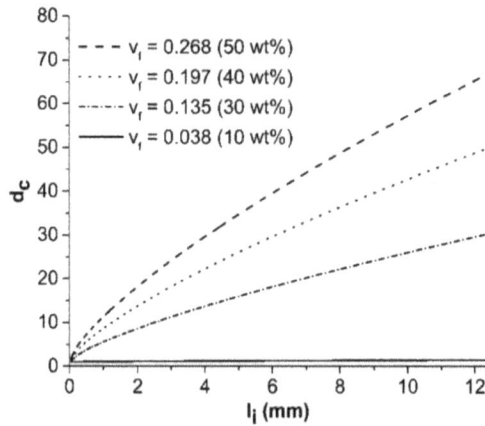

**Figure 9.** Predicted fiber bundles' size as a function of fiber length with various fiber concentrations by fitting the predictions using EMT method to experimental tensile data.

## 5. Conclusions

The stiffness properties have been studied in this work for 10 wt%, 30 wt%, 40 wt%, and 50 wt% injection molded LFTs (glass/PP). Experimental measurements of fiber length distribution and fiber orientation were obtained to calculate the transverse modulus ($\theta$ direction) using both Halpin-Tsai (HT) model and the Mori-Tanaka model based on Eshelby's equivalent inclusion (EMT). It has been shown that the EMT method generates slightly larger values than does the HT model. The accuracy of the various closure approximations for predicting the elastic properties has also been evaluated. The IBOF and ORW3 approximations turned out to be the best ways of describing the fourth order tensor in terms of the second order tensor. An important finding from this work is that the models over-predicted the modulus for higher fiber content injection molded LFTs. An empirical model was developed to correct the fibers' aspect ratio in the computations obtained for the actual as-formed LFTs with fiber bundles under high fiber content. After the correction, the analytical predictions matched well with the experimental stiffness values from tensile tests on the LFTs. The leveling off of the elastic properties of the injection molded LFTs as fiber content increases (>30 wt%) is due to the existence of the fiber bundles. In this study, we assume the adhesion is perfect to simplify the problem, which most likely not be true. However, the 10 wt%, 30 wt%, 40 wt%, and 50 wt% materials have the same surface treatments for the fibers. Therefore, it is legitimate to only include the effects of the clumping of fibers for comparison purpose. Moreover, the calculated bundle sizes from the proposed empirical model can be further applied to predict the strength in future work.

**Author Contributions:** H.C. performed the work on the fiber orientation, fiber length analysis, and tensile test. H.C. also did all the elastic properties predictions and the development of the proposed empirical bundles' size model. D.G.B. had oversight in all aspects of the project and were involved in all stages of the research.

**Acknowledgments:** We would like to thank the SABIC Innovative Plastics for supplying all the materials in this study. The authors would also like to thank the Material Science and Engineering department and the Sustainable Energy laboratory at Virginia Tech for the use of laboratory equipment. This work was funded, in part, by the American Chemical Council.

**Conflicts of Interest:** The authors declare no conflict of interest. The sponsors had no role in the design of the study; in the collection, analyses, or interpretation of data; in the writing of the manuscript, and in the decision to publish the results.

## References

1. Henning, F.; Ernst, H.; Brüssel, R. Lfts for automotive applications. *Reinf. Plast.* **2005**, *49*, 24–33. [CrossRef]
2. Ortman, K.; Baird, D.; Wapperom, P.; Whittington, A. Using startup of steady shear flow in a sliding plate rheometer to determine material parameters for the purpose of predicting long fiber orientation. *J. Rheol.* **2012**, *56*, 955–981. [CrossRef]
3. Von Turkovich, R.; Erwin, L. Fiber fracture in reinforced thermoplastic processing. *Polym. Eng. Sci.* **1983**, *23*, 743–749. [CrossRef]
4. Advani, S.G.; Tucker, C.L., III. The use of tensors to describe and predict fiber orientation in short fiber composites. *J. Rheol.* **1987**, *31*, 751–784. [CrossRef]
5. Garesci, F.; Fliegener, S. Young's modulus prediction of long fiber reinforced thermoplastics. *Compos. Sci. Technol.* **2013**, *85*, 142–147. [CrossRef]
6. Tucker Iii, C.L.; Liang, E. Stiffness predictions for unidirectional short-fiber composites: Review and evaluation. *Compos. Sci. Technol.* **1999**, *59*, 655–671. [CrossRef]
7. Hine, P.J.; Rudolf Lusti, H.; Gusev, A.A. Numerical simulation of the effects of volume fraction, aspect ratio and fibre length distribution on the elastic and ther moelastic properties of short fibre composites. *Compos. Sci. Technol.* **2002**, *62*, 1445–1453. [CrossRef]
8. Ingber, M.S.; Papathanasiou, T.D. A parallel-supercomputing investigation of the stiffness of aligned, short-fiber-reinforced composites using the boundary element method. *Int. J. Numer. Methods Eng.* **1997**, *40*, 3477–3491. [CrossRef]
9. Takao, Y.; Taya, M. The effect of variable fiber aspect ratio on the stiffness and thermal expansion coefficients of a short fiber composite. *J. Compos. Mater.* **1987**, *21*, 140–156. [CrossRef]
10. Halpin, J.C.; Jerine, K.; Whitney, J.M. The laminate analogy for 2 and 3 dimensional composite materials. *J. Compos. Mater.* **1971**, *5*, 36–49. [CrossRef]
11. Hine, P.J.; Lusti, H.R.; Gusev, A.A. On the possibility of reduced variable predictions for the thermoelastic properties of short fibre composites. *Compos. Sci. Technol.* **2004**, *64*, 1081–1088. [CrossRef]
12. Bumm, S.H.; White, J.L.; Isayev, A.I. Glass fiber breakup in corotating twin screw extruder: Simulation and experiment. *Polym. Compos.* **2012**, *33*, 2147–2158. [CrossRef]
13. Thomason, J.L. The influence of fibre length and concentration on the properties of glass fibre reinforced polypropylene. 6. The properties of injection moulded long fibre pp at high fibre content. *Compos. Part A Appl. Sci. Manuf.* **2005**, *36*, 995–1003. [CrossRef]
14. Thomason, J.L. Structure-property relationships in glass-reinforced polyamide, part 1: The effects of fiber content. *Polym. Compos.* **2006**, *27*, 552–562. [CrossRef]
15. Halpin, J. Stiffness and expansion estimates for oriented short fiber composites. *J. Compos. Mater.* **1969**, *3*, 732–734. [CrossRef]
16. Affdl, J.; Kardos, J. The halpin-tsai equations: A review. *Polym. Eng. Sci.* **1976**, *16*, 344–352. [CrossRef]
17. Hill, R. A self-consistent mechanics of composite materials. *J. Mech. Phys. Solids* **1965**, *13*, 213–222. [CrossRef]
18. Dray, D.; Gilormini, P.; Régnier, G. Comparison of several closure approximations for evaluating the thermoelastic properties of an injection molded short-fiber composite. *Compos. Sci. Technol.* **2007**, *67*, 1601–1610. [CrossRef]
19. Taya, M.; Mura, T. On stiffness and strength of an aligned short-fiber reinforced composite containing fiber-end cracks under uniaxial applied stress. *J. Appl. Mech.* **1981**, *48*, 361–367. [CrossRef]
20. Camacho, C.W.; Tucker, C.L.; Yalvaç, S.; McGee, R.L. Stiffness and thermal expansion predictions for hybrid short fiber composites. *Polym. Compos.* **1990**, *11*, 229–239. [CrossRef]
21. Hyer, M.W. *Stress Analysis of Fiber-reinforced Composite Materials*; DEStech Publications Inc.: Lancaster, PA, USA, 2009.
22. Standard, A. Standard Test Method for Tensile Properties of Polymer Matrix Composite Materials. Available online: http://file.yizimg.com/175706/2012061422194947.pdf (accessed on 6 August 2018).
23. Kunc, V.; Frame, B.J.; Nguyen, B.N.; Tucker, C.L., III; Velez-Garcia, G. Fiber Length Distribution Measurement for Long Glass and Carbon Fiber Reinforced Injection Molded Thermoplastics. Available online: https://www.researchgate.net/profile/Gregorio_Velez-Garcia/publication/237431694_FIBER_LENGTH_DISTRIBUTION_MEASUREMENT_FOR_LONG_GLASS_AND_CARBON_FIBER_REINFORCED_INJECTION_MOLDED_THERMOPLASTICS/links/00b49531712e6067b3000000/

FIBER-LENGTH-DISTRIBUTION-MEASUREMENT-FOR-LONG-GLASS-AND-CARBON-FIBER-REINFORCED-INJECTION-MOLDED-THERMOPLASTICS.pdf (accessed on 3 August 2018).

24. Vélez-García, G.; Wapperom, P.; Kunc, V.; Baird, D.; Zink-Sharp, A. Sample preparation and image acquisition using optical-reflective microscopy in the measurement of fiber orientation in thermoplastic composites. *J. Microsco.* **2012**, *248*, 23–33. [CrossRef] [PubMed]

25. Vélez-García, G.M.; Wapperom, P.; Baird, D.G.; Aning, A.O.; Kunc, V. Unambiguous orientation in short fiber composites over small sampling area in a center-gated disk. *Compos. Part A Appl. Sci. Manuf.* **2012**, *43*, 104–113. [CrossRef]

26. Chung, D.H.; Kwon, T.H. Improved model of orthotropic closure approximation for flow induced fiber orientation. *Polym. Compos.* **2001**, *22*, 636–649. [CrossRef]

27. Chung, D.H.; Kwon, T.H. Invariant-based optimal fitting closure approximation for the numerical prediction of flow-induced fiber orientation. *J. Rheol.* **2002**, *46*, 169–194. [CrossRef]

28. Houshyar, S.; Shanks, R.A.; Hodzic, A. The effect of fiber concentration on mechanical and thermal properties of fiber-reinforced polypropylene composites. *J. Appl. Polym. Sci.* **2005**, *96*, 2260–2272. [CrossRef]

29. Little, J.E.; Yuan, X.; Jones, M.I. Characterisation of voids in fibre reinforced composite materials. *NDT E Int.* **2012**, *46*, 122–127. [CrossRef]

30. Lee, D.J.; Oh, H.; Song, Y.S.; Youn, J.R. Analysis of effective elastic modulus for multiphased hybrid composites. *Compos. Sci. Technol.* **2012**, *72*, 278–283. [CrossRef]

31. Swolfs, Y.; Gorbatikh, L.; Romanov, V.; Orlova, S.; Lomov, S.V.; Verpoest, I. Stress concentrations in an impregnated fibre bundle with random fibre packing. *Compos. Sci. Technol.* **2013**, *74*, 113–120. [CrossRef]

32. Pratten, N.A. The precise measurement of the density of small samples. *J. Mater. Sci.* **1981**, *16*, 1737–1747. [CrossRef]

33. Berger, R. Comparison of the gompertz and logistic equations to describe plant disease progress. *Phytopathology* **1981**, *71*, 716–719. [CrossRef]

34. Richard, V.T.; Lewis, E. Fiber fracture in reinforced thermoplastic processing. *Polym. Eng. Sci.* **1983**, *23*, 743–749.

35. Cieslinski, M.J.; Wapperom, P.; Baird, D.G. Influence of fiber concentration on the startup of shear flow behavior of long fiber suspensions. *J. Non-Newton. Fluid Mech.* **2015**, *222*, 163–170. [CrossRef]

36. Phelps, J.H.; Abd El-Rahman, A.I.; Kunc, V.; Tucker, C.L. A model for fiber length attrition in injection-molded long-fiber composites. *Compos. Part A Appl. Sci. Manuf.* **2013**, *51*, 11–21. [CrossRef]

Journal of
*Composites Science*

MDPI

*Article*

# Process-Induced Fiber Orientation in Fused Filament Fabrication

Tom Mulholland [1,*], Sebastian Goris [1], Jake Boxleitner [1], Tim A. Osswald [1] and
Natalie Rudolph [2]

[1]  Department of Mechanical Engineering, University of Wisconsin-Madison, Madison, WI 53706, USA;
    sgoris@wisc.edu (S.G.); boxleitner@wisc.edu (J.B.); tosswald@wisc.edu (T.A.O.)
[2]  AREVO, Inc., Santa Clara, CA 95054, USA; natalie.rudolph@wisc.edu
*  Correspondence: tmulholland@wisc.edu; Tel.: +1-608-265-2405

Received: 2 July 2018; Accepted: 27 July 2018; Published: 2 August 2018

**Abstract:** As the applications for additive manufacturing have continued to grow, so too has
the range of available materials, with more functional or better performing materials constantly
under development. This work characterizes a copper-filled polyamide 6 (PA6) thermoplastic
composite designed to enhance the thermal conductivity of fused filament fabrication (FFF) parts,
especially for heat transfer applications. The composite was mixed and extruded into filament
using twin screw extrusion. Because the fiber orientation within the material governs the thermal
conductivity of the material, the orientation was measured in the filament, through the nozzle, and in
printed parts using micro-computed tomography. The thermal conductivity of the material was
measured and achieved 4.95, 2.38, and 0.75 W/(m·K) at 70 °C in the inflow, crossflow, and thickness
directions, respectively. The implications of this anisotropy are discussed using the example of an air-
to-water crossflow heat exchanger. The lower conductivity in the crossflow direction reduces thermal
performance due to the orientation in thin-walled parts.

**Keywords:** thermal conductivity; fiber orientation; composite; filler; heat exchanger; copper

## 1. Introduction

Additive manufacturing (AM) is publicized as a new, exciting technology that is on the cusp
of revolutionizing manufacturing everywhere and changing the way we think about the products we
buy, whether it be custom designing one's own clothing or producing spare parts at home (or in space).
Of course, there are elements of truth and exaggerations with these ideas. Although there are now
many examples of metal AM end-use products in high-cost, high-value industries like aerospace [1],
polymer end-use parts are less prominent. Polymer AM examples are more often related to prototyping
or tool-making; however, the correct combination of material, process, and product can lead to more
successful applications.

Functionalized polymer materials can increase the scope of available applications for polymer AM
parts. Fused filament fabrication (FFF), a material extrusion process that melts and deposits polymer
filament through a heated nozzle, is well-suited for AM with composite materials [2]. FFF filament
is produced by extrusion, a mature process that lends itself to the easy incorporation of high loads
of fillers [3]. FFF filaments have been produced with up to 40% vol fillers or higher [4] to enhance
different properties, such as strength, stiffness, electrical conductivity, or thermal conductivity.

While electrically conductive FFF materials have been available for some time, increased thermal
conductivity requires the incorporation of high loads of fillers or generally larger filler sizes.
Although some increase in thermal conductivity can be achieved at low filler content [5], the largest
gains are seen as conductive chains are formed along filler particles in contact, leading to increasing
gains in the compound thermal conductivity [6]. These conductive chains are more easily formed by

fillers with a higher aspect ratio [7]. While a sphere has an aspect ratio of 1, a square platelet 15 μm wide and 2 μm thick would have an aspect ratio of 7.5, and a fiber that is 500 μm long and 30 μm in diameter (as is used in this work) has an aspect ratio of 16.7. These high aspect ratio fillers lead to higher thermal conductivity but also exhibit a highly anisotropic thermal conductivity due to the process-induced alignment of the fillers [7]. This is well-studied in shear flows [8,9] and injection molding [7,10] but is little understood for FFF processing.

The process-induced fiber alignment in injection molding of discontinuous fiber-reinforced composites has been studied by several groups in the past [11–15]. However, little research work has been published on fiber alignment in FFF using discontinuous fiber-reinforced materials. Some simulation work has been done on understanding fiber orientation in a composite material as it passes through a nozzle. Heller found, using the Folgar-Tucker model [9], that the alignment will increase through the constriction section of the nozzle, and then may decrease again through the land at the nozzle tip, if the land is long enough. The orientation can decrease substantially following the nozzle exit due to die swell, which acts perpendicularly to the shear flow, however this analysis was based on an unrestricted flow through the nozzle into space [16]. In reality, material exiting the nozzle is immediately deposited on the platform or previously deposited beads, changing the flow direction by 90 degrees from vertical to its final horizontal position. Additionally, different flow lines in the deposited bead undergo different deformations. Therefore, the actual flow in FFF at the nozzle exit is more complex than simple extrusion into space with die swell.

In any case, the thermal conductivity can be seven times higher or more in the flow direction compared to the through-plane direction, as shown in injection molding [7] and later in this work. This may be advantageous or may present problems in different areas of an FFF part. This work explores FFF filament produced with copper fillers in order to print parts with thermal applications: in this case, a crossflow air-cooled heat exchanger (HX). It will be shown that the flow fields in the production of the filament and the deposition during printing govern the fiber orientation, in turn governing the thermal conductivity, which is one of the main factors in the overall performance of the printed HX. A better understanding of the fiber orientation will help improve the performance of the HX and possibly lead to the prediction of the thermal conductivity in different sections of a printed part.

In addition, measuring the orientation of fibers is essential in the understanding of the physics that govern their movement during processing, but it remains a challenging task [17]. Characterization methodologies are limited for copper fibers as the measurements by polishing samples and inspecting the cross-sectional footprint of fibers are not feasible due to copper fiber's irregular shape.

In this article, we show the results of a make-measure-model concept for 3D printing of copper filled polyamide to manufacture air-cooled heat exchangers. This work comprises the compounding and extrusion of the filament, the production and characterization of the samples, and the modeling of the thermal properties that can be achieved.

## 2. Materials and Methods

### 2.1. Filament Extrusion

Polyamide 6 (PA6, Ultramid® B33-01, BASF®, Ludwigshafen, Germany) was chosen as the polymer matrix in this work due to its ability to resist the HX service temperature of 70 °C. This was mixed with copper fibers with a 30 μm nominal diameter and 500 μm length (Deutsches Metallfaserwerk, Neidenstein, Germany) or copper spheres with a nominal diameter of 45 μm (Chemical Store, Clifton, NJ, USA). These were compounded on a Leistritz® (Nürnberg, Germany) ZSE 27 HPe corotating twin screw extruder with a screw diameter of 27 mm and L/D ratio of 36. The screw configuration, shown in Figure 1 below, was based on previous work by Amesöder [7] on thermally conductive composites for injection molding. The extruder temperature was controlled in eight zones along the barrel and another for the die, increasing along the length as indicated in Figure 1.

These temperatures produced a final melt temperature for the compounded PA6 between 248 and 251 °C. Initial trials used higher temperatures for extrusion, but these led to yellowing of the PA6, indicating degradation, so they were lowered to the given temperatures. The screw speed and the side stuffer speed were 100 RPM and 140 RPM, respectively. The resin feed rate was kept at a constant 4 kg/h, and the filler feed rate was varied to achieve the desired fill ratio. For example, 4 kg/h of PA6 with 10.5 kg/h of copper makes a filler mass fraction of 72.4% wt, or a volume fraction of approximately 25% vol. This speed was chosen to run the line slowly while still achieving a steady filler mass flow rate from the gravimetric feeders.

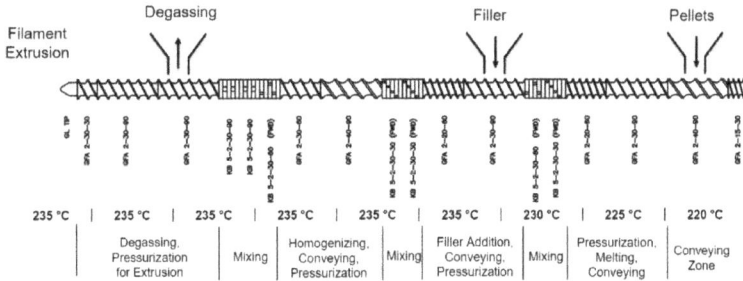

**Figure 1.** The screw configuration included two mixing zones for compounding the fillers.

The extruded filament passes from the die into a vacuum cooling tank, a laser micrometer, a belt puller, and finally a winder. A strand die with a 3.5 mm orifice was used to produce the 1.75 mm filament. The strand leaves the die and enters the Conair® (Cranberry Township, PA, USA) MT104-13-3 vacuum cooling tank. This 4 m (13 ft) stainless steel tank is filled with water to a level higher than the top of the product. A light vacuum less than 1.2 kPa (5 inH$_2$O) keeps the water at the entrance flowing inwards, improving the circularity. The strand then passes through a laser micrometer (LaserLinc$^{TM}$ Triton331, Fairborn, OH, USA), which measures the diameter in three locations and calculates the ovality of the product. After the laser micrometer, the strand passes through a Conair® 3-20 precision belt puller, which is in a closed feedback loop with the laser micrometer to control the filament diameter. Finally, at the very end of the line is a winder, which winds the filament onto a large spool for easy handling.

Because of the weight of the copper-filled compound, startup of the downstream operations was very difficult; with the low melt viscosity of PA6, the weight would cause the strand to sag before the strand could be passed into the vacuum cooling tank. Therefore, the line had to be started with just PA6 resin. Once the line was in full operation, the side stuffer was started and the filler was added to the extruder. The maximum filler content was limited to 25% vol due to issues such as die drool causing line breakage.

## 2.2. Fused Filament Fabrication

Fused filament fabrication is an additive manufacturing process that is classified as material extrusion. A solid circular filament of plastic is fed into a heated nozzle, in which it is melted. The molten plastic is extruded through a small diameter nozzle. The material is then deposited onto the part in a layer-by-layer fashion. The molten plastic cools and solidifies in a very short time as it is deposited.

The printer used in this work was the Aon 3D$^{TM}$ (Montreal, QC, Canada), which is equipped with dual hot ends capable of printing at temperatures up to 450 °C. This printer is built in a gantry style, where the extruders move in the x- and y-directions, and the print bed only moves in the z-direction. One interesting feature of the Aon, which should become more common, is its ability to use two nozzles to print a part in two locations on the bed simultaneously. This essentially reduces the machine cost by half in a manufacturing setting.

Filaments selected for printing in this work were PA6 with 20% vol copper fiber (PA6-CuF-20), 25% vol copper fiber (PA6-CuF-25), or neat PA6. Some copper sphere-filled samples were prepared by compression molding at approximately 10, 20, and 30% vol for comparison.

*2.3. Laser Flash Analysis*

The thermal diffusivity can be measured by laser flash analysis (LFA). The instrument, in this work a Netzsch® Instruments (Selb, Germany) Nanoflash 447, analyzes the transient heat flow in a material by firing an energy pulse at the bottom side of a sample using a xenon flash lamp. The top side of the sample is monitored with a liquid nitrogen-cooled infrared detector, which measures a signal proportional to temperature rise on the top surface. A thermal diffusivity model can then be fit to the time-dependent temperature rise. The thermal conductivity ($k$) can be calculated by multiplying the thermal diffusivity ($\alpha$) with the mass density ($\rho$) and specific heat capacity ($c_p$), as:

$$k = \alpha \rho c_p. \tag{1}$$

This study analyzed the thermal diffusivity in a range from 25 to 100 °C, as the working temperature for the materials was intended to be 70 °C.

The minimum testable sample size on the Nanoflash 447 LFA is a 6-mm round disc, which is much larger than an FFF filament strand of 1.75 mm or 3 mm diameter. Thus, the filament itself is not easily tested, and the thermal diffusivity must be measured on a prepared sample by printing a test bar that can be used to cut 10 mm × 10 mm × 1 mm square samples. In this case, the test bar is 60 mm × 10 mm × 1 mm. This bar is printed with all tool paths parallel to the long dimension using SciSlice, an open-source slicer developed in-house and available on GitHub [18]. The ends of the printed bar were cut and discarded, as they contain turns in the toolpath that may disrupt the fiber orientation within each printed bead. The printed bar was polished with 1000 grit sandpaper to produce a flat surface, then cut to the correct sample size. Finally, the sample was spray-coated with graphite to promote absorption of the LFA energy pulse.

Samples produced in this way measure the thermal diffusivity perpendicular to the printhead direction (in the layer direction), therefore perpendicular to the primary fiber orientation direction. This is referred to as the through-plane (TP) direction, which is the direction through the layers of a print. To study the diffusivity parallel to the fiber orientation direction, which is referred to as the in-plane direction, the samples need to be prepared differently. Using the printed 60 mm × 10 mm × 1 mm bar, strips of dimensions 1 mm × 10 mm × 1 mm are cut with a razor blade. No additional polishing is done. Each strip is rotated 90 degrees, then placed in a sample holder with the printing direction facing the detector. The overall sample thickness was measured with a digital micrometer with the strips already in the sample holder. Now, the energy pulse travels along the length of the printed beads, and therefore along the principal fiber orientation direction. These samples are referred to as laminate (LAM) samples. Sample preparation is illustrated in Figure 2. The samples were printed with a 1-mm nozzle with an extruder temperature of 270 °C, a bed temperature of 110 °C, a layer height of 0.2 mm, an extrusion ratio of 101%, and a printhead speed of 60 mm/s. Finally, additional samples were printed to measure the crossflow thermal conductivity. Freestanding walls 20 mm long × 1 mm wide × 60 mm tall were printed with a 1-mm nozzle with an extruder temperature of 270 °C, a bed temperature of 110 °C, a layer height of 0.2 mm, an extrusion ratio of 101%, and a printhead speed of 19 mm/s. The 10 mm section of wall closest to the build plate was cut and discarded, since the build plate temperature affects this region. The remaining section was used to cut LFA samples which can be used to measure the thermal diffusivity across the width of a bead. Samples are printed this way, instead of cutting the TP samples as with LAM, because these walls are more representative of the conductivity in a thin-walled part.

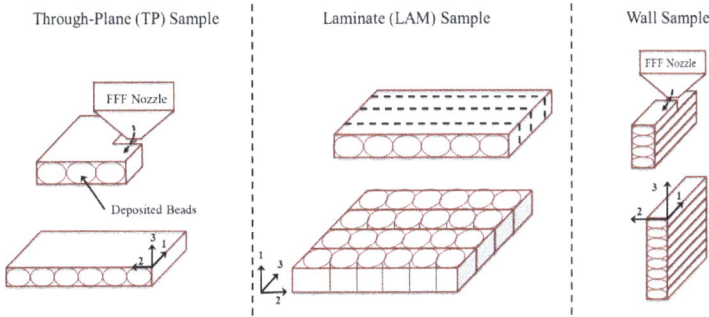

**Figure 2.** Illustration of laser flash analysis (LFA) sample preparation. From left to right, the sample is printed and a through-plane (TP) sample is scanned through the thickness. The same sample may be cut, and each strip is rotated, making a laminate (LAM) sample, which is also scanned through the thickness. Other samples are printed as walls to scan across the bead width. The 1-direction is the principal fiber orientation direction and the printing direction. The 2-direction is the crossflow, and the 3-direction is the through-plane or build direction.

## 2.4. Fiber Orientation Analysis Using Microcomputed Tomography

Microcomputed tomography (μCT) is a nondestructive testing (NDT) method which allows one to obtain the internal material structure of an object and evaluate its micro-structural properties. In general, the system consists of an X-ray source, a rotating platform, and a detector. The basic principle of μCT is to irradiate a sample with penetrating X-rays, which are attenuated and captured downstream of the object with a detector system creating radiographs. At defined energy levels, the X-ray source irradiates the specimen, which is placed on a rotating platform to achieve a full scan of the sample. The detector records the attenuated X-rays as radiographs at each increment of angle during the rotation of the sample. Each captured projection (radiograph) is a two-dimensional intensity distribution of the attenuated X-rays. The intensity distribution is directly related to the material's atomic density. A phase of high atomic density within the specimen absorbs more energy than low-density materials. The 3D reconstruction of the scanned sample is generated from all captured radiographs using tomographic reconstruction. The μCT data set can be processed using image processing algorithms for both qualitative and quantitative analyses.

The μCT scans in this work were performed with a Metrotom® 800 μCT system (Carl Zeiss® AG, Oberkochen, Germany) and the scan settings are summarized in Table 1. The fiber orientation analysis was performed using VG StudioMAX® 3.0 (Volume Graphics® GmbH, Heidelberg, Germany), which computes the fiber orientation based on the structure tensor approach [19].

**Table 1.** Zeiss Metrotom® 800 scan parameters.

| Parameter | Value |
|---|---|
| Voltage (kV) | 50 |
| Current (μA) | 80 |
| Integration Time (ms) | 1000 |
| Gain (-) | 8.0 |
| Spot Size (μm) | 5.0 |
| Voxel Size (μm) | 5.0 |

In this work, we use tensor representation proposed by Advani and Tucker [20] to describe the fiber orientation. The orientation of a single rigid fiber in three-dimensional (3D) space is described by the angle pair $(\theta, \phi)$, or by the unit vector $p(\theta, \phi)$ directed along the fiber axis as shown in Figure 3.

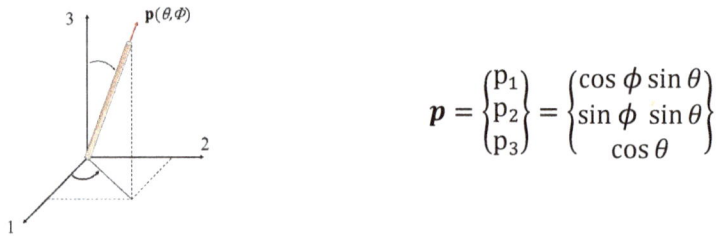

$$p = \begin{Bmatrix} p_1 \\ p_2 \\ p_3 \end{Bmatrix} = \begin{Bmatrix} \cos \phi \sin \theta \\ \sin \phi \sin \theta \\ \cos \theta \end{Bmatrix}$$

**Figure 3.** Representation of the orientation of a single rigid fiber by the vector $p(\theta, \phi)$.

For a fiber population in a defined volume, a complete description of the orientation is a probability density function $\psi(p)$, which describes the probability of a fiber oriented between the angles $\theta_i$ and $(\theta_i + d\theta)$, and between $\phi_i$ and $(\phi_i + d\phi)$ [9]. Advani and Tucker [20] proposed a more concise orientation of the fiber population by using a tensorial description. The Advani orientation tensor $a_{ij}$ computes the average orientation property of all fibers in a unit volume. The components of the symmetric second-order orientation tensor are defined as:

$$a_{ij} = \oint p_i p_j \, \psi(p) \, dp \tag{2}$$

where $p_i p_j$ is the dyadic or tensor product of the fiber orientation vector $p$ with itself, and $\oint (\ldots) \psi(p) \, dp$ denotes an integral over all possible fiber orientations. In terms of the orientation distribution and the angle pair $(\theta, \phi)$, the tensor components can be calculated as follows [17]:

$$
\begin{array}{lll}
a_{11} = \langle \cos^2 \phi \sin^2 \theta \rangle & a_{12} = \langle \cos \phi \sin \phi \sin^2 \theta \rangle & a_{13} = \langle \cos \phi \sin \theta \cos \theta \rangle \\
a_{21} = a_{12} & a_{22} = \langle \sin^2 \phi \sin^2 \theta \rangle & a_{23} = \langle \sin \phi \sin \theta \cos \theta \rangle \\
a_{31} = a_{13} & a_{23} = a_{23} & a_{33} = \langle \cos^2 \theta \rangle
\end{array}
$$

where the angle brackets $\langle \cdot \rangle$ indicate the average of all fibers in a volume.

The diagonal components of the second order orientation tensor ($a_{11}$, $a_{22}$ and $a_{33}$) describe the degree of orientation with respect to the defined coordinate system. Conventionally, the reference coordinates are defined so that the 1-direction represents the in-flow direction, the 2-direction is the crossflow direction and the 3-direction is the thickness direction. The off-diagonal components of the orientation tensor show the tilt of the orientation tensor from the coordinate axes. Hence, they are zero only if the coordinate axes align with the principal directions of the orientation tensor [20]. The physical interpretation of the tensor components focuses mainly on the diagonal components of the tensor, illustrated in Figure 4.

Scans of different samples were made using μCT to capture the change in orientation during processing:

- Strands of filament, which defines the initial orientation of the fibers;
- Strands extruded from the 1 mm nozzle illustrating the change in fiber orientation due to shear deformations in the nozzle;
- Single bead samples deposited on the print bed with a layer height of 0.2 mm;
- LFA samples in the through-plane (TP, $a_{33}$), wall ($a_{22}$), and laminate (LAM, $a_{11}$) directions. Wall samples were not tested at 25% vol copper fiber due to issues with producing quality samples at the higher filler content.

Three samples each of filament and through-plane, laminate, and wall samples were scanned for the 20% vol copper fiber material in order to understand the measurement and material variability.

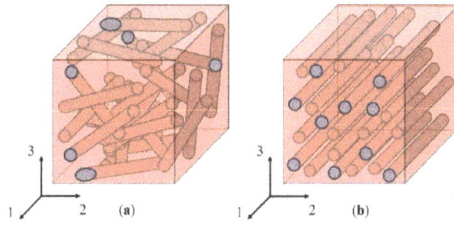

**Figure 4.** Orientation of a fiber population within a volume: (**a**) randomly oriented fibers and (**b**) aligned fibers.

## 3. Results

### 3.1. Fiber Orientation Measurements

The orientation of the copper fibers was successfully obtained from the μCT scans. Due to their isotropy and aspect ratio of 1, the orientation in the sphere-filled material was not analyzed. Figure 5a shows the 3D reconstruction of a 20% vol copper fiber-filled PA6 filament. The results of the orientation analysis are summarized in Figure 5b, which shows the degree of orientation in the extrusion direction ($a_{11}$ value) and in the cross-section ($a_{22}$ and $a_{33}$) along the diameter of the filament. The measurements suggest a fairly constant orientation along the filament diameter with a high degree of orientation in the extrusion direction and an average tensor component value for $a_{11}$ of 0.8. The orientation in the cross-section is low and random with average values for $a_{22}$ and $a_{33}$ of 0.1. The results correspond with the idea of fiber alignment as a function of the deformation experienced by the material. The fibers align in the shear direction (extrusion direction) during the manufacturing process. The repeatability of the measurements is very high, shown by a small standard deviation indicated as error bars in Figure 5b. Hence, the initial orientation of the copper fibers before the FFF process is known and constant throughout the compounded filament.

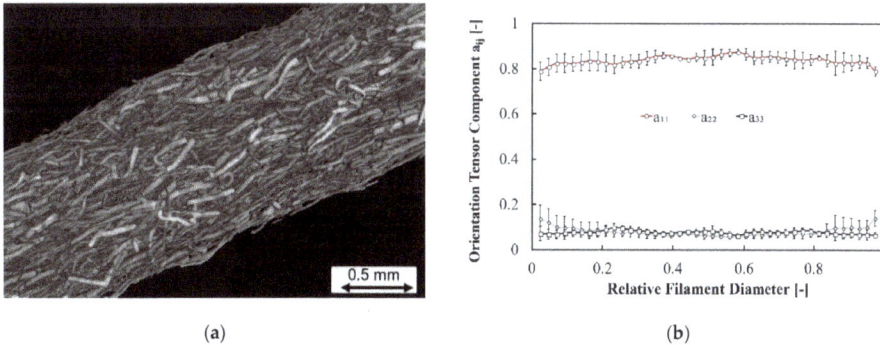

(a)    (b)

**Figure 5.** (**a**) 3D reconstruction of a PA6 and 20% vol copper fiber filament strand; (**b**) the measured fiber orientation distribution where $a_{11}$ is the orientation in the extrusion direction, and $a_{22}$ and $a_{33}$ are the orientation in the cross-section.

The 3D reconstruction and 2D slices from the μCT scan of a 20% vol copper fiber-filled PA6 through-plane (TP) sample is shown in Figure 6. The printing direction is associated with the 1-direction. As seen in the cross-section view in Figure 6b, the copper fibers can be clearly identified and qualitatively show a preferential orientation in the printing direction.

(a)  (b)

**Figure 6.** Illustration of fiber alignment in the through-plane (TP) samples: (**a**) 3D reconstruction and (**b**) 2D cross-sectional views.

Figure 7 summarizes the quantitative results of the fiber orientation through the thickness of the through-plane (TP), laminate (LAM), and wall samples. The measurements of the TP sample suggest a preferential orientation of the fibers in the 1-direction with an average value of 0.68. The through-thickness fiber alignment ($a_{33}$) is low with an average value of 0.08, and the fiber alignment perpendicular to the print direction ($a_{22}$) has an average value of 0.24. These results indicate that the main fiber direction in the filament is maintained through the printing process, but it is slightly reduced from 0.8 to 0.68. The reorientation of the fibers during printing occurs mainly in the printing plane (1-2 plane) since the orientation in the through-thickness direction ($a_{33}$) is reduced, the $a_{11}$ is also reduced, and the $a_{22}$ is increased, compared to the filament orientation. Overall, the orientation of the printed samples is uniform through the thickness.

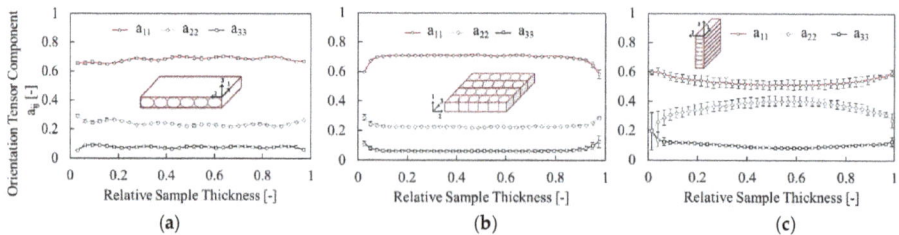

(a)  (b)  (c)

**Figure 7.** Fiber orientation for printed samples: (**a**) through-plane (TP) sample, (**b**) laminate (LAM) sample, and (**c**) wall sample.

Maintaining the coordinate system defined in Figure 2 for the prepared LAM samples, fiber orientation measurements are shown in Figure 7b. The LAM samples show fiber alignment in the thickness direction of the sample with an average value of 0.70 for $a_{11}$. This shows that the fiber orientation is maintained through the preparation of the LAM by the rotation of TP sample strips. While the orientation is fairly constant throughout the sample thickness, there are some edge effects at the surfaces of the LAM sample, which can be attributed to the sample preparation leading to not perfectly uniform planes. The wall samples varied significantly from the other two types, with the $a_{11}$ and $a_{22}$ orientation tensor components nearly equal towards the center of the sample. Unlike the other two types, these samples are completely unconstrained at the edges, allowing the polymer to flow more into the crossflow direction and ultimately moving fibers from the 1- into the 2-direction.

The average orientation in the three axes is presented in Figure 8. Data for the filament and LFA samples were averaged from relative thicknesses between 0.2 and 0.8 to eliminate edge effects. The fiber orientation for the filament before and after passing through the 1-mm nozzle are shown on the left. The orientation in the extrusion direction decreases slightly. That principal orientation decreases further when measuring one bead deposited on the print bed. The change in fiber orientation appears higher in the 25% vol fiber material, which could be related to higher fiber-fiber interaction. However, this small difference could be attributed to scan artifacts like beam hardening caused by the fiber concentration, which results in a skewed fiber orientation calculation. The 20% vol copper fiber LAM and TP samples are nearly equal, as expected, since a LAM sample is just cut from a TP sample and turned. The standard deviations are small, from 3% in the filament to 0.7% in the laminate samples. Likewise, the orientation in the $a_{22}$ direction increases from the filament to the printed TP sample, from 0.07 to 0.23 for the 20% vol copper fiber, as plotted in Figure 8. This means that the squeezing flow that occurs as material is extruded from the nozzle and flattened onto the previous layer has reoriented the fibers to some degree in this direction. This is supported by the lower $a_{11}$ and higher $a_{22}$ values for the wall samples. Since there are no other beads to constrain the flow, the material moves more in the crossflow direction.

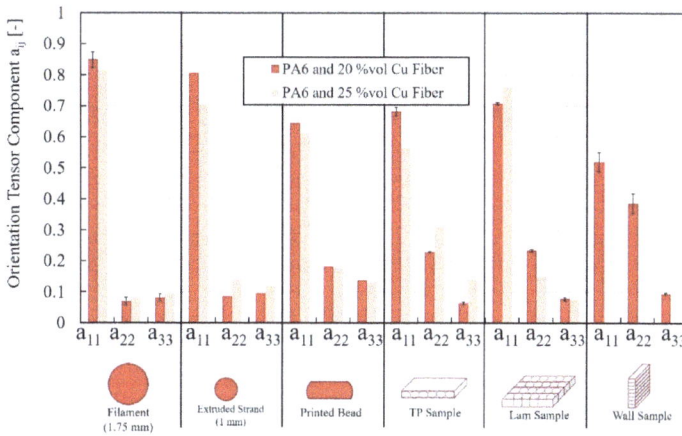

**Figure 8.** Average orientation components in material after processing for PA6 with 20 and 25% vol copper fiber.

The change in orientation through processing is consistent with Heller's and Skrabala's results [16,21], showing that orientation will decrease during the expansion due to die swell occurring after the nozzle. However, the actual flow field is more complex than simple die swell, since the material turns 90° and is deformed as it is deposited. Unfortunately, there is not much published information about the actual flow field as material is deposited on an FFF part. However, any reorientation that occurs, as suggested by these results, is positive for the wall conductivity, as discussed in the following section. Geometric changes to the nozzle could increase this effect; for example, Heller's work suggests that lengthening the land at the nozzle exit could lead to higher perpendicular orientation [16].

*3.2. Thermal Diffusivity Measurements*

Figure 9 shows the thermal diffusivity for 20% vol copper fiber PA6 in the through-plane, wall, and laminate samples. As predicted, the fiber orientation in the samples has a significant effect on the thermal diffusivity. The diffusivity decreases linearly with temperature, and the laminate sample has a diffusivity about seven times larger than the through-plane sample. The same samples used for fiber orientation analysis were measured (three laminate, three wall, and three through-plane samples), and the standard deviations at different temperatures were below 4.5%, 3.3%, and 1.1%, respectively, indicating good repeatability. Since the variation in fiber orientation was very small, the larger variation in laminate sample diffusivity is probably due to variations in the cutting of the strips, leading to thickness variations.

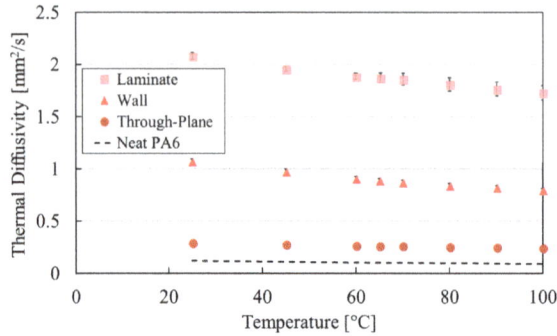

**Figure 9.** The thermal diffusivity decreases with temperature in the 20% vol copper fiber material. Standard deviations are imperceptibly small in most cases.

The thermal conductivity of the materials as a function of filler content is plotted in Figure 10 for copper fiber-filled and sphere-filled PA6. As seen with the thermal diffusivity, the thermal conductivity in the fiber-filled material is highly anisotropic. For example, at 20% vol copper fiber, the conductivity in the flow direction was 5.0 W/(m·K), 2.4 W/(m·K) across the flow, and 0.75 W/(m·K) in the build direction. The large differences in thermal conductivity for different directions can lead to important design challenges for thermal products, which are discussed in more detail below.

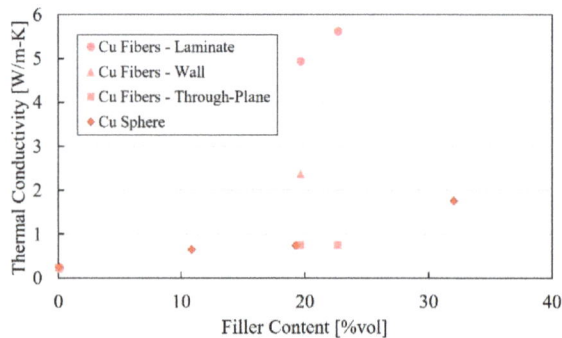

**Figure 10.** Thermal conductivity at 70 °C in the flow direction is many times higher than the through-plane direction. Sphere-filled PA6 performs about the same as the fibers through-plane.

### 3.3. Orientation in FFF Prints

The overall goal of this project was to produce FFF heat exchangers (HXs) printed with thermally conductive plastics. In general, air-cooled HX performance is dominated by thermal resistance on the air side, from the air into the solid. With additive manufacturing, airfoil fin arrays can be added to the air channel to enhance the heat transfer, allowing polymer HXs to compete with traditional metal designs. The HX considered here consists of air and water channels in cross-flow and can be viewed at two scales: the microstructure and the macrostructure. The air channels are filled with airfoil-shaped fins, which are separated from the water channel by a wall, as illustrated on the left in Figure 11. That microstructure is repeated many times to make the macrostructure of the HX, displayed on the right.

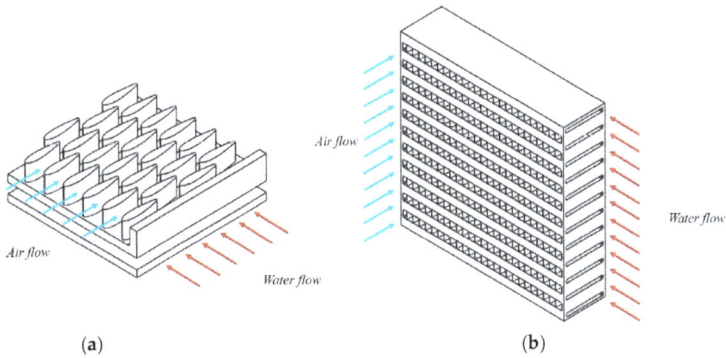

**Figure 11.** Cross-flow HX configuration displaying (**a**) airfoil pins in the microstructure; (**b**) the overall crossflow HX macrostructure.

The process-induced fiber orientation can affect the performance of printed HXs as the toolpath will govern the principle fiber orientation direction. This presents a fundamental challenge in HX printing; the low conductivity of plastics (even filled plastics) drives the HX design to smaller and smaller features, but small features greatly constrain the possible tool paths. The two most important features in the design are the fins and the walls dividing the two fluids. The fins will ideally conduct heat from the fluid to the walls (in-plane) at a very high rate. The walls will ideally conduct heat from one fluid channel into the other (crossflow), but not conduct along the length or width of the wall (in-plane). However, the walls should be as thin as possible to minimize material use and reduce the conductive resistance, but this unfortunately leads to a lower thermal conductivity through the walls. Thin walls with a thickness of only a few beads are printed parallel to the wall, meaning the fibers orient along, not through, the wall. As illustrated in Figure 12a, the fibers are oriented in the ideal direction along the axis of the pin, but they are oriented perpendicular to the ideal direction in the walls.

A HX microstructure performance model was developed in order to simulate and optimize heat transfer. The model carries out standard HX analysis in which geometry and material properties are used to calculate thermal resistances associated with conduction resistances through the walls separating air and water channels and convection resistances in the air channel (convection resistance in the water channel is negligible). The convection heat transfer coefficient associated with flow across airfoil shaped fins was determined using computational fluid dynamics. The overall performance was calculated by finding the overall resistance to heat transfer and using the effectiveness-number of transfer units (ε-NTU) method [22]. This method employs correlations based on the flow orientation, the fluid properties, flow rates, and temperatures, and the thermal resistances to find the overall heat transfer rate.

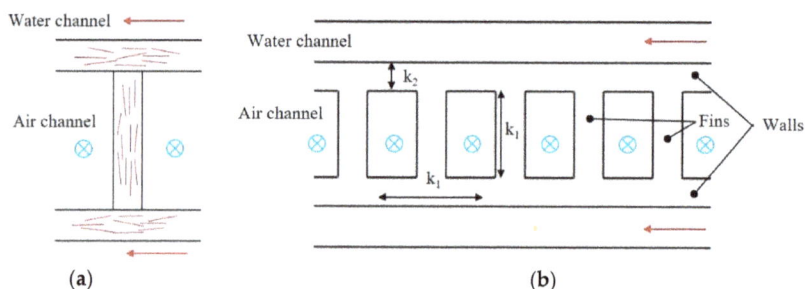

Figure 12. (a) Illustration of a fin between two walls showing the fiber orientation. (b) Cross-section of plate fin geometry with anisotropic conductivity orientations $k_1$ (aligned with fibers) and $k_2$ (perpendicular to fibers).

When modeling conduction resistances and fin efficiencies, print orientation must be considered because of the fiber orientation and resulting anisotropic conductivity. The most practical way of printing the HX consists of printing the walls as perimeters and the fins as bridges. This toolpath results in the conductivities shown in Figure 12b; the conductivity parallel to bead print direction is $k_1$ and conductivity perpendicular to bead print direction is $k_2$.

Using that thermal model, Figure 13 shows the simultaneous effect that wall thickness, isotropic conductivity, and anisotropic conductivity have on the performance of the HX. As a function of wall thickness, the figure compares the behavior of the anisotropic conductivity of PA6-CuF-20 (4.95 W/(m·K) in-plane, 2.38 W/(m·K) crossflow, and 0.75 W/(m·K) through-plane) to three materials: a material with an isotropic conductivity of 0.75 W/(m·K), one with an isotropic conductivity of 2.38 W/(m·K), and one with an isotropic conductivity of 4.95 W/(m·K). The performance is normalized with respect to the best case. The figure demonstrates that increasing thermal conductivity leads to better performance. However, as wall thickness decreases, the performance of the anisotropic material and the lower bound isotropic material become more alike. This demonstrates the importance of the conduction resistance through the wall and how the conductivity perpendicular to bead print direction limits performance. The similarity between the 2.38 W/(m·K) case and the 4.95 W/(m·K) case demonstrates the performance-limiting effect of air-side convection resistance.

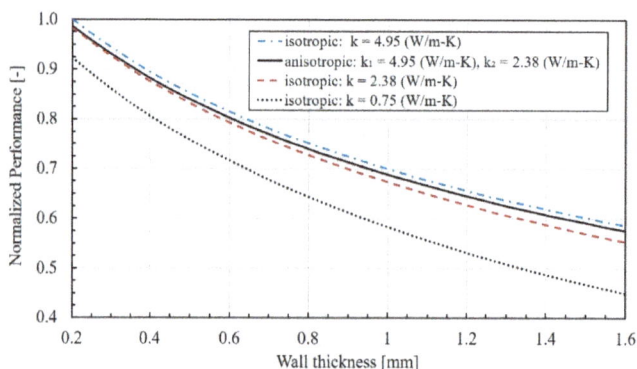

Figure 13. Effect of wall thickness, conductivity, and anisotropy on normalized performance.

## 4. Conclusions

The fiber orientation in FFF filament was shown to be highly preferential in the extrusion direction. This could be analyzed and quantified with μCT analysis, displaying an average orientation as high as 0.8 in filament and 0.7 in printed parts in the principal flow direction. Consequentially, the orientation was lower in the two perpendicular directions. This anisotropy was reflected in the thermal conductivity, where the flow direction conductivity was about seven times larger than the build direction conductivity. Although the addition of fibers can increase the conductivity 19 times higher than the base resin conductivity at only 20% vol, the geometry of the printed part and the toolpath have an equally important role in the final performance, since the toolpath ultimately determines the fiber orientation in the part. While high fiber orientation leads to effective cooling fins in the analyzed air-cooled heat exchanger, the lower orientation in the walls between water and air channels also governs the performance.

The crossflow conductivity can be increased by raising the fiber content; however, a 1-mm FFF nozzle already experienced clogging at 25% vol copper fiber content, implying that higher fiber content is not possible for this filler. The relationships between filler size, shape, and content on FFF nozzle clogging and on the composite thermal conductivity is essential for optimizing the HX performance with the given manufacturing restrictions.

As additive manufacturing slowly starts to become feasible for serial production, continued development and functionalization of materials is essential. Fiber-filled plastics have significant benefits, including thermal and electrical conductivity, so it is essential to advance machine design and manufacturing techniques to take advantage of the materials for new applications.

**Author Contributions:** Conceptualization—T.M., T.A.O. and N.R.; Investigation—T.M., S.G. and J.B.; Writing—original draft, T.M., S.G. and J.B.; Writing—review & editing—T.M., S.G., J.B., T.A.O. and N.R.

**Funding:** The research was done at the University of Wisconsin-Madison. The information, data, or work presented herein was funded in part by the Advanced Research Projects Agency-Energy (ARPA-E), U.S. Department of Energy, under Award Number DE-AR0000573. The views and opinions of authors expressed herein do not necessarily state or reflect those of the United States Government or any agency thereof.

**Conflicts of Interest:** The authors declare no conflict of interest.

## References

1. Murr, L.E. A metallographic review of 3D printing/additive manufacturing of metal and alloy products and components. *Metallogr. Microstruct. Anal.* **2018**, *7*, 103–132. [CrossRef]
2. Brenken, B.; Favaloro, A.; Barocio, E.; Denardo, N.M.; Kunc, V.; Pipes, R.B. Fused deposition modeling of fiber-reinforced thermoplastic polymers: Past progress and future needs. In Proceedings of the American Society for Composites 31st Technical Conference and ASTM Committee D30 Meeting, Williamsburg, VA, USA, 19–22 September 2016; pp. 1–14.
3. Osswald, T.A.; Menges, G. *Materials Science of Polymers for Engineers*, 3rd ed.; Carl Hanser Verlag: Munich, Germany, 2012.
4. Nikzad, M.; Masood, S.H.; Sbarski, I. Thermo-mechanical properties of a highly filled polymeric composites for fused deposition modeling. *Mater. Des.* **2011**, *32*, 3448–3456. [CrossRef]
5. T'Joen, C.; Park, Y.; Wang, Q.; Sommers, A.; Han, X.; Jacobi, A. A review on polymer heat exchangers for HVAC&R applications. *Int. J. Refrig.* **2009**, *32*, 763–779.
6. Agari, Y.; Uno, T. Thermal conductivity of polymer filled with carbon materials: Effect of conductive particle chains on thermal conductivity. *J. Appl. Polym. Sci.* **1985**, *30*, 2225–2235. [CrossRef]
7. Amesöder, S. Wärmeleitende Kunststoffe für das Spritzgießen. Ph.D. Thesis, Universität Erlangen-Nürnberg, Erlangen, Germany, 2010.
8. Jeffrey, G.B. The motion of ellipsoidal particles immersed in a viscous fluid. *Proc. R. Soc. Lond. A* **1922**, *102*, 161–179. [CrossRef]
9. Folgar, F.; Tucker, C.L. Orientation behavior of fibers in concentrated suspensions. *J. Reinf. Plast. Compos.* **1984**, *3*, 98–119. [CrossRef]

10. Heinle, C. Simulationsgestätzte Entwicklung von Bauteilen aus Wärmeleitfähigen Kunststoffen. Ph.D. Thesis, Universität Erlangen-Nürnberg, Erlangen, Germany, 2012.

11. Goris, S.; Osswald, T.A. Fiber orientation measurements using a novel image processing algorithm for micro-computed tomography scans. In Proceedings of the 15th SPE Automotive Composites Conference & Exhibition, Novi, MI, USA, 9 September 2015.

12. Foss, P.H.; Tseng, H.-C.; Snawerdt, J.; Chang, Y.-J.; Yang, W.-H.; Hsu, C.-H. Prediction of fiber orientation distribution in injection molded parts using Moldex3D simulation. *Polym. Compos.* **2014**, *35*, 671–680. [CrossRef]

13. Vélez-Garcia, G.M.; Wapperom, P.; Baird, D.G.; Aning, A.O.; Kunc, V. Unambiguous orientation in short fiber composites over small sampling area in a center-gated disk. *Compos. Part A Appl. Sci. Manuf.* **2012**, *43*, 104–113. [CrossRef]

14. Sun, X.; Lasecki, J.; Zeng, D.; Gan, Y.; Su, X.; Tao, J. Measurement and quantitative analysis of fiber orientation distribution in long fiber reinforced part by injection molding. *Polym. Test.* **2015**, *42*, 168–174. [CrossRef]

15. Gandhi, U.; Sebastian, D.B.; Kunc, V.; Song, Y. Method to measure orientation of discontinuous fiber embedded in the polymer matrix from computerized tomography scan data. *J. Thermoplast. Compos. Mater.* **2016**, *29*, 1696–1709. [CrossRef]

16. Heller, B.P.; Smith, D.E.; Jack, D. Effect of extrudate swell, nozzle shape, and convergence zone on fiber orientation in fused deposition modeling nozzle flow. In Proceedings of the Solid Freeform Fabrication, Austin, TX, USA, 10 August 2015; pp. 1220–1236.

17. Bernasconi, A.; Cosmi, F.; Hine, P.J. Analysis of fibre orientation distribution in short fibre reinforced polymers: A comparison between optical and tomographic methods. *Compos. Sci. Technol.* **2012**, *72*, 2002–2008. [CrossRef]

18. Van Hulle, L. "SciSlice", GitHub. 2017. Available online: https://github.com/VanHulleOne/SciSlice. (accessed on 21 June 2017).

19. Krause, M.; Hausherr, J.M.; Burgeth, B.; Herrmann, C.; Krenkel, W. Determination of the fibre orientation in composites using the structure tensor and local X-ray transform. *J. Mater. Sci.* **2010**, *45*, 888–896. [CrossRef]

20. Advani, S.G.; Tucker, C.L. The use of tensors to describe and predict fiber orientation in short fiber composites. *J. Rheol.* **1987**, *31*, 751–784. [CrossRef]

21. Skrabala, O. Wärmeleitfähige Kunstoffe: Verarbeitungsinduzierte Eigenschaftsbeeinflussung und Deren Numerische Vorhersage. Ph.D. Thesis, Universität Stuttgart, Stuttgart, Germany, 2016.

22. Nellis, G.F.; Klein, S.A. *Heat Transfer*, 1st ed.; Cambridge University Press: New York, NY, USA, 2009.

*Journal of*
*composites science*

MDPI

*Article*

# Prediction of the Fiber Orientation State and the Resulting Structural and Thermal Properties of Fiber Reinforced Additive Manufactured Composites Fabricated Using the Big Area Additive Manufacturing Process

**Timothy Russell, Blake Heller, David A. Jack * and Douglas E. Smith**

Baylor University, Waco, TX 76706, USA; Timothy_Russell@baylor.edu (T.R.); Blake_Heller@baylor.edu (B.H.); Douglas_E_Smith@baylor.edu (D.E.S.)
* Correspondence: David_Jack@baylor.edu; Tel.: +1-254-710-3347

Received: 31 December 2017; Accepted: 26 March 2018; Published: 10 April 2018

**Abstract:** Recent advances in Fused Filament Fabrication (FFF) include large material deposition rates and the addition of chopped carbon fibers to the filament feedstock. During processing, the flow field within the polymer melt orients the fiber suspension, which is important to quantify as the underlying fiber orientation influences the mechanical and thermal properties. This paper investigates the correlation between processing conditions and the resulting locally varying thermal-structural properties that dictate both the final part performance and part dimensionality. The flow domain includes both the confined and unconfined flow indicative of the extruder nozzle within the FFF deposition process. The resulting orientation is obtained through two different isotropic rotary diffusion models, the model by Folgar and Tucker and that of Wang et al., and a comparison is made to demonstrate the sensitivity of the deposited bead's spatially varying orientation as well as the final processed part's thermal-structural performance. The results indicate the sensitivity of the final part behavior is quite sensitive to the choice of the slowness parameter in the Wang et al. model. Results also show the need, albeit less than that of the choice of fiber interaction model, to include the extrudate swell and deposition within the flow domain.

**Keywords:** additive manufacturing; short-fiber reinforcement; fiber orientation modeling; fiber interactions

## 1. Introduction

Fused Filament Fabrication (FFF) is an Additive Manufacturing (AM) technology that extrudes beads of thermoplastic materials onto a moving substrate, layer by layer, based on a digital three-dimensional model. Since its introduction nearly three decades ago, FFF (sometimes called Fused Deposition Modeling or FDM$^{TM}$) has become one of the fastest growing forms of AM. An attractive feature of FFF as compared to other AM systems is material selection (see e.g., [1]). While this advantage is responsible, in part, for the growth of FFF, the lack of structural materials and scalability to industrially relevant volumes are limiting factors that must be overcome in order to realize widespread application of AM systems. Recently, large scale FFF processes such as the Big Area Additive Manufacturing (BAAM) process developed at the Manufacturing Demonstration Facility (MDF) at Oak Ridge National Laboratories has significantly increased build volume. In addition, fiber reinforcements have recently been introduced in FFF systems to reduce distortion and improve the mechanical performance of the final part, particularly in the Z- or through-thickness direction where they are weak. This work seeks to quantify the intra-layer structural and thermal performance of the deposited bead. The inter-layer performance is another significant consideration, and there are a variety of authors who have presented methods to improve the inter-layer bonding strength such as tamping (see e.g., [2,3]), infrared preheating (see e.g., [4]), and plasma treating (see e.g., [5]).

Fiber reinforced matrix (FRM) composites are used in many industries using multiple manufacturing processes. For example, FRM composites have been used in construction for many years (see e.g., [6]). They have also been used in the aerospace and medical industries (see e.g., [7–9]). In addition, the use of short-fiber reinforced composites is widely prevalent throughout the injection and compression molding community and is the most closely related field to the present work. The significant difference between the presented work and previous modeling efforts of molding processes is that FFF systems have a closed cavity flow followed by an unconfined expansion with a subsequent confined boundary from the moving plate.

With the recent advances of large scale FFF, the use of fiber reinforcements in the AM process is a viable option for increasing the performance of AM parts (see e.g., [10,11]). The process of blending short, chopped fibers into a polymer melt coupled with the freeform nature of AM allows for the fabrication of complex and intricate parts with the potential for improved structural performance and with a geometric complexity previously unavailable. A key aspect to realize this vision is understanding the impact of the nozzle, both its geometry and position relative to the moving platen, on the internal fiber orientation state within the deposited bead and, by extension, the part's final stiffness. The thermo-mechanical behavior of the bead is also quite critical as the deposition process induces large thermal gradients during fabrication and can impact the final part's dimensionality. To quantify a-priori the distortion of the entire fabricated product, the spatially and directionally varying anisotropic stiffness and coefficient of thermal expansion tensors would be required. This work seeks to present a method to predict the anisotropic stiffness and thermal expansion tensors due to local variations in the fiber microstructure. The results from this paper can be used in a full scale simulation of the macroscopic fabricated product, and the methodology presented in this work can be used by designers in establishing new nozzle designs for the extruder to tailor the properties of the deposited bead.

Fiber microstructure predictions in a short-fiber reinforced polymer composite require an accurate understanding of the relationship between the fiber kinetics and the flow kinematics. There is an extensive body of literature to address the relationship between the local fiber orientation kinetics and the surrounding velocity field (see e.g., [12–19]). Several authors have developed methods for predicting part stiffness and thermal conductivity as a function of local fiber orientation in the hardened polymer melt (see e.g., [13,20–22]). Most fiber kinetics models are based on the motion of individual ellipsoids in a dilute Newtonian solvent as constructed by Jeffery [23]. It is unfeasible to track the motion of every fiber within a suspension, thus the fiber orientation distribution function (ODF) is often more desirable. The full ODF has been solved using control volumes (see e.g., Bay [24]) or the computationally efficient and numerically exact spherical harmonics (see e.g., Montgomery-Smith et al. [25]). Flow kinematics are coupled to the fiber orientation state in densely packed flows (see e.g., [26,27]) and several authors have performed numerical solutions for fully coupled flows (see e.g., [28,29]). In the present study, flow kinematics are decoupled from the fiber orientation state similar to what is often done in injection molding simulations (see e.g., [17,29–31]). This work focuses on the impact that the swell of the extrudate has on the final orientation state.

The kinetic equation for the ODF may be used to describe the concentrated suspension behavior of interacting fibers through the addition of a rotary diffusivity term $D_r$ as generalized by Bird et al. [32] for the related system of polymer chain alignment. The Folgar and Tucker [12] isotropic rotary diffusion (IRD) model to represent fiber interactions has been widely used for more than three decades. Recent studies demonstrate the IRD model predicts fiber alignment occurs at a rate faster than that observed experimentally and several authors have sought to address this limitation (see e.g., [15,17,18,33,34]). The strain-reduction factor model of Huynh [16] sought to reduce the alignment rate by a scaling parameter pre-multiplied on the equation of motion. The reduced strain closure model (IRD-RSC) of Wang et al. [17] expanded the Huynh [16] approach by reducing the rate of fiber alignment by operating exclusively on the eigenvalues of the orientation tensors and thus is invariant with respect to the chosen coordinate frame. Recently Phelps and Tucker [18] introduced the anisotropic rotary diffusion model (ARD) which introduces a directional dependence to fiber interactions and has been shown to be quite effective for select long-fiber systems. Unfortunately, the general behavior of the five ARD empirical coefficients as a function of the polymer melt and fiber packing is not yet well understood in the literature. Strautins and Latz [35] introduced a computationally efficient form for flexible fibers using the moments of the orientation distribution, and Ortman et al. [19] extended their work to allow fiber interactions using a strain-reduction extension of the IRD model based on the work of Sepehr et al. [34]. The current paper focuses on short, rigid fibers within a polymer melt and compares results between the classical IRD model and the recent IRD-RSC model. Particular focus is given to the sensitivity of the processed orientation state to the interaction coefficient $C_I$ within both models and the slow-down parameter $\kappa$ in the IRD-RSC model.

One objective of predicting the fiber orientation is to determine the mechanical performance of the deposited bead. Advani and Tucker [13] linked the stiffness of the solidified part to the local fiber microstructure using an orientation homogenization approach. In Jack and Smith [22] closed form expressions for the stiffness expectation and variance were derived using complex spherical harmonics. Nguyen et al. [36] performed a comparison between the $E_1$ and $E_2$ predicted by the IRD-RSC model and tensile tests. Recently Nguyen et al. [31] used the ARD model to predict the fiber orientation and the subsequent stress response for an injection molded long fiber thermoplastic. This work was extended by Agboola et al. [37] to investigate the impact the choice of fiber interaction model has on the expected macroscopic stiffness of a simple two dimensional part. Recently Stair and Jack [38] demonstrated the effectiveness of taking the measured properties of the neat polymer and fibers to predict both the locally varying stiffness and Coefficient of Thermal Expansion (CTE) tensors and then used this information to successfully predict the part warpage of laminated composites due to processing.

The current paper presents a computational approach that extends the work of Agboola et al. [37] and Stair and Jack [38] for related composite systems to predict the stiffness and CTE tensors for the deposited bead of a fiber reinforced polymer bead fabricated using the large volume FFF process. Results are presented based on predictions of the spatially varying fiber orientation obtained from process simulations that include the nozzle, die-swell and subsequent deposition and the constitutive properties of the neat polymer and individual fibers. A method is presented to predict the spatially varying, bulk stiffness tensors and CTE tensors for the beads for results from the IRD and the IRD-RSC fiber interaction models to determine the sensitivity of the final part performance to variations in fiber interaction parameters.

## 2. Fiber Orientation Modeling

The foundational work of Jeffery [23] describes the orientation change of a rigid and mass-less ellipsoidal particle in a fluid. The particle direction is defined by **p**, a unit vector along the fiber longitudinal axis. While Jeffery's equation is effective for the motion of a single fiber in a Newtonian solvent, it cannot adequately capture the motion of interacting fibers. Folgar and Tucker [12] applied the probability density distribution function for the orientation of fibers, labeled as $\psi$ (**p**). For a dense suspension of interacting fibers the equation of motion is cast in a form, which is similar to the expression initially used for polymer chain alignment developed by Bird et al. [32], as

$$\frac{D\psi}{Dt} = -\frac{1}{2}\nabla_{\mathbf{p}} \cdot (\mathbf{\Omega} \cdot \mathbf{p} + \lambda(\mathbf{\Gamma} \cdot \mathbf{p} - \lambda\mathbf{\Gamma} : \mathbf{ppp})\,\psi) + \nabla_{\mathbf{p}} \cdot \nabla_{\mathbf{p}}(D_r\psi) \tag{1}$$

where the response to fiber interactions is contained within the rotary diffusivity term $D_r$, often expressed as a function of $\psi$ (**p**) and the cartesian gradient of the suspension velocity **v** (see, e.g., [12,15,18]). In Equation (1), $\nabla_{\mathbf{p}}$ is the gradient operator in orientation space, and $\lambda$ relates to the fiber aspect ratio (see e.g., Zhang et al. [39]). In Equation (1) $\mathbf{\Omega}$ is the vorticity tensor $\mathbf{\Omega} = [(\nabla\mathbf{v}) - (\nabla\mathbf{v})^T]$, $\mathbf{\Gamma}$ is the rate of deformation tensor $\mathbf{\Gamma} = [(\nabla \mathbf{v}) + (\nabla \mathbf{v})^T]$, and $\frac{D}{Dt}$ is the material derivative defined as $\frac{D}{Dt} = \frac{\partial}{\partial t} + \mathbf{v} \cdot \nabla$, where $\nabla$ is the gradient operator in cartesian space. Solutions of $\psi(\mathbf{p}, t)$ often take minutes to hours of computational time for a single streamline using control volume methods, but spherical harmonic methods (see e.g., Montgomery-Smith et al. [25]) reduce the computational time by several orders of magnitude. Unfortunately neither approach is easily mapped to a finite element form, thus precluding their use in complex geometries. This later issue is often addressed using the moments of $\psi(\mathbf{p})$, called the orientation tensors, as used by Advani and Tucker [13] who define the second-order orientation tensor $\mathbf{A}$ and the fourth-order orientation tensor $\mathbb{A}$ as, respectively,

$$\mathbf{A} = \int_{\mathbb{S}} \mathbf{pp}\psi(\mathbf{p})d\mathbb{S}, \quad \mathbb{A} = \int_{\mathbb{S}} \mathbf{pppp}\psi(\mathbf{p})d\mathbb{S} \tag{2}$$

where $\mathbb{S}$ is the surface of the unit sphere. The moments in orientation space have the interpretation similar to the moments of a one-dimensional probability distribution function, thus the first moment (called the expectation in one dimension) has the analogy of the mean and the second moment would be similar to the variance or the spread. As the fiber orientation distribution function is symmetric, all odd ordered moments will yield the zero tensor of the respective order, we will limit the discussion to the even ordered orientation tensors. The orientation equation of motion for $\psi(\mathbf{p})$ from Equation (1) can be recast in terms of the orientation tensors as

$$\frac{D\mathbf{A}}{Dt} = -\frac{1}{2}(\mathbf{\Omega} \cdot \mathbf{A} - \mathbf{A} \cdot \mathbf{\Omega}) + \frac{1}{2}\lambda(\mathbf{\Gamma} \cdot \mathbf{A} + \mathbf{A} \cdot \mathbf{\Gamma} - 2\mathbb{A} : \mathbf{\Gamma}) + \mathcal{D}[\mathbf{A}] \tag{3}$$

where $\mathcal{D}[\mathbf{A}]$ is the diffusion component of Equation (1) describing fiber interactions. The orientation tensor form allows solutions for the spatially varying fiber orientation to be readily incorporated into industrial finite element codes with the drawback that the higher-order orientation tensor is contained within the equation of motion for the lower ordered orientation tensor. This is observed in Equation (3) for the equation of motion for $\mathbf{A}$ that contains $\mathbb{A}$. This necessitates the need for a closure by which the higher order orientation tensor is approximated as a function of a lower order tensor, i.e., $\mathbb{A} = f(\mathbf{A})$. There exist many closure approximations in the literature of $\mathbb{A}$ in terms of $\mathbf{A}$ (see e.g., [13,14,28,40–46] to list just a few) and several for the sixth-order orientation tensor (see e.g., [30,47,48]). Montgomery-Smith et al. [45] demonstrated that the orthotropic fitted (ORT) closure implemented by VerWeyst and Tucker [28] and the fast exact closure (FEC) [45] are the most accurate. In the current paper, we present orientation results obtained using an ORT closure.

## 2.1. Fiber Interactions: Isotropic Rotary Diffusion (IRD)

One of the most widely used models to predict the behavior of short fiber interactions is the isotropic rotary diffusion (IRD) model of Folgar and Tucker [12] where $\mathcal{D}[\mathbf{A}]$ of Equation (3) is expressed as

$$\mathcal{D}[\mathbf{A}] = C_I \gamma (2\mathbf{I} - 6\mathbf{A}) \tag{4}$$

where $\gamma$ is the scalar magnitude of the rate of deformation $\gamma = \sqrt{\mathbf{\Gamma} : \mathbf{\Gamma}/2}$ and $C_I$ is an empirically measured parameter termed the interaction coefficient. The isotropic rotary diffusion model assumes that all fibers within the melt interact with neighboring fibers in the same way, regardless of the surrounding orientation state of the fibers. The interaction coefficient is often found by matching the orientation observations of the steady state orientation in a pure shearing flow with those predicted by solutions of Equation (3). The value for the interaction coefficient is often obtained through the use of either measuring the resulting orientation state using sectioning (see e.g., [49]) and fitting the flow prediction results to that of the measured orientation state or through transient shear rheology and coupling the orientation state with that of the effective shear stress solution for concentrated suspensions (see e.g., [17]). The resulting value of the interaction coefficient will be different for different polymers and different packing densities. Regardless, authors typically present values for $C_I$ between $10^{-4}$ and $10^{-2}$ and their results show that $C_I$ is an increasing function of fiber packing density, the fiber aspect ratio, and the viscosity of the polymer matrix. The IRD results predict the steady state orientation with reasonable accuracy with the proper selection of $C_I$, and for several decades this model was considered the standard for use in industrial simulations of fiber motion. Recent work, such as that of Huynh's thesis [16], Sepehr et al. [34] and Wang et al. [17], demonstrated that the alignment rate predicted by the IRD was faster than that observed experimentally. Of note in the present context is that the velocity gradients are continually changing within the FFF nozzle and subsequent deposition and then the flow immediately ceases prior to a steady state orientation being attained.

### 2.2. Fiber Interactions: Isotropic Rotary Diffusion—Reduced Strain Closure (IRD-RSC)

The isotropic rotary diffusion model with the reduced strain closure (IRD-RSC) developed by Wang et al. [17] controls the rate of change of the predicted fiber orientation through an empirical scaling parameter $0 \leq \kappa \leq 1$ in an objective manner by providing mathematical diffusion exclusively on the eigenvalues of the orientation tensor $\mathbf{A}$ as [17]

$$\mathcal{D}^{\text{IRD-RSC}} = -2\left(1-\kappa\right)\left(\mathbb{L}-\mathbb{M}:\mathbb{A}\right):\Gamma + 2\kappa C_I \gamma\left(\mathbf{I}-3\mathbf{A}\right) \tag{5}$$

where $\mathbb{L}$ and $\mathbb{M}$ are expressed in terms of the eigenvalues $\alpha_i$ and the unit eigenvectors $\mathbf{e}_i$ of the second order orientation tensor defined as [17]

$$\mathbb{L} = \sum_{i=1}^{3} \alpha_i \mathbf{e}_i \mathbf{e}_i \mathbf{e}_i \mathbf{e}_i, \quad \mathbb{M} = \sum_{i=1}^{3} \mathbf{e}_i \mathbf{e}_i \mathbf{e}_i \mathbf{e}_i \tag{6}$$

While the RSC is termed the reduced strain closure, it is not a closure in the context of Equation (3) as it is not an approximation of the fourth order orientation tensor. The fourth order orientation tensor remains explicit in Equation (3), both through the Jeffery component and the fiber interaction component and a closure is still required to solve $D\mathbf{A}/Dt$. The IRD and IRD-RSC for the same choice of $C_I$ achieve a similar steady state orientation so methods to capture $C_I$ from experimental data for the IRD model remain applicable. The parameter $\kappa$ can be fit to experimental observations during the transient alignment state and has been determined using rheological studies (see e.g., [17]) with values reported in the literature ranging between $\kappa = 1/30$ and $\kappa = 1$ for different fiber-polymer blends.

### 2.3. Mechanical Stiffness Tensor Predictions from Orientation

The homogenization approach for stiffness first proposed by Advani and Tucker [13] translates the constitutive behavior of the fiber and the matrix along with the spatially varying orientation distribution information to locally varying effective mechanical properties. This approach is derived in detail in [22] and was shown to extend to orthotropic reinforcement materials. The local stiffness $\left\langle C_{ijkl} \right\rangle$ may be expressed as [13]

$$\left\langle C_{ijkl} \right\rangle = B_1 A_{ijkl} + B_2(A_{ij}\delta_{kl} + A_{kl}\delta_{ij}) + B_3(A_{ik}\delta_{jl} + A_{il}\delta_{jk}\delta_{jl} + A_{jl}\delta_{ik}\delta_{jl} + A_{jk}\delta_{il})$$
$$+ B_4(\delta_{ij}\delta_{kl}) + B_5(\delta_{ik}\delta_{jl} + \delta_{il}\delta_{jk}) \tag{7}$$

where $\delta_{ij}$ is the Kronecker delta, $A_{ij}$ is $\mathbf{A}$ in component form, $A_{ijkl}$ is $\mathbb{A}$ in component form, and the $B_m$ terms are given as [13]

$$
\begin{aligned}
B_1 &= \overline{C}_{1111} + \overline{C}_{2222} - 2\overline{C}_{1122} - 4\overline{C}_{1212} & B_4 &= \overline{C}_{2233} \\
B_2 &= \overline{C}_{1122} - \overline{C}_{2233} & B_5 &= \tfrac{1}{2}(\overline{C}_{2222} + \overline{C}_{2233}) \\
B_3 &= \overline{C}_{1212} + \tfrac{1}{2}(\overline{C}_{2233} - 2\overline{C}_{2222})
\end{aligned}
\tag{8}
$$

where $\overline{C}_{ijkl}$ in Equation (8) indicates the $(i, j, k, l)$ component of the underlying unidirectional composite stiffness tensor defined in its principal material reference frame. In Equation (7), it can be seen that the material stiffness is a function of the second and the fourth-order orientation tensors as well as the underlying stiffness tensor of the unidirectional composite. The homogenization equation is constructed for dilute and semi-dilute suspensions as stress concentrations due to the fiber inclusions [13,22] are neglected. However, it was demonstrated in Caselman [50] that this approach yields reasonable results for both semi-concentrated and the low range of concentrated suspensions as well.

There are many options for the choice of micromechanics model used to calculate the unidirectional composite stiffness $\overline{C}_{ijkl}$ for the homogenization in Equation (7). The Halpin-Tsai micromechanics model [51], which is based on laminate theory, is a concise approach for aligned short fiber composites. However, as indicated in the work of Tucker and Liang [52], this approach is ineffective for the aspect ratios found in typical short-fiber composites, whereas the Tandon-Weng [53] and Lielens [54] models are quite effective (see e.g., [52]). Using the closed form suggested by Tucker and Liang [52], as expressed in Zhang [55], the Tandon-Weng micromechanics theory yields one of the most accurate set of solutions over the range of aspect ratios found in short-fiber composites and is used in the present study.

### 2.4. CTE Tensor Predictions from Orientation

The second order, homogenized CTE tensor $\alpha_{ij}$ may be found from Camacho et al. [20] as

$$
\langle \alpha_{ij} \rangle = \langle C_{ijkl}\alpha_{kl} \rangle \langle C_{ijkl} \rangle^{-1}
\tag{9}
$$

In this equation, $\langle C_{ijkl} \rangle^{-1}$ is the inverse of $\langle C_{ijkl} \rangle$ from Equation (7), which therefore must be determined prior to solving Equation (9). Calculating $\langle C_{ijkl} \rangle^{-1}$ involves converting $\langle C_{ijkl} \rangle$ into its contracted form, a $6 \times 6$ matrix, inverting the contracted stiffness, and then converting the inverted $6 \times 6$ stiffness matrix back into a full fourth order tensor. Equation (9) also involves the term $\langle C_{ijkl}\alpha_{kl} \rangle$ which may be found from

$$
\langle C_{ijkl}\alpha_{kl} \rangle = D_1 A_{ij} + D_2 \delta_{ij}
\tag{10}
$$

where

$$
\begin{aligned}
D_1 &= A_1(B_1 + B_2 + 4B_3 + B_5) + A_2(B_1 + 3B_2 + 4B_3) \\
D_2 &= A_1(B_2 + B_4) + A_2(B_2 + 3B_4 + B_5)
\end{aligned}
\tag{11}
$$

In Equation (11), the $B_i$ quantities come from Equation (8) which requires $\overline{C}_{ijkl}$. The $A_i$ quantities in Equation (11) come from

$$
\begin{aligned}
A_1 &= \alpha_{11} - \alpha_{22} \\
A_2 &= \alpha_{22}
\end{aligned}
\tag{12}
$$

where $\alpha_1$ and $\alpha_2$ are components of the transversely isotropic CTE tensor $\overline{\alpha}_{ij}$. Thus, $\overline{\alpha}_{ij}$ must also be predicted before Equation (9) can be calculated and this is done so using the method of Schapery [56] using the form expressed in Stair and Jack [38].

## 3. Flow Modeling

The planar deposition fluid flow model, defining the flow domain in this paper, is generated using the die-swell method suggested by the authors in Heller et al. [57]. A low nozzle configuration is used in this paper, where the deposited bead height is actually higher than that of the separation distance between the nozzle tip and moving platen as depicted in Figure 1a,b. The dimensions of the nozzle are taken to be similar to that of the Strangpresse Model 19 large scale FFF extrusion nozzle. The flow domain is confined within the nozzle region and experiences a no-slip boundary on the nozzle walls. Upon exiting the nozzle, the flow expands and the boundary is that of either air or the moving platen. The flow domain is modeled as a Newtonian fluid undergoing Stokes flow where the inertial effects in the flow domain are neglected as is typically done when modeling polymer flows (see e.g., [58]). The Newtonian fluid values used in the present study are those for a typical extrusion grade ABS polymer with a density, $\rho$, of 1040 kg/m$^3$ and a dynamic viscosity, $\mu$, of 3200 Pa-s. Under the Newtonian fluid assumption, the density and viscosity only effect the pressure drop through the nozzle and do not change the velocity field within the flow domain and thus have no impact upon the resulting fiber orientation within the fluid. The modulus, CTE, and Poisson's ratio of the ABS matrix were taken to be 2.25 GPa, $90 \times 10^{-6}$ (mm/mm)/$^\circ$C, and 0.35, respectively. The Young's modulus, CTE, Poisson's ratio, and density of the carbon fiber were taken to be 230 GPa, $-2.6 \times 10^{-6}$ (mm/mm)/$^\circ$C, 0.2, and 1700 kg/m$^3$, respectively. The weight fraction was taken to be 13% to match that of the composite system studied in Duty et al. [3] for a carbon fiber reinforced FFF bead with similar bead deposition geometry to that considered here. The shape of the fiber was taken into account using a geometric aspect ratio of 20, within the range of the carbon fiber reinforcement in [3].

The boundary conditions for the model are summarized in Figure 1b. The fiber-filled polymer enters the nozzle at the inlet with an average velocity of 24 mm/s, which would correspond to a mass flow rate of 12.2 kg/h for a circular nozzle with the same dimensions as that of the nozzle depicted in Figure 1. The nozzle walls are defined to be no-slip, meaning the velocity is zero all along the nozzle walls. The end of the bead, called the "outlet" in Figure 1b, is constrained to have zero pressure. The flat, bottom boundary of the printed bead is considered to be a moving wall with a velocity of 101.6 mm/s in the negative $x_1$ direction and zero velocity in the $x_2$ direction.

The shape of the leading and trailing free extrudate surfaces after the nozzle exit are defined using a zero surface tension minimization method presented by Heller et al. [57]. Free surfaces can be specified as zero penetration where the normal velocity is set equal to zero along the surface or zero surface tension where the normal stress along the surface is set to zero (see e.g., [59]). The selection of the curve to define the extrudate surfaces was complex due to a need for smoothness as well as having enough variability to accurately model the extrudate surface and reach an acceptable minimum. The curve that was found to give acceptable smoothness and variability for the given model is an $n^{th}$ order Bezier curve ranging from 15 to 25 terms to define the polynomial, and the full details are provided in the companion paper by Heller et al. [57].

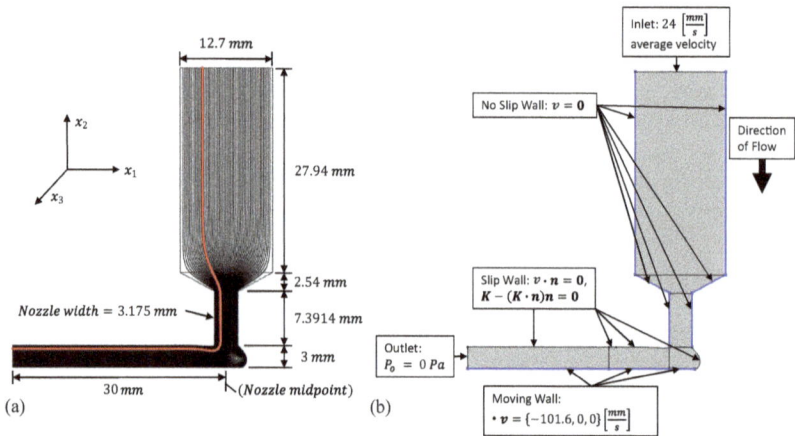

**Figure 1.** (**a**) The flow domain and dimensions; streamline 15 is highlighted in red. (**b**) The flow domain with boundary conditions.

## 4. Results

In this section, the method of obtaining the effective longitudinal modulus and CTE is illustrated. The results between those of the full nozzle, die-swell and deposition problem are contrasted with those from a flow domain that neglects the orientational changes that occur outside of the nozzle by treating the problem as just that of an injection molded part. The flow domain of this simplified model, which is also planar, can be seen in Figure 2. The boundary conditions for the flow domain are also shown in Figure 2 and it has the same dimensions as the nozzle in the previous flow domain which was shown in Figure 1. For the new flow domain of just the nozzle, the pressure at the nozzle outlet was constrained to be 0 and the flow at the outlet was constrained to be in the normal, or vertical, direction. The flow simulations in both flow domains used an initial random orientation state, i.e., $A_{ij} = \frac{1}{3}\delta_{ij}$, taken to be well up-stream of the nozzle tip. In addition, the velocities and velocity gradients for both of the flow simulations were found along 61 streamlines, numbered from left to right starting from the top of the nozzle. Results are provided in this section for a comparison between the IRD and IRD-RSC models for both the full die-swell process and the flow domain that neglects the swell. After this, a final results table with the longitudinal property predictions of all the flow conditions considered will be used for conclusion purposes.

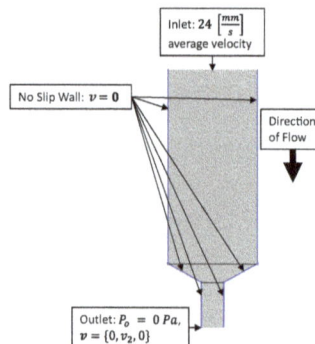

**Figure 2.** The flow domain of the nozzle alone with boundary conditions. (Nozzle dimensions are the same as those shown in Figure 1).

## 4.1. Results for a Single Flow Type—IRD, $C_I = 10^{-2}$

Beginning in COMSOL (Version 5.3; COMSOL Inc., Berlin, Germany, 2017), the flow domain was defined and the velocities in the $x_1$ and $x_2$ directions, that is $v_1$ and $v_2$, respectively, as well as the velocity gradients $\frac{\partial v_1}{\partial x_1}$, $\frac{\partial v_1}{\partial x_2}$, $\frac{\partial v_2}{\partial x_1}$, and $\frac{\partial v_2}{\partial x_2}$ were found along 61 streamlines. These were exported to MATLAB (R2016a; MathWorks, 2016) which, using an in-house implementation of the IRD model, obtained solutions for the orientation tensors for $C_I = 10^{-2}$ along each streamline. From the orientation tensors, Equations (7)–(12) were used to predict $\langle C_{ijkl} \rangle$ and $\langle \alpha_{ij} \rangle$ across cross sections of the flow domain.

### 4.1.1. Orientation Tensors along a Streamline

The second order orientation tensor $A_{ij}$ was calculated using a custom code within MATLAB along streamline 15 (shown in Figure 1a) and some of the components of $A_{ij}$ are shown in Figure 3. The $A_{ij}$ results for the IRD model with $C_I = 10^{-2}$ are shown in Figure 3. It is worth noting that the choice of a random initial condition is effectively meaningless for the IRD model as the orientation attains a steady state just prior to the nozzle's tapered zone. This will not be the case for the IRD-RSC results shown in Figure 8, where the final orientation state is clearly a function of the initial alignment state due to the slow rate of orientational changes. In the case of the IRD-RSC, a proper identification of the orientation state prior to the nozzle would need to be performed prior to industrial implementations.

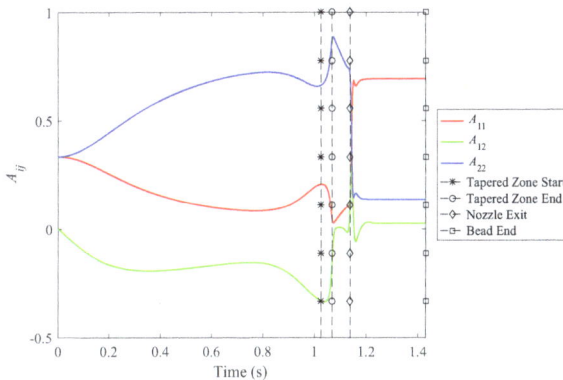

**Figure 3.** Components of $A_{ij}$, predicted with the IRD model with $C_I = 10^{-2}$, along streamline 15.

There are two important cross sections of the flow that are worth highlighting. One is where the flow reaches the end of the nozzle and the other is where the flow reaches the end of the flow domain which is indicative of the final processed bead and will exhibit properties that represent the properties of the final 3D printed bead. At the end of the nozzle, the IRD model predicts a relatively high value of $A_{22}$ and a relatively low value of $A_{11}$. Since $A_{11}$ and $A_{22}$ are representative of the overall fiber alignment in the $x_1$ and $x_2$ directions, respectively, this means that the IRD model predicts that the nozzle induces a relatively high and low fiber alignment in the $x_2$ and $x_1$ directions, respectively. At the end of the flow domain, however, the IRD predicts higher $x_1$ alignment than $x_2$ alignment. Thus, the direction of highest fiber alignment tends to follow the flow direction.

### 4.1.2. Orientation Tensors at the Exit and End

More information about the orientation state at the exit of the 3D printer nozzle was gained by calculating the second and fourth order orientation tensors $A_{ij}$ and $A_{ijkl}$ along 61 streamlines in the flow domain. This permitted $A_{ij}$ and $A_{ijkl}$ to be obtained across cross sections of the flow domain, such as at the nozzle exit. $A_{ij}$ as a function of $x_1$ at the nozzle exit was calculated (using the IRD model with $C_I = 10^{-2}$) and select components are shown in Figure 4a. In addition, to predict $A_{ij}$ as a function of $x_2$ at the end of the flow domain (the bead end, which is representative of the deposited bead), the flow domain as seen in Figure 1 was used. Components of $A_{ij}$ are shown in Figure 4b.

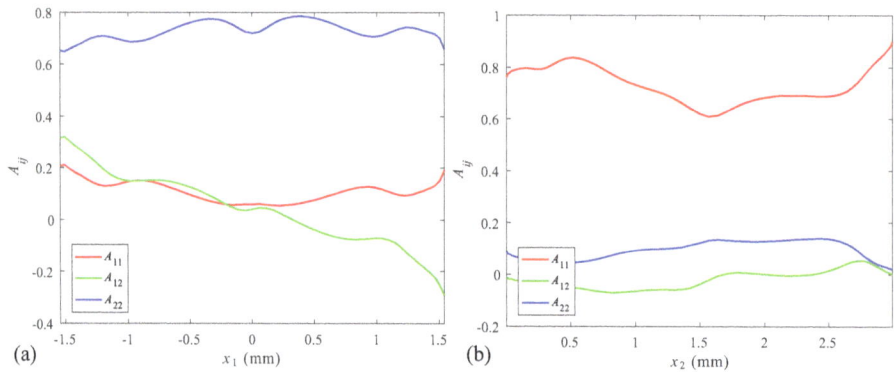

**Figure 4.** Orientation tensors calculated using the IRD model with $C_I = 10^{-2}$. (**a**) $A_{ij}$ as a function of $x_1$ at the nozzle exit, calculated using the model in Figure 2, and (**b**) $A_{ij}$ as a function of $x_2$ within the deposited bead, calculated using the model in Figure 1.

Since $A_{22}$ was found to be the dominant component of the orientation tensor along streamline 15 at the nozzle exit (see Section 4.1.1), it comes as no surprise that $A_{22}$ was the dominant component of the orientation tensor across the width of the nozzle at the nozzle exit. In addition, since the nozzle is symmetrical about the $x_2 = 0$ axis and the inlet condition for the flow was an average laminar inflow velocity, $A_{11}$ and $A_{22}$ are symmetrical about the $x_2 = 0$ axis as well and this can be seen in Figure 4a. Since $A_{12}$ contains the product of $p_1$ and $p_2$ according to Equation (2), $A_{12}$ is not symmetric about the $x_2 = 0$ axis. Furthermore, as $A_{11}$ was the dominant term at the end of streamline 15, it comes as no great surprise that this was the dominant $A_{ij}$ component within the deposited bead. This can be seen in Figure 4b. It can also be noted that $A_{11}$ and $A_{22}$ are not symmetrical within the deposited bead because the extruded melt turns over and experiences non-symmetrical shears as the bead is deposited.

### 4.1.3. Stiffness and CTE at the Exit and End

In this section, we discuss the results for the stiffness and CTE at the nozzle exit and the deposited bead. Again, these results use the orientation tensors predicted using the IRD model with $C_I = 10^{-2}$, along with Equations (7)–(12). Select components of the fourth order, homogenized stiffness tensor across the nozzle exit and across the deposited bead are shown in Figure 5a,b, respectively.

There are striking similarities between Figures 4 and 5. Figure 5a shows some of the components of the homogenized, fourth order stiffness tensor and shows that $\langle C_{2222} \rangle$ dominates across the width of the nozzle at the nozzle exit. Since $\langle C_{2222} \rangle$ is indicative of the stiffness in the $x_2$ direction, this result implies that the composite has greater stiffness in the vertical (or $x_2$) direction. It can be seen that $\langle C_{2222} \rangle$ and $\langle C_{1111} \rangle$ are symmetrical about $x_2 = 0$ at the nozzle exit and follow the trends of $A_{22}$ and $A_{11}$, respectively. Figure 5b shows similar trends to that of Figure 4b for the stiffness in the deposited bead. Again we see that $\langle C_{2222} \rangle$ tends to follow the trend of $A_{22}$ and $\langle C_{1111} \rangle$ follows the trend of $A_{11}$. Thus, the stiffness of the composite tends to be highest in the direction of highest fiber alignment and lower in the transverse direction. Since the direction of highest fiber alignment generally follows the direction of the flow, the direction of highest stiffness also tends to be in the flow direction.

The second order, homogenized CTE $\langle \alpha_{ij} \rangle$, like $\langle C_{2222} \rangle$, could also be found at any arbitrary point in the flow domain. Figure 6 shows some of the components of $\langle \alpha_{ij} \rangle$ as a function of $x_1$ at the nozzle exit and as a function of $x_2$ within the deposited bead.

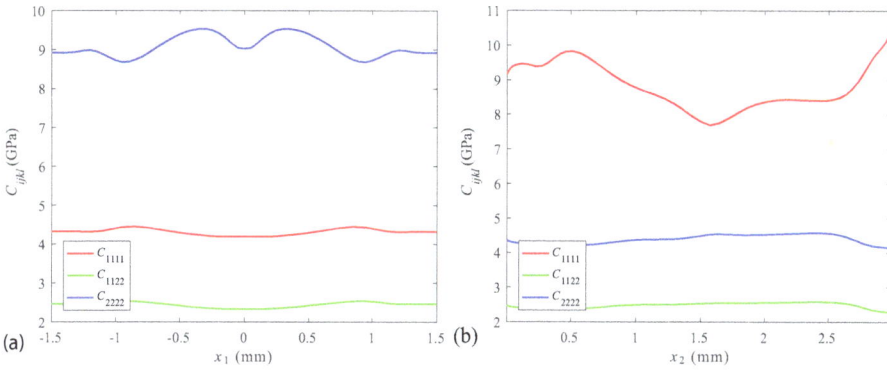

**Figure 5.** Components of $\left\langle C_{ijkl} \right\rangle$ from using the IRD model with $C_I = 10^{-2}$ (**a**) as a function of $x_1$ at the nozzle exit, and (**b**) as a function of $x_2$ within the deposited bead.

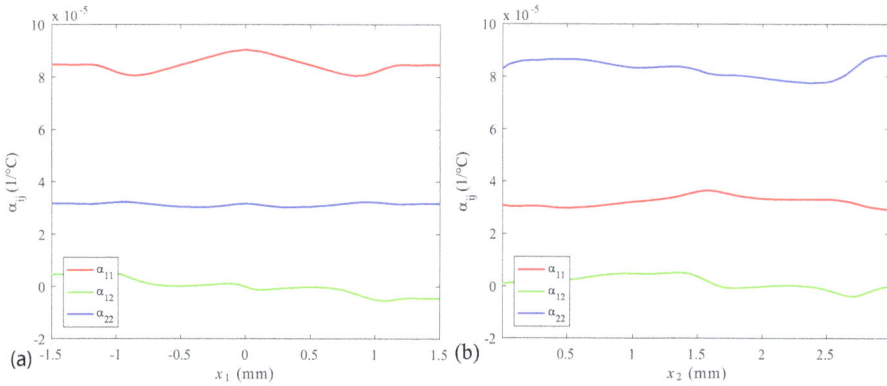

**Figure 6.** Components of $\left\langle \alpha_{ij} \right\rangle$ from using the IRD model with $C_I = 10^{-2}$ (**a**) as a function of $x_1$ at the nozzle exit, and (**b**) as a function of $x_2$ at the end of the bead.

At first glance, it is perhaps harder to notice trends in the CTE that were set by the fiber orientation state, at least while using the IRD model with $C_I = 10^{-2}$. However, it can be seen that $\alpha_1$ is actually the dominant component of the CTE tensor at the nozzle exit and $\alpha_2$ is the dominant component in the deposited bead. Thus, the direction of highest fiber alignment is not the direction of highest CTE. This is due to the low CTE of carbon fibers which counteracts the higher CTE of the ABS matrix and decreases the overall CTE of the composite in the direction of highest fiber alignment. On the other hand, the direction transverse to the direction of highest fiber alignment has a relatively high CTE.

### 4.1.4. Effective Longitudinal Properties

After the stiffness and CTE tensors were found in the deposited bead, the stiffness tensor values were used to define the anisotropic stiffness tensor values across the width of a simulation tensile sample in COMSOL. MATLAB and COMSOL were linked via LiveLink so that the stiffness tensors, which were predicted at the end of each streamline, could be accessed by COMSOL. For COMSOL, the MATLAB function linearly interpolates the stiffness tensor values between streamlines and sets the stiffness tensor values in the region between the outermost streamlines and the boundary of the flow domain to be equivalent to the stiffness tensor values of the outermost streamlines. The simulation involved a displacement-prescribed tensile test after which the bulk, effective, longitudinal Young's modulus was derived using the reaction stress and strain. The simulation tensile sample with boundary conditions as shown in Figure 7a represents an equivalent tensile test for obtaining the longitudinal modulus. The sample before deformation was 1.5 mm wide and 3 mm tall, the same height as the bead in Figure 1a. A 20 × 60 rectangular element mesh was used (20 elements in $x_1$, 60 elements in $x_2$), the entire left side of the sample had a prescribed displacement of 0.01 mm in the $-x_1$ direction, the entire right side was fixed in the $x_1$ direction with the center point fixed also in the $x_2$ direction, and the top and bottom sides were free. After the displacement-prescribed tensile simulation was computed, the bulk effective longitudinal Young's modulus in the $x_1$ direction, $E_1$, was derived using a line average method. This involved dividing the average $x_1$-reaction stress across the entire left side of the sample by the strain, where strain is the average $x_1$-displacement across the entire left side of the sample divided by the original width of the sample. In mathematical terms, this can be expressed as $E_1 = \sigma_{11}/(u_1/w_o)$, where $E_1$ is the bulk, effective, longitudinal Young's modulus, $\sigma_{11}$ is the reaction stress on the left side of the sample, $u_1$ is the average displacement of the entire left side of the sample in the $x_1$ direction, and $w_o$ is the original $x_1$ dimension of the sample. Solving this equation yields a Young's modulus at the end of the bead of $E_1 = 6.82$ GPa for the IRD model with $C_I = 10^{-2}$, not too dissimilar from those results provided in Duty et al. [3] for a similar system. A similar procedure can be done using the stiffness tensor values at the nozzle exit and this yields a Young's modulus of $E_2 = 7.10$ GPa at the nozzle exit for the IRD model with $C_I = 10^{-2}$.

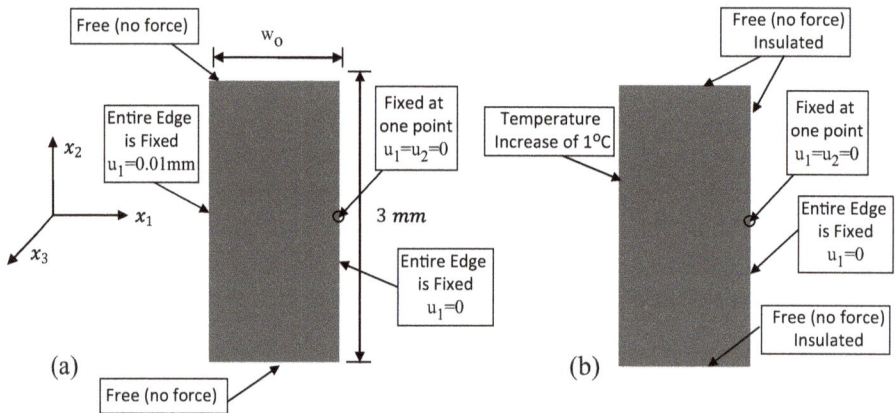

**Figure 7.** Boundary conditions for finite element analysis: (**a**) the tensile sample boundary conditions, and (**b**) the TMA boundary conditions.

A thermomechanical analyzer (TMA) machine is often used to measure the linear CTE of a material sample by heating the sample in a furnace and measuring the displacement (expansion or contraction) of the sample along a particular direction due to the changing temperature (see, e.g., http://www.tainstruments.com/q400/). If the displacement is measured in the $x_1$ direction, for example, then the linear CTE in the $x_1$ direction can be calculated using the equation $\alpha_1 = 1/L_1 \times dL_1/dT$, where $L_1$ is the original $x_1$-length of the sample, $dL_1$ is the change in the $x_1$-length, and $dT$ is the change in temperature of the sample. In this study, the boundary conditions for the representative domain for that of an equivalent test are shown in Figure 7b to calculate the bulk, effective, longitudinal CTE which is $\alpha_1$. The TMA sample was identical in dimensions to the tensile sample, the finite element mesh used was identical to that used in the tensile simulation, and the displacement boundary conditions were identical to those used in the tensile simulation. However, in this simulation, no loads were put on the sample. The sample was initially at a uniform temperature and a temperature increase of $\Delta T = 1\ °C$ was applied to the entire left side of the sample while the other sides were insulated. The bulk, effective, longitudinal CTE in the $x_1$

direction is calculated when thermal equilibrium is attained using the equation $\alpha_1 = -u_1/w_0/\Delta T$, where the negative sign ensures $\alpha_1$ will be positive due to expansion in the $-x_1$ direction, $u_1$ is the average displacement in the $x_1$ direction of the entire left side of the sample, and $w_0$ is the initial $x_1$-length of the sample. After the TMA simulation, the bulk, effective, longitudinal CTE evaluated is $\alpha_1 = 32.2 \times 10^{-6}/°C$ within the deposited bead for the IRD model with $C_I = 10^{-2}$. A similar procedure can be done for the CTE at the nozzle exit and this yields $\alpha_2 = 31.3 \times 10^{-6}/°C$ at the nozzle exit for the IRD model with $C_I = 10^{-2}$.

## 4.2. Contrasting Results between IRD and RSC

The IRD-RSC model decreases the influence of strain on the fiber orientation state and thus predicts different values for the orientation tensors along the flow domain than the IRD model predicts and thus inevitably leads to different predictions for the stiffness and CTE tensors as well. Figure 8 shows components of $A_{ij}$ along streamline 15 in the low nozzle flow domain calculated using the IRD-RSC model with $C_I = 10^{-2}$ and $\kappa = 1/30$. The IRD-RSC model predicted a much lower rate of alignment than the IRD model (compare Figure 8 with Figure 3), although the IRD-RSC model predicted a similar trend in the changes in the components of $A_{ij}$ along streamline 15. That is, $A_{22}$ is highest at the nozzle exit but $A_{11}$ is highest at the bead end, indicating that the fibers tend to align most in the flow direction.

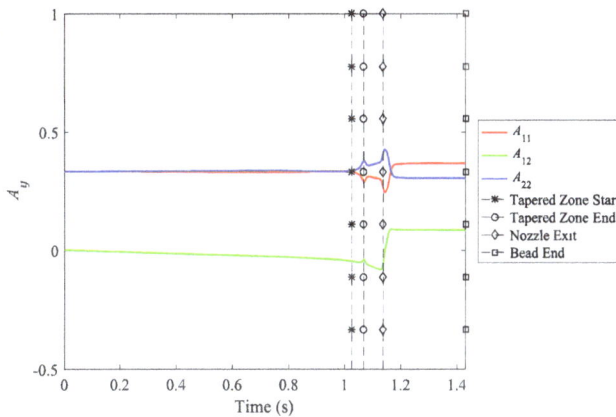

**Figure 8.** $A_{ij}$ components along streamline 15—results are from using the IRD-RSC model with $C_I = 10^{-2}$ and $\kappa = 1/30$.

Components of $A_{ij}$ across the nozzle exit and within the deposited bead, calculated using the IRD-RSC model with $C_I = 10^{-2}$ and $\kappa = 1/30$, are shown in Figure 9. Once again, $A_{11}$ and $A_{22}$ are symmetrical about the $x_2 = 0$ axis at the nozzle exit as one would expect. Figure 9a shows that the IRD RSC model predicts much higher $x_2$ alignment along the sides of the nozzle. This is due to the higher shear along the sides of the nozzle. Figure 9b also shows that the fiber alignment is highest near the sides of the flow domain in the deposited bead.

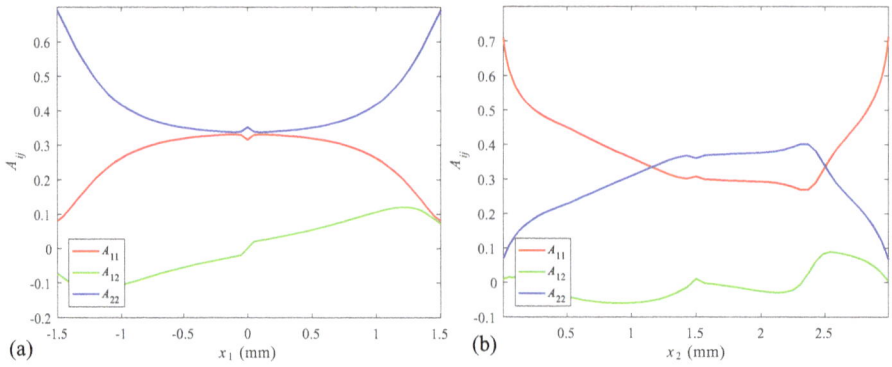

**Figure 9.** Orientation tensors calculated from the IRD-RSC model with $C_I = 10^{-2}$ and $\kappa = 1/30$. (a) $A_{ij}$ as a function of $x_1$ at the nozzle exit and (b) $A_{ij}$ as a function of $x_2$ within the deposited bead.

The homogenized stiffness and CTE tensors were also calculated using the orientation tensors from the IRD-RSC model with $C_I = 10^{-2}$ and $\kappa = 1/30$. The results for these quantities at the end of the bead for the low nozzle domain are shown in Figure 10. Once again, it can be seen that the stiffness tends to follow a similar trend as the orientation tensors and that the CTE, in a sense, follows the opposite trend. Regardless, when comparing the results from that of the IRD-RSC with $\kappa = 1/30$ to that of the IRD model with the same interaction coefficient $C_I$, there is a clear difference in the orientation, the stiffness, and the resulting coefficient of thermal expansion. Performing the same finite element analysis as in the preceding section within the deposited bead, the bulk longitudinal stiffness is $E_1 = 4.82$ GPa and the bulk coefficient of thermal expansion is $\alpha_1 = 45.5 \times 10^{-6}/°C$. Both of these numbers are significantly different than those of the IRD results.

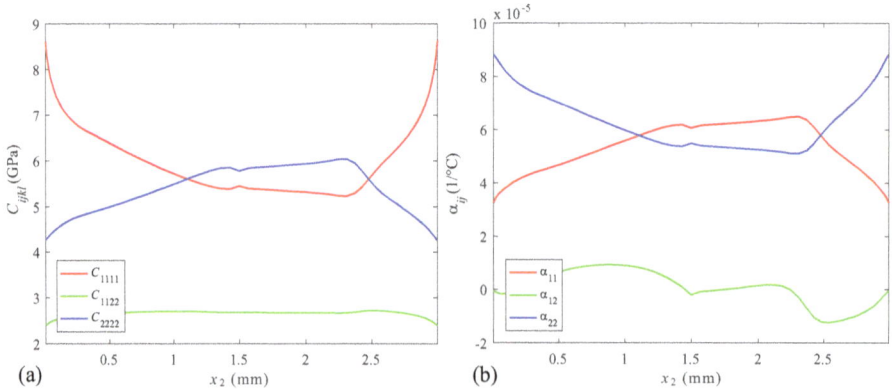

**Figure 10.** Results in the deposited bead from the IRD-RSC model with $C_I = 10^{-2}$ and $\kappa = 1/30$. (a) $\langle C_{ijkl} \rangle$ as a function of $x_2$ in the deposited bead, and (b) $\langle \alpha_{ij} \rangle$ as a function of $x_2$.

### 4.3. Effective Longitudinal Properties—All Flow Conditions

Using the methodology discussed above, the effective, longitudinal stiffness and CTE properties are studied as functions of the interaction model, the degree of fiber interaction $C_I$, the amount of slowness $\kappa$, and whether or not the die swell is included in the analysis. A fiber interaction coefficient of $C_I = 3 \times 10^{-3}$ is chosen to represent melt with a lower degree of interaction, and slowness parameters of $\kappa = 1/10$ and $1/30$ are presented. Table 1 shows the results from the complete study. Comparisons of the results of the deterministic models between the various parameters is made through the use of the percent relative difference, defined as

$$PRD = (|value_1 - value_2|)/(|value_1 - value_2|/2) \times 100\% \tag{13}$$

**Table 1.** Summary of Effective Longitudinal Properties for All Flow Conditions.

| Location | Model | $C_I$ | $\kappa$ | $E_{long}$ (GPa) | $\alpha_{long} \times 10^{-6}/°C$ |
|---|---|---|---|---|---|
| Deposited Bead | IRD | 0.01 | 1 | 6.82 | 32.2 |
| Deposited Bead | RSC | 0.01 | 1/10 | 4.82 | 45.5 |
| Deposited Bead | RSC | 0.01 | 1/30 | 4.18 | 53.3 |
| Deposited Bead | IRD | 0.003 | 1 | 7.69 | 30.3 |
| Deposited Bead | RSC | 0.003 | 1/10 | 4.88 | 45.1 |
| Deposited Bead | RSC | 0.003 | 1/30 | 4.19 | 53.2 |
| Nozzle Exit | IRD | 0.01 | 1 | 7.10 | 31.3 |
| Nozzle Exit | RSC | 0.01 | 1/10 | 5.07 | 42.4 |
| Nozzle Exit | RSC | 0.01 | 1/30 | 4.40 | 49.5 |
| Nozzle Exit | IRD | 0.003 | 1 | 7.89 | 29.9 |
| Nozzle Exit | RSC | 0.003 | 1/10 | 5.20 | 41.7 |
| Nozzle Exit | RSC | 0.003 | 1/30 | 4.45 | 49.2 |

The first study is of the relative importance of the full die-swell model and bead deposition versus that of model domain that stops at the nozzle end. Taking the bulk stiffness and CTE results from the IRD model, the PRD for the longitudinal modulus $E_{long}$ would be 4.02% for $C_I = 10^{-2}$ and 2.56% for $C_I = 3 \times 10^{-3}$, whereas for the same interaction coefficients and a $\kappa = 1/10$ the PRD in the longitudinal modulus is respectively, 5.05% and 6.34% with similar percentages for a $\kappa = 1/30$. The trend in the PRD in the various CTEs ranges between 1.3% and 7.8%. Thus, it would appear that if the ultimate objective is the bulk behavior of the composite, the inclusion of the full model domain with die swell and deposition on the moving platen will refine the results on the order of 5%.

Focusing on the results from the deposited bead and looking at the results as a function of the interaction coefficient for the IRD model, there is a 12% PRD for the longitudinal modulus and a 6% PRD for the longitudinal CTE as the interaction coefficient changes from $C_I = 3 \times 10^{-3}$ to $10^{-2}$. The change as a function of interaction coefficient is nearly non-existent when the IRD-RSC model is used with either value of $\kappa$, with the greatest PRD of 1.3% occurring in the longitudinal modulus for $\kappa = 1/10$.

Comparing the results of the IRD model to that of the IRD-RSC model with $\kappa = 1/10$, the PRD in the longitudinal modulus for $C_I = 10^{-2}$ is 34% and for $C_I = 3 \times 10^{-3}$ it is nearly 45%, with similar numbers for the CTE values. Similarly, when looking at the longitudinal modulus for $\kappa = 1/10$ as compared to that of $\kappa = 1/30$, the PRD is 14% for the higher $C_I$ and 15% for the lower $C_I$, with similar values for the PRD of the CTEs.

Thus, of all the comparisons, the results are most sensitive to the choice of whether or not to use the IRD-RSC model or the original IRD model. It is very interesting to note, that in this flow scenario, where the melt experiences a rapid change in the nozzle region and then the velocity gradients go to zero, the choice of the interaction coefficient has very little bearing on the final part behavior for the newer IRD-RSC model. This observation is in contrast to results within an injection molded product as there is a large period of the melt time that the fibers are subjected to a shear after they have left the mold inlet. It is also worth noting, that although the die-swell and deposition have a substantial bearing on the internal orientation state, the bulk material response is less sensitive to that effect than to issues effecting the interaction coefficient and the slowness parameter. It would also be appropriate to note the sensitivity of the choice of initial conditions on the orientation tensor for the IRD-RSC model. The initial condition can be neglected for the IRD model as the orientation reaches a steady state prior to the tapered nozzle region, but for the IRD-RSC results this is not the case and the final orientation maintains some history based on the somewhat arbitrary choice of the initial conditions.

## 5. Conclusions

In this study, we have developed a methodology for predicting the fiber orientation state within the flow of an FFF 3D printer and subsequently predicting the effective, longitudinal stiffness $E_{long}$ and CTE $\alpha_{long}$. We found that incorporating the die swell into the FFF flow domain did not have a substantial impact on the predicted moduli. However, there was a significant difference between the IRD predictions and the IRD-RSC predictions. When the interaction coefficient $C_I$ varies within the range 0.003–0.01, it has moderate effects on the IRD predictions but the effects are insignificant on the IRD-RSC predictions. Finally, varying the value of the slowness factor $\kappa$, which appears in the IRD-RSC model, does indeed appear to significantly affect the $E_{long}$ and $\alpha_{long}$ predictions. A future study will focus on experimental validation of the computational methodology used in this paper.

**Acknowledgments:** We would like to acknowledge the financial support of Oak Ridge National Laboratories RAMP-UP 40001455134 and Baylor University during this study. We would also like to thank Strangpresse for the donation of the Model 19 extruder which allows us to evaluate large scale deposition both through lab experiments and fluid modeling approaches.

**Author Contributions:** T.R. and D.A.J. performed the work on the orientation analysis and the resulting structural predictions. B.H. was instrumental in the development of flow domain and in finding the stress free surface for the proper accounting for the die swell. D.E.S. and D.A.J. had oversight in all aspects of the project and were involved in all stages of the research.

**Conflicts of Interest:** The authors declare no conflict of interest. The sponsors had no role in the design of the study; in the collection, analyses, or interpretation of data; in the writing of the manuscript, and in the decision to publish the results.

## References

1. Kannan, S.; Senthilkumaran, D.; Elangovan, K. Development of Composite Materials by Rapid Prototyping Technology using FDM Method. In Proceedings of the IEEE International Conference on Current Trends in Engineering and Technology (ICCTET), Coimbatore, India, 3 July 2013; pp. 281–283.
2. Lind, R.; Post, B.; Blue, C.; Love, L. *Enhanced Additive Manufacturing with a Reciprocating Leveling and Force Sensing Platen*; Technical Report 201403257; Oak Ridge National Laboratory: Oak Ridge, TN, USA, 2014.
3. Duty, C.; Kunc, V.; Compton, B.; Post, B.; Erdman, D.; Smith, R.; Lind, R.; Lloyd, P.; Love, L. Structure and Mechanical Behavior of Big Area Additive Manufacturing (BAAM) Materials. *Rapid Prototyp. J.* **2017**, *23*, 181–189.
4. Kishore, V.; Ajinjeru, C.; Nycz, A.; Post, B.; Lindahl, J.; Kunc, V.; Duty, C. Infrared preheating to improve interlayer strength of big area additive manufacturing (BAAM) components. *Addit. Manuf.* **2017**, *14*, 7–12.
5. Narahara, H.; Shirahama, Y.; Koresawa, H. Improvement and Evaluation of the Interlaminar Bonding Strength of FDM Parts by Atmospheric-Pressure Plasma. *Proc. CIRP* **2016**, *42*, 754–759.
6. Lau, K.; Hung, P.; Zhu, M.; Hui, D. Properties of natural fibre composites for structural engineering applications. *Compos. Part B Eng.* **2018**, *136*, 222–233.
7. Soutis, C. Fibre reinforced composites in aircraft construction. *Prog. Aerosp. Sci.* **2015**, *41*, 143–151.
8. Sfondrini, M.; Cacciafesta, V.; Scribante, A. Shear bond strength of fibre-reinforced composite nets using two different adhesive systems. *Eur. J. Orthod.* **2011**, *33*, 66–70.
9. Cacciafesta, V.; Sfondrini, M.; Lena, A.; Scribante, A.; Vallittu, P.; Lassila, L. Flexural strengths of fiber-reinforced composites polymerized with conventional light-curing and additional postcuring. *Am. J. Orthodont. Dent. Orthop.* **2007**, *132*, 524–527.
10. Love, L.; Kunk, V.; Rios, O.; Duty, C.; Elliott, A.; Post, B.; Smith, R.; Blue, C. The Importance of Carbon Fiber to Polymer Additive Manufacturing. *J. Mater. Res.* **2014**, *29*, 1893–1898.
11. Tekinalp, H.; Kunc, V.; Velez-Garcia, G.; Duty, C.; Love, L.; Naskar, A.; Blue, C.; Ozcan, S. Highly Oriented Carbon Fiber-Polymer Composites via Additive Manufacturing. *Compos. Sci. Technol.* **2014**, *105*, 144–150.

12. Folgar, F.; Tucker, C. Orientation Behavior of Fibers in Concentrated Suspensions. *J. Reinf. Plast. Compos.* **1984**, *3*, 98–119.

13. Advani, S.G.; Tucker, C. The Use of Tensors to Describe and Predict Fiber Orientation in Short Fiber Composites. *J. Rheol.* **1987**, *31*, 751–784.

14. Cintra, J.S.; Tucker, C. Orthotropic Closure Approximations for Flow-Induced Fiber Orientation. *J. Rheol.* **1995**, *39*, 1095–1122.

15. Koch, D. A Model for Orientational Diffusion in Fiber Suspensions. *Phys. Fluids* **1995**, *7*, 2086–2088.

16. Huynh, H.M. Improved Fiber Orientation Predictions for Injection-Molded Composites. Master's Thesis, University of Illinois at Urbana-Champaign, Champaign, IL, USA, 2001.

17. Wang, J.; O'Gara, J.; Tucker, C. An Objective Model for Slow Orientation Kinetics in Concentrated Fiber Suspensions: Theory and Rheological Evidence. *J. Rheol.* **2008**, *52*, 1179–1200.

18. Phelps, J.; Tucker, C. An Anisotropic Rotary Diffusion Model for Fiber Orienation in Short- and Long Fiber Thermoplastics. *J. Non-Newt. Fluid Mech.* **2009**, *156*, 165–176.

19. Ortman, K.; Baird, D. Using Startup of Steady Shear Flow in a Sliding Plate Rheometer to Determine Material Parameters for the Purpose of Predicting Long Fiber Orientation. *J. Rheol.* **2012**, *56*, 955–981.

20. Camacho, C.; Tucker, C.; Yalvac, S.; McGee, R. Stiffness and Thermal Expansion Predictions for Hybird Short Fiber Composites. *Polym. Compos.* **1990**, *11*, 229–239.

21. Gusev, A.; Heggli, M.; Lusti, H.; Hine, P. Orientation Averaging for Stiffness and Thermal Expansion of Short Fiber Composites. *Adv. Eng. Mater.* **2002**, *4*, 931–933.

22. Jack, D.; Smith, D. Elastic Properties of Short-Fiber Polymer Composites, Derivation and Demonstration of Analytical Forms for Expectation and Variance from Orientation Tensors. *J. Compos. Mater.* **2008**, *42*, 277–308.

23. Jeffery, G. The Motion of Ellipsoidal Particles Immersed in a Viscous Fluid. *Proc. R. Soc. Lond. A* **1923**, *102*, 161–179.

24. Bay, R.; Tucker, C. Fiber orientation in simple injection moldings. Part I: Theory and numerical methods. *Polym. Compos.* **1992**, *13*, 317–331.

25. Montgomery-Smith, S.J.; Jack, D.; Smith, D. A Systematic Approach to Obtaining Numerical Solutions of Jeffery's Type Equations using Spherical Harmonics. *Compos. Part A* **2010**, *41*, 827–835.

26. Dinh, S.; Armstrong, R. A Rheological Equation of State for Semiconcentrated Fiber Suspensions. *J. Rheol.* **1984**, *28*, 207–227.

27. Lipscomb, G.I.; Denn, M.; Hur, D.; Boger, D. Flow of Fiber Suspensions in Complex Geometries. *J. Non-Newt. Fluid Mech.* **1988**, *26*, 297–325.

28. Verweyst, B.; Tucker, C.L., III. Fiber Suspensions in Complex Geometries: Flow-Orientation Coupling. *Can. J. Chem. Eng.* **2002**, *80*, 1093–1106.

29. Chung, D.; Kwon, T. Numerical Studies of Fiber Suspensions in an Axisymmetric Radial Diverging Flow: The Effects of Modeling and Numerical Assumptions. *J. Non-Newt. Fluid Mech.* **2002**, *107*, 67–96.

30. Advani, S.; Tucker, C. Closure Approximations for Three-Dimensional Structure Tensors. *J. Rheol.* **1990**, *34*, 367–386.

31. Nguyen, N.; Bapanapalli, S.; Kunc, V.; Phelps, J.; Tucker, C. Prediction of the Elastic-Plastic Stress/Strain Response for Injection-Molded Long-Fiber Thermoplastics. *J. Compos. Mater.* **2009**, *43*, 217–246.

32. Bird, R.B.; Curtiss, C.; Armstrong, R.C.; Hassager, O. *Dynamics of Polymeric Liquids. Vol. 2: Kinetic Theory*, 2nd ed.; John Wiley & Sons, Inc.: New York, NY, USA, 1987.

33. Phan-Thien, N.; Fan, X.J.; Tanner, R.; Zheng, R. Folgar-Tucker Constant for a Fibre Suspension in a Newtonian Fluid. *J. Non-Newt. Fluid Mech.* **2002**, *103*, 251–260.

34. Sepehr, M.; Carreau, P.; Grmela, M.; Ausias, G.; Lafleur, P. Comparison of Rheological Properties of Fiber Suspensions with Model Predictions. *J. Polym. Eng.* **2004**, *24*, 579–610.

35. Strautins, U.; Latz, A. Flow-Driven Orientation Dynamics of Semiflexible Fiber Systems. *Rheol. Acta* **2007**, *46*, 1057–1064.

36. Nguyen, N.; Bapanapalli, S.; Holbery, J.; Smith, M.; Kunc, V.; Frame, B.; Phelps, J.; Tucker, C. Fiber Length and orientation in Long-Fiber Injection-Molded Thermoplastics—Part I: Modeling of Microstructure and Elastic Properties. *J. Compos. Mater.* **2008**, *42*, 1003–1029.

37. Agboola, B.; Jack, D.; Montgomery-Smith, S. Effectiveness of Recent Fiber-interaction Diffusion Models for Orientation and the Part Stiffness Predictions in Injection Molded Short-fiber Reinforced Composites. *Compos. Part A* **2012**, *43*, 1959–1970.

38. Stair, S.; Jack, D. Comparison of Experimental and Modeling Results for Cure Induced Curvature of a Carbon Fiber Laminate. *Polym. Compos.* **2017**, *38*, 2488–2500.
39. Zhang, D.; Smith, D.; Jack, D.; Montgomery-Smith, S. Numerical Evaluation of Single Fiber Motion for Short-Fiber-Reinforced Composite Materials Processing. *J. Manuf. Sci. Eng.* **2011**, *133*, 051002.
40. Doi, M. Molecular Dynamics and Rheological Properties of Concentrated Solutions of Rodlike Polymers in Isotropic and Liquid Crystalline Phases. *J. Polym. Sci. Part B Polym. Phys.* **1981**, *19*, 229–243.
41. Hand, G. A Theory of Anisotropic Fluids. *J. Fluid Mech.* **1962**, *13*, 33–46.
42. Chung, D.; Kwon, T. Improved Model of Orthotropic Closure Approximation for Flow Induced Fiber Orientation. *Polym. Compos.* **2001**, *22*, 636–649.
43. Chung, D.; Kwon, T. Invariant-Based Optimal Fitting Closure Approx. for the Numerical Prediction of Flow-Induced Fiber Orientation. *J. Rheol.* **2002**, *46*, 169–194.
44. Jack, D.; Schache, B.; Smith, D. Neural Network Based Closure for Modeling Short-Fiber Suspensions. *Polym. Compos.* **2010**, *31*, 1125–1141.
45. Montgomery-Smith, S.J.; Jack, D.; Smith, D. The Fast Exact Closure for Jeffery's Equation with Diffusion. *J. Non-Newt. Fluid Mech.* **2011**, *166*, 343–353.
46. Montgomery-Smith, S.J.; He, W.; Jack, D.; Smith, D. Exact Tensor Closures for the Three Dimensional Jeffery's Equation. *J. Fluid Mech.* **2011**, *680*, 321–335.
47. Jack, D.; Smith, D. An Invariant Based Fitted Closure of the Sixth-order Orientation Tensor for Modeling Short-Fiber Suspensions. *J. Rheol.* **2005**, *49*, 1091–1116.
48. Jack, D.; Smith, D. Sixth-order Fitted Closures for Short-fiber Reinforced Polymer Composites. *J. Thermoplast. Compos.* **2006**, *19*, 217–246.
49. Velez-Garcia, G.; Mazahir, S.; Hofmann, J.; Wapperom, P.; Barid, D.; Zink-Sharp, A.; Kunc, V. Improvement in Orientation Measurement for Short and Long Fiber Injection Molded Composites. In Proceedings of the 10th Annual SPE Automotive Composites Conference and Exposition, Detroit, MI, USA, 15–16 September 2010.
50. Caselman, E. Elastic Property Prediction of Short Fiber Composites Using a Uniform Mesh Finite Element Method. Master's Thesis, University of Missouri, Columbia, MO, USA, 2007.
51. Halpin, J. Stiffness and Expansion Estimates for Oriented Short Fiber Composites. *J. Compos. Mater.* **1969**, *3*, 732–734.
52. Tucker, C.; Liang, E. Stiffness Predictions for Unidirectional Short-Fiber Composites: Review and Evaluation. *Compos. Sci. Technol.* **1999**, *59*, 655–671.
53. Tandon, G.; Weng, G. The Effect of Aspect Ratio of Inclusions on the Elastic Properties of Unidirectionally Aligned Composites. *Polym. Compos.* **1984**, *5*, 327–333.
54. Lielens, G.; Pirotte, P.; Couniot, A.; Dupret, F.; Keunings, R. Prediction of Thermo-Mech Properties for Compression Moulded Composites. *Compos. Part A* **1998**, *29*, 63–70.
55. Zhang, C. Modeling of Flexible Fiber Motion and Prediction of Material Properties. Master's Thesis, Baylor University, Waco, TX, USA, 2011.
56. Schapery, R. Thermal Expansion Coefficients of Composite Materials Based on Energy Principles. *J. Compos. Mater.* **1968**, *2*, 380–404.
57. Heller, B.; Smith, D.; Jack, D. Simulation of Planar Deposition Polymer Melt Flow and Fiber Orientation in Fused Filament Fabrication. In Proceedings of the Solid Freeform Fabrication Symposium, Austin, TX, USA, 7–9 August 2017; pp. 1096–1111.
58. Tadmor, Z.; Gogos, C. *Principles of Polymer Processing*, 2nd ed.; Wiley Interscience: Hoboken, NJ, USA, 2006.
59. Athanasopulos, I.; Makrakis, G.; Rodrigues, J.F. *Free Boundary Problems: Theory and Applications*; Chapman and Hill/CRC Press: Boca Raton, FL, USA, 1999.

*Journal of*
*composites science*

MDPI

*Article*

# Rheology Effects on Predicted Fiber Orientation and Elastic Properties in Large Scale Polymer Composite Additive Manufacturing

Zhaogui Wang and Douglas E. Smith *

Department of Mechanical Engineering, Baylor University, Waco, TX 76798, USA; Zhaogui_Wang@baylor.edu
* Correspondence: Douglas_E_Smith@baylor.edu; Tel.: +1-254-710-6830

Received: 1 January 2018; Accepted: 13 February 2018; Published: 16 February 2018

**Abstract:** Short fiber-reinforced polymers have recently been introduced to large-scale additive manufacturing to improve the mechanical performances of printed-parts. As the short fiber polymer composite is extruded and deposited on a moving platform, velocity gradients within the melt orientate the suspended fibers, and the final orientation directly affects material properties in the solidified extrudate. This paper numerically evaluates melt rheology effects on predicted fiber orientation and elastic properties of printed-composites in three steps. First, the steady-state isothermal axisymmetric nozzle melt flow is computed, which includes the prediction of die swell just outside the nozzle exit. Simulations are performed with ANSYS-Polyflow, where we consider the effect of various rheology models on the computed outcomes. Here, we include Newtonian, generalized Newtonian, and viscoelastic rheology models to represent the melt flow. Fiber orientation is computed using Advani–Tucker fiber orientation tensors. Finally, elastic properties in the extrudate are evaluated based from predicted fiber orientation distributions. Calculations show that the Phan–Thien–Tanner (PTT) model yields the lowest fiber principal alignment among considered rheology models. Furthermore, the cross section averaged elastic properties indicate a strong transversely isotropic behavior in these composites, where generalized Newtonian models yield higher principal Young's modulus, while the viscoelastic fluid models result in higher shear moduli.

**Keywords:** short fiber-reinforced polymer; large-scale additive manufacturing; rheology effect; die swell; fiber orientation; elastic properties

---

## 1. Introduction

Recently, extrusion-based Additive Manufacturing (AM) (otherwise known as fused filament fabrication or fused deposition modeling) has moved rapidly from small scale rapid prototyping to the manufacture of large-scale parts and tooling such as the Big Area Additive Manufacturing (BAAM) system developed by Oak Ridge National Laboratories (ORNL) (Oak Ridge, TN, USA) [1]. Typical thermoplastic polymer extrusion-based AM is a process where polymer feedstock materials are melted and deposited on a heated platform, layer-by-layer, to form three-dimensional (3D) objects [2]. To achieve a relatively high dimensional accuracy and superior mechanical performances in large-scale parts, carbon fiber filled polymers are employed. Duty et al. show that adding short carbon fibers into the neat Acrylonitrile Butadiene Styrene (ABS) polymer yields a composite with improved elastic properties, especially along the printing direction, and less distortion in the printed part following the bead deposition process of the BAAM system [3,4].

A key factor in the polymer composite deposition AM process is the flow-induced fiber orientation within the printed composite (cf. Figure 1) since material properties of solidified parts depend on the fiber alignment within the printed bead [5]. Therefore, the prediction of fiber orientation during the polymer extrusion process, and the subsequent evaluation of mechanical properties in the short fiber polymer composite extrudate is of great importance.

Evans et al. [6] and Lipscomb et al. [7] considered a fully coupled approach where the motion of suspended fibers depend on the flow field, and the fiber orientation influences flow kinematics, typically through the suspension melt viscosity. A computationally more efficient approach that is often employed is a one-way weakly coupled formation that ignores the effect of the fiber suspension on viscosity. This approach has been effective in applications having shear dominant narrow gap flows such as injection and compression modelling simulations [8–10], and is the approach we use in this study.

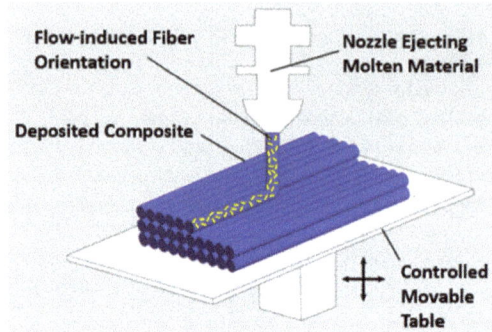

**Figure 1.** Flow-induced fiber orientation occurs during the process of large-scale additive manufacturing.

Fiber orientation studies in polymer composite AM applications have recently become of interest. Nixon et al. [11] simulated fiber orientation in three Fused Deposition Modeling (FDM) nozzle geometries (convergent, straight and divergent) using Moldflow (Moldflow Corporation, Framingham, MA, USA) and the Folgar–Tucker Isotropic Rotary Diffusion (IRD) model [12]. Their work, which ignored die swell, showed that a converging geometry yielded the highest principal fiber alignment and the divergent geometry resulted in the lowest. Additionally, at the exit of the straight and the converging nozzle, a higher alignment was predicted near the center than at the edge, unlike the experimental result reported by Kunc [13]. Heller et al. [14] computed the fiber orientation tensors in a conventional small scale FDM nozzle and extruded filament. In their work, die swell was computed by minimizing the integrated normal stress on the free surface using COMSOL Multiphysics (Comsol, Inc., Burlington, MA, USA). Their approach modeled the molten polymer as an isothermal Newtonian fluid in a creeping flow, and assumed an axisymmetric velocity field. Orientation tensors (cf. e.g., Advani and Tucker [15]) were computed along streamlines within the flow domain from velocity and velocity gradient information. Their results showed that fiber alignment reached its peak at the outer edge of the nozzle, and then decreased towards the center of the flow.

Extrudate swell occurs in many extrusion-based polymer processing applications and is known to be highly influenced by the non-Newtonian behavior of the melt. Crochet et al. [16] theoretically analyzed the die swell of an upper-convected Maxwell fluid based on the mixed finite element method for fluids with implicit constitutive equations. Luo and Tanner [17] applied the Streamline Finite Element Method (SFEM) to the die swell problem, which avoided the numerical instability in high Weissenberg number problems. Luo and Mitsoulis [18] extended the SFEM by adding in a particle-tracking scheme along the streamlines with a Picard iterative scheme. Béraudo et al. [19] applied a finite-element-based method to investigate the extrudate swell of Linear Low-Density Polyethylene (LLDPE) and Low-Density Polyethylene (LDPE) melts using a multi-mode Phan–Thien–Tanner (PTT) model [20]. Their approach provided an accurate die swell prediction for die geometries of a 2D slit die and a 2D axisymmetric capillary die in low and intermediate shear conditions. Ganvir et al. [21] applied an Arbitrary Lagrangian Eulerian (ALE) algorithm to calculate the extrudate free surface, which enabled the die swell simulations to be performed in both steady state and transient problems. Alternatively, Limtrakarn et al. [22] employed the Simplified Viscoelastic (SV) model implemented in ANSYS-Polyflow (ANSYS, Inc., Canonsburg, PA, USA) to predict die swell of a 3D circular die flow of LDPE. A good agreement between the numerical results and the experimental data was achieved. Clemeur et al. [23] found that the SV model was a cost-effective approach for evaluating the flow-viscoelasticity as compared to conventional viscoelastic fluid flow models including the Oldroyd-B and the Phan–Thien–Tanner models.

This paper presents a numerical approach to study the effect of assumed polymer melt rheology on predicting fiber orientation and elastic properties of short fiber polymer composites extruded in large-scale AM. The weakly coupled formation is used to compute the fiber orientation within the polymer melt flow where we first obtain the flow kinematic in an isothermal axisymmetric large-scale AM nozzle. Our flow model is created in two dimensions and solved with the finite element suite ANSYS-Polyflow (version 17.1, ANSYS, Inc.) [24], and includes melt flow within the nozzle in addition to a short section of post-nozzle extrudate, which enables the prediction of die swell at the nozzle exit. We consider a Newtonian fluid model, a Power law model, a Carreau–Yasuda model, a multi-mode Phan–Thien–Tanner (PTT) model, and a Simplified Viscoelastic (SV) model in separate flow simulations. Secondly, the fiber orientation along streamlines within the flow domain is computed from the velocity field computed within the melt flow domain. The Advani–Tucker fiber orientation tensor evaluation equation [15] and the Folgar–Tucker Isotropic Rotary Diffusion (IRD) model [12] are employed to solve the fiber orientation problem. In addition, the Orthotropic Closure (ORT) [5] is used to address the closure problem encountered in the fiber orientation computation. Finally, elastic properties are computed from fiber orientation predictions for each melt rheology using the Tandon–Wang approach with fiber orientation averaging [25,26]. This paper uses computational methods alone to gain useful insight into the effect of assumed rheology on properties of an extruded composite bead. It does not include the difficult, if not impossible, experimental procedures that would be required to validate these results.

## 2. Materials

In this study, we consider the material rheology of Acrylonitrile Butadiene Styrene (ABS) fabricated by the PolyOne Corporation (Avon Lake, OH, USA). The rheological properties, including the complex shear viscosity $\eta$, storage shear modulus $G'$ and loss shear modulus $G''$, are measured using a HAAKE MARS 40 rheometer (Thermo Fisher Scientific, Waltham, MA, USA) at 210 °C. Once the experimental data is obtained, curve-fitting for the various rheology models is performed using ANSYS-Polymat (version 17.1, ANSYS, Inc.) [27].

The experimental data appears in Figure 2, which shows apparent shear shinning behavior as expected for ABS. The shear thinning behavior of polymer melts can be expressed through various Generalized Newtonian Fluid (GNF) models. Here, we consider the Power law model written as [28]

$$\eta(\dot{\gamma}) = K(\dot{\gamma})^{n-1},\tag{1}$$

and Carreau–Yasuda model given as [28]

$$\eta(\dot{\gamma}) = \eta_\infty + (\eta_0 - \eta_\infty)\left(1 + (\kappa\dot{\gamma})^a\right)^{\frac{n-1}{a}}.\tag{2}$$

In the above, K is the consistency index, n is the power-law index, $\eta_\infty$ is the infinite-shear-rate viscosity, $\eta_0$ is the zero-shear-rate viscosity, and $\kappa$ is the natural time. In Equation (2), the constant a controls the transition from the Newtonian plateau to the Power-law region in the Carreau–Yasuda model. For measured data appearing in Figure 2, the fitted rheology model data for the Power law model is K = 16761 Pa·s$^n$ and n = 0.4503. For the Carreau–Yasuda model, we obtained

$\eta_0 = 204064$ Pa·s$^{\frac{n-1}{a}}$, $\eta_\infty = 0$, $\mu = 0.3333$ s$^{-1}$, $n = 0.000001455$, and $a = 0.2398$. For comparison, we also consider the Newtonian fluid model, also appearing in Figure 2, having a viscosity $\mu = 3200$ Pa·s, which corresponds to a shear rate of ~30 s$^{-1}$ from our measured rheology. This shear rate was used to determine a Newtonian viscosity value based on a typical shear rate of 30~40 s$^{-1}$ given by Duty et al. [29] for Big Area Additive Manufacturing (BAAM) systems.

We also consider a multi-mode PTT rheology model, which is a differential-type viscoelastic fluid model. The exponential form of the PTT model is expressed as [20]

$$\exp\left[\frac{\varepsilon\lambda}{\eta_1}tr(\mathbf{T}_1)\right]\mathbf{T}_1 + \lambda\left[\left(1 + \frac{\xi}{2}\right)\mathbf{T}_1^\nabla + \frac{\xi}{2}\mathbf{T}_1^\Delta\right] = 2\eta_1\mathbf{D}, \tag{3}$$

with

$$\mathbf{T}_1{}^\Delta = \frac{D\mathbf{T}_1}{Dt} + \mathbf{T}_1\cdot(\nabla\mathbf{v})^T + \nabla\mathbf{v}\cdot\mathbf{T}_1, \tag{4}$$

and

$$\mathbf{T}_1{}^\nabla = \frac{D\mathbf{T}_1}{Dt} - \mathbf{T}_1\cdot\nabla\mathbf{v} - (\nabla\mathbf{v})^T\cdot\mathbf{T}_1, \tag{5}$$

and

$$\mathbf{D} = \frac{1}{2}\left[(\nabla\mathbf{v}) + (\nabla\mathbf{v})^T\right]. \tag{6}$$

Here, $\mathbf{T}_1$ and $\eta_1$ are the stress tensor and the viscosity component associated with the viscoelasticity, $\mathbf{D}$ is the strain rate tensor, $\mathbf{v}$ is the velocity tensor, $\lambda$ is the mode relaxation time, $\eta$ is the mode viscosity, $\xi$ controls the shear viscosity behavior, and $\varepsilon$ controls the elongational behavior. The fitting results in Figure 3 are in good agreement with the experimental rheology data. The fitted parameters of the PTT model appear in Table 1.

Alternatively, ANSYS-Polyflow includes the Simplified Viscoelastic (SV) model that reduces computational expense when predicting die swell in viscoelastic flows. In the SV formulation, it is understood that extrudate swell in polymer extrusion is associated with the first normal stress difference in the fluid. Hence, the SV model extends the Generalized Newtonian Fluid (GNF) model, where the total stress tensor is given as [24]

$$\mathbf{T}_1 = \begin{bmatrix} \Psi\mu(\dot{\chi})\dot{\chi} & \eta(\dot{\chi})\dot{\chi} & 0 \\ \eta(\dot{\chi})\dot{\chi} & 0 & 0 \\ 0 & 0 & 0 \end{bmatrix}, \tag{7}$$

where the off-diagonal terms given as $\eta(\dot{\gamma})\dot{\gamma}$ are the shear stress components. In this form, $\eta(\dot{\gamma})$ is expressed by a typical generalized Newtonian model, and $\dot{\gamma}$ is the magnitude of the strain rate tensor $\mathbf{D}$ (cf. Equation (6)). In the above, $\Psi\mu(\dot{\chi})\dot{\chi}$ represents the first normal stress component, in which $\mu(\dot{\chi})$ is described in a similar fashion as is done for the shear strain rate. In Equation (7), $\dot{\chi}$ is the specialized viscoelastic variable, which is evaluated with the transport equation

$$\theta(\dot{\gamma})\frac{D\dot{\chi}}{Dt} + \dot{\chi} = \dot{\gamma}, \tag{8}$$

where $\theta(\dot{\gamma})$ is the relaxation time of the melt which controls the development of the extrudate swell diameter once the melt flow exits the nozzle. In addition, $\Psi$ appearing in Equation (7) is an artificial weighting factor, which controls the swelling enhancement versus the input flow rate [24]. To use Equation (7), a description of the shear viscosity is required. Since the Power law model exhibits an unbounded viscosity at a near-zero shear rate, it is not realistic, especially when free surface prediction is of primary interest. Hence, we employ the Carreau–Yasuda law to represent the behavior of shear viscosity. It has been shown that defining an independent law for the normal stress viscosity increases computational cost, but does not greatly enhance the accuracy of die swell calculation. Therefore, the same Carreau–Yasuda model form is used to describe the first normal stress viscosity term.

The SV model is an empirical construction defined in terms of $\theta$ and $\Psi$, each of which is typically defined to obtain a known flow domain property such as die swell. In our simulation, different sets of parameters $\theta$ and $\Psi$ are attempted and values are selected to provide a predicted die swell profile that is in good agreement with results obtained using the PTT viscoelastic fluid model. In this study, $\theta$ and $\Psi$ are thus defined as 0.26 and 0.47, respectively.

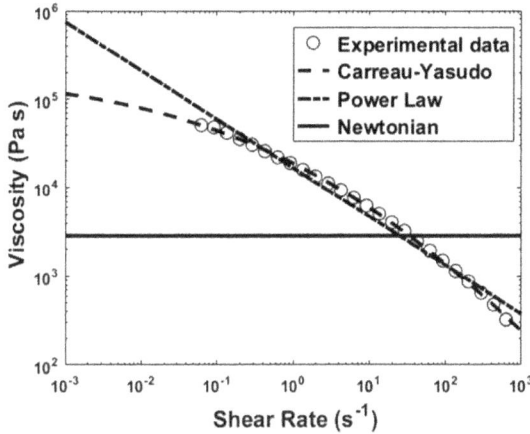

**Figure 2.** Shear viscosity curve fitting using Generalized Newtonian Fluid models.

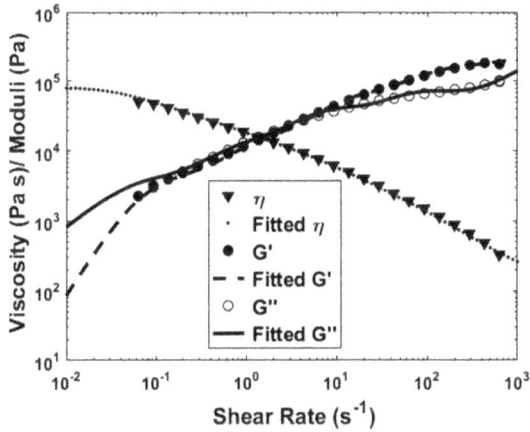

**Figure 3.** Curve fitting the experimental rheology data using the Phan–Thien–Tanner model.

**Table 1.** Phan–Thien–Tanner model parameters for Acrylonitrile Butadiene Styrene.

| Mode No. ($i$) | $\lambda_i$ (s) | $\eta_i$ (Pa s) | $\varepsilon$ | $\xi$ |
|---|---|---|---|---|
| 1 | 0.00022 | 131.7 | 0.75 | 0.18 |
| 2 | 0.0022 | 44.7 | 0.75 | 0.18 |
| 3 | 0.012 | 1180.8 | 0.75 | 0.18 |
| 4 | 0.12 | 6286.4 | 0.75 | 0.18 |
| 5 | 1.14 | 13065.7 | 0.75 | 0.18 |
| 6 | 13.82 | 61917.7 | 0.75 | 0.18 |

## 3. Methods

### 3.1. Flow Kinematics and Die Swell Evaluation

We use ANSYS-Polyflow [24] to evaluate the flow kinematics in the polymer melt flow domain based on conservation of momentum

$$-\nabla p + \nabla \cdot \mathbf{T} + \mathbf{f} = \rho \mathbf{a}, \tag{9}$$

and conservation of mass

$$\nabla \cdot \mathbf{v} = 0, \tag{10}$$

where we have assumed an isothermal incompressible fluid represents the polymer melt as often appeared in extrusion die flow numerical studies [19,21,30]. In the above, p is the pressure, **T** is the total stress tensor, **f** is the body force, ρ is the density of the fluid and **a** is acceleration. Note that non-isothermal effects such as the temperature gradients within the free extrudate or viscous heating in the melt flow may result in nonuniform melt rheology properties such as K in Equation (1), which is often addressed with an Arrhenius-type temperature dependency (e.g., the Williams–Landel–Ferry equation) [28]. Consequently, we understand that our isothermal assumption may yield some inaccuracy in the predicted numerical data. However, it is expected, as suggested by others [19,21,30], that any temperature-related variation in the rheology properties would not significantly alter the trends that appear in our result section.

Moreover, for a Generalized Newtonian Fluid (GNF) model, the total stress tensor may be written as

$$\mathbf{T} = \mathbf{T}_2 = 2\eta_2 \mathbf{D}, \tag{11}$$

and for a viscoelastic fluid model

$$\mathbf{T} = \mathbf{T}_1 + \mathbf{T}_2, \tag{12}$$

with

$$\eta = \eta_1 + \eta_2, \tag{13}$$

where $\mathbf{T}_1$ and $\eta_1$ are the non-viscous contributions, and $\mathbf{T}_2$ and $\eta_2$ are related to the viscous effect of the flow and η is the total viscosity.

The geometry of the flow domain in our study is based on the large-scale Additive Manufacturing Strangpresse Model-19 extruder nozzle appearing in Figure 4. In addition, we include a 1-inch section of free extrudate beyond the nozzle exit in the simulation to capture die swell. Due to the axisymmetry of the nozzle geometry and assumed flow, we are able to simplify the flow domain as a 2D axisymmetric model, which saves significant computational expense. Furthermore, our axisymmetric assumption ignores any swirling motion in the flow that may result from the extruder screw. Since creeping flow is assumed, the inertia contribution "ρa" in Equation (9) is ignored in our simulations. In addition, since only a short section of the free extrudate material is considered, gravitational effects are ignored, which results in the body force **f** in Equation (9) being zero as well.

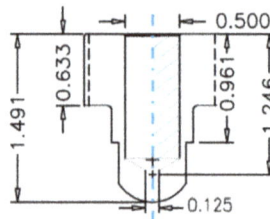

**Figure 4.** Geometric dimensions of a Strangpresse Model-19 extruder nozzle.

The boundary conditions of the flow domain appear in Figure 5, in which

- $\Gamma_1$: Flow domain inlet, where the prescribed volumetric flow rate Q is specified. In addition, a fully developed velocity profile is computed and imposed at the inlet by ANSYS-Polyflow based on Q and the selected rheology model.
- $\Gamma_2$: No slip wall boundary, where $v_s = v_n = 0$.
- $\Gamma_3$: Axis of symmetry, where $F_s = v_n = 0$.
- $\Gamma_4$: No slip wall boundary, where $v_s = v_n = 0$.
- $\Gamma_5$: Free surface, where $\mathbf{v} \cdot \mathbf{n} = 0$.
- $\Gamma_6$: Flow domain exit, where $F_n = v_s = 0$.

In the above, $F_s$ is the tangential force, $F_n$ is the normal force, $v_s$ is the tangential velocity, $v_n$ is the normal velocity, $\mathbf{v}$ is the velocity vector at the free surface, and $\mathbf{n}$ is a unit vector normal to the free surface [24]. The die swell of the free surface is predicted using the methods of spines in ANSYS-Polyflow, which is an efficient remeshing rule often applied to 2D free surface problem [31]. The finite element domain is discretized into 704 nodes and 630 elements using 4-node quadrilateral elements as shown in Figure 5. The mesh size is reduced near the flow boundary as well as the nozzle exit to avoid potential singularity issues. Additionally, results obtained using a coarse mesh (448 nodes, 378 elements) and a fine mesh (960 nodes, 882 elements) were compared with those obtained with the model in Figure 5. Elastic moduli predictions appearing in the results section below were obtained using the model in Figure 5 and are within 1% absolute relative difference to the fine mesh model output. We therefore use the model in Figure 5 in the remainder of the paper to avoid the extra computational expense that would be required for a model having a finer mesh.

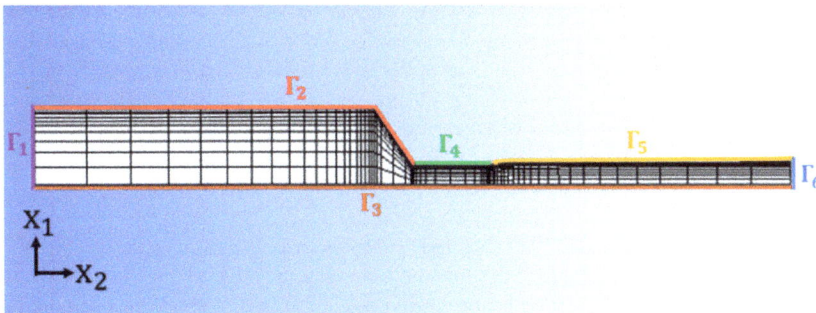

**Figure 5.** Mesh and boundary condition of the flow domain.

*3.2. Fiber Orientation Distribution Prediction*

The direction of a single rigid fiber within a polymer matrix is commonly described by a unit vector $\mathbf{p}(\varphi, \phi)$, as shown in Figure 6, with coordinates [15]

$$\mathbf{p}(\varphi, \phi) = \left\{ \begin{array}{c} \sin\varphi\cos\phi \\ \sin\varphi\sin\phi \\ \cos\varphi \end{array} \right\}. \tag{14}$$

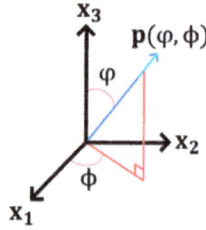

**Figure 6.** Coordinates of the vector $\mathbf{p}(\varphi, \phi)$ defining by the angles $\varphi$ and $\phi$.

Advani and Tucker [15] considered the statistical behavior of the fibers using the computationally efficient orientation tensor approach where the orientation tensor evolution equation is written as

$$\frac{D\mathbf{A}}{Dt} = (\mathbf{A} \cdot \mathbf{W} - \mathbf{W} \cdot \mathbf{A}) + \beta(\mathbf{D} \cdot \mathbf{A} + \mathbf{A} \cdot \mathbf{D} - 2\mathbb{A} : \mathbf{D}) + 2C_I\dot{\gamma}(\mathbf{I} - 3\mathbf{A}), \tag{15}$$

which assumes isotropic rotary diffusion first proposed by Folgar and Tucker [12]. Here, $\mathbf{A}$ and $\mathbb{A}$ are the second and fourth order orientation tensors, respectively, written as

$$\mathbf{A} = A_{ij} = \oint_\mathbb{S} p_i p_j \delta(\varphi, \phi) d\mathbb{S}, \tag{16}$$

and

$$\mathbb{A} = A_{ijkl} = \oint_\mathbb{S} p_i p_j p_k p_l \delta(\varphi, \phi) d\mathbb{S}, \tag{17}$$

where $\delta(\varphi, \phi)$ is a probability distribution function and $\mathbb{S}$ is unit sphere surface. Note that, due to the normalization condition, the integral of $\delta(\varphi, \phi)$ over the surface $\mathbb{S}$ equates to unity, making the trace of $\mathbf{A}$ equal to 1 (see e.g., [25,26]). It can also be shown that $\mathbf{A}$ is symmetric, yielding just five independent components in Equation (15).

In additional, the vorticity tensor $\mathbf{W}$ appearing in Equation (15) is given as

$$\mathbf{W} = \frac{1}{2}\left[(\nabla\mathbf{v}) - (\nabla\mathbf{v})^\mathrm{T}\right], \tag{18}$$

and the rate of deformation tensor $\mathbf{D}$ is evaluated by Equation (6). We evaluate the tensors $\mathbf{W}$ and $\mathbf{D}$, from the velocity vector $\mathbf{v}$ computed along streamlines within the polymer melt flow field obtained from our ANSYS-Polyflow simulation result. The constant $\beta$ in Equation (15) depends on the fiber aspect ratio as

$$\beta = \frac{\alpha^2 - 1}{\alpha^2 + 1}, \tag{19}$$

where $\alpha$ is the fiber aspect ratio. The interaction coefficient $C_I$ is used to capture the effect of fiber–fiber interaction, and $\dot{\gamma}$ represents the scalar magnitude of the rate of deformation tensor $\mathbf{D}$. The last term in Equation (15) written as $2C_I\dot{\gamma}(\mathbf{I} - 3\mathbf{A})$ results from the the Folgar–Tucker Isotropic Rotary Diffusion (IRD) model [12]. Fu et al. [32] experimentally observed that molten short fiber polymer composite exhibited an asymmetric profile of fiber length distribution with a peak skewed toward small values of fiber length. In addition, we experimentally measured the fiber aspect ratio of the sample prepared by performing a burn-off test on the 13 wt % carbon fiber filled ABS (manufactured by Polyone, Avon Lake, OH, USA) and found that the values of the fiber aspect ratio are in a range of 10 to 60. Following Fu et al., we define $\beta = 0.9802$ (corresponding to $\alpha = 10$ in Equation (19)). Bay and Tucker [9] defined an empirical formula for evaluating $C_I$, which depends on the values of the fiber volume fraction and aspect ratio as

$$C_I = 0.0184 \exp(-0.7148 v_f \alpha), \tag{20}$$

where $v_f$ is the fiber volume fraction. Equation (20) has been effective in concentrated suspensions flows, which are typical for short fiber polymer composites in large-scale deposition processes. We also assume a fiber volume fraction of 13% following that used by ORNL in prior study during the development of the Big Area Additive Manufacturing (BAAM) [3]. Therefore, we use $C_I = 0.0073$, which is computed with Equation (20) for $v_f = 0.13$ and $\alpha = 10$.

The fourth order fiber orientation tensor, $\mathbb{A}$, is typically computed with a closure approximation in fiber orientation simulations. Prior studies have focused on the natural–type closure [33,34] and the orthotropic-type closure [35,36]. In this paper, we employ the Orthotropic Closure (ORT) to compute for $\mathbb{A}$ from **A** as defined in [5].

Note that the second order orientation tensor **A** has seen widespread use for statistically describing the orientation of suspended fibers in narrow gap shear dominant applications [8–10,13]. In this study, flow within the nozzle has a significant shear component; however, there is considerable extensional flows within the converging section of the nozzle, and also just outside the nozzle exit where die swell begins to form. Here, we have chosen to demonstrate the use of this common orientation tensor formulation in large-scale AM flows realizing that the limitations of the model in these flows is still to be determined. It is important to note that the orientation tensor approach does not track each individual fiber, but instead provides an indication of the degree of alignment through the nine components of tensor **A**. The second order orientation tensor **A** is the second moment of the orientation distribution function $\delta(\varphi, \phi)$. Specifically, Figure 7 gives two important examples of **A**, in which a diagonal component of **A** having a value of one represents full alignment in the corresponding direction, and three diagonal components all equal to $1/3$ represents the case of a uniformly random orientation.

In addition, the assumption of the initial fiber orientation state at the nozzle inlet directly influences the fiber orientation throughout the flow domain. We assume that the fiber orientation state prior to entering the nozzle has attained a fully developed steady state as the flow reaches the nozzle inlet. It is important to note that the initial condition of the fiber orientation has been found to have an influence on predicted fiber orientation in injection molding processes by Baird et al. [37]. We note that the complexity in the melt flow during the extrusion process before the flow reaches the nozzle will indeed influence the fiber orientation state as the melt enters the nozzle. To better understand the effect of inlet fiber orientation on computed outputs, we performed other simulations using a uniformly random fiber orientation at the nozzle inlet. In this case, we found that using the alternate inlet fiber orientation condition had little effect on the trends in predicted extrudate fiber orientation and mechanical properties shown below. In addition, the steady state fiber orientation tensor is obtained by setting the left-hand side of Equation (15) to zero and then solving for components of **A**.

$$A_{ij} = \begin{bmatrix} 0 & 0 & 0 \\ 0 & 1 & 0 \\ 0 & 0 & 0 \end{bmatrix} \qquad A_{ij} = \begin{bmatrix} 1/3 & 0 & 0 \\ 0 & 1/3 & 0 \\ 0 & 0 & 1/3 \end{bmatrix}$$

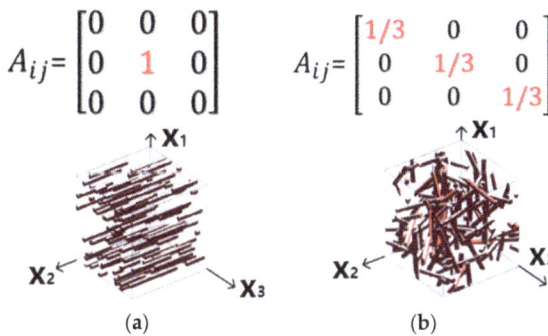

(a)                          (b)

**Figure 7.** Fiber orientation tensors of (**a**) fully alignment along $x_2$ direction and (**b**) uniformly random alignment.

*3.3. Printed Bead Elastic Properties*

Advani and Tucker [15] first computed the volume-averaged stiffness tensor **C** ($C_{ijkl}$) using the fiber orientation tensors **A** ($A_{ij}$) and $\mathbb{A}$ ($A_{ijkl}$), where components of **C** ($C_{ijkl}$) are given as

$$C_{ijkl} = M_1 A_{ijkl} + M_2 \left( A_{ij}\delta_{kl} + A_{kl}\delta_{ij} \right) + M_3 \left( A_{ik}\delta_{jl} + A_{il}\delta_{jk} + A_{jl}\delta_{ik} + A_{jk}\delta_{il} \right) +$$
$$M_4 A_{ij}\delta_{kl} + M_5 \left( A_{ik}\delta_{jl} + A_{il}\delta_{jk} \right), \tag{21}$$

where the constants $M_i$ is computed from the five independent components of the related unidirectional composite stiffness tensor $\widetilde{C}_{ijkl}$ and $\delta_{il}$ is the Kronecker delta [38]. We follow the work of Jack et al. [25] who used the Tandon–Wang analytical equations [39] to evaluate the unidirectional elastic moduli of the discontinuous fiber-reinforced polymer. The elastic properties of the fiber and the matrix used in our calculations below are given in Table 2, where we assume that the matrix and fiber are both isotropic materials. In addition, as in Section 3.2, we use a fiber aspect ratio of 10 and the fiber volume fraction is assumed to be 13% when calculating the unidirectional elastic moduli with the Tandon–Wang approach (see e.g., [26]).

**Table 2.** Elastic properties of the fiber and the matrix of the composite.

| Material | Young's Modulus, E (GPa) | Shear Modulus, G (GPa) | Poisson's Ratio, $\nu$ |
|---|---|---|---|
| Carbon fiber | 230 | 95.83 | 0.2 |
| ABS matrix | 2.5 | 0.93 | 0.35 |

## 4. Results and Discussions

The objective of this paper is to demonstrate how different representations of the melt rheology properties affects the predicted fiber orientation in the polymer melt and also the elastic properties of the resulting extrudate material. In this section, we first consider the melt flow die swell predictions computed with ANSYS-Polyflow. Then, fiber orientation tensors are obtained along streamlines within the polymer melt flow from the computed velocity field. Furthermore, the elastic properties within the solidified extrudate are evaluated from the fiber orientation tensors. We also evaluate the average elastic properties in the extrudate by numerical integration of the data points over the cross-section.

*4.1. Predicted Die Swell*

Ajinjeru and Duty showed that the typical wall shear rate appearing in Big Area Additive Manufacturing (BAAM) systems is between 30 and 40 s$^{-1}$, and reaches a peak value near 100 s$^{-1}$ at the nozzle exit [29]. In our simulations, the average wall shear rate using the PTT model with material constants from Table 1 and an inlet flow rate of Q = 100 mm$^3$/s is calculated to be 36 s$^{-1}$, with a peak value of 87 s$^{-1}$ at nozzle exit, which agrees well with the literature data [29]. Here, the average wall shear rate $\bar{\dot{\gamma}}_w$ is computed from the wall shear rate $\dot{\gamma}_w$ on $\Gamma_4$ (cf. Figure 5) as

$$\bar{\dot{\gamma}}_w = \frac{1}{L} \int_{\Gamma_4} \dot{\gamma}_w \, dx_2, \tag{22}$$

where L is the length of the nozzle exit tube defined by $\Gamma_4$ in the $x_2$ direction.

Die swell profiles for the nozzle flow problem defined in Section 2 using each of the rheology models defined above appear in Figure 8. The die swell just downstream of the nozzle exit is assessed using the apparent swell ratio B defined as

$$B = \frac{d}{d_0},$$ (23)

where d is the steady state swell flow diameter evaluated along the free surface downstream of the die exit (length of $\Gamma_6$ appearing in Figure 5), and $d_0$ is the nozzle exit diameter. The computed data for B at the $\Gamma_6$ surface is given in Table 3. The apparent die swell ratio B = 1.133 computed using the Newtonian fluid model agrees with the swell ratio of 1.13 proposed by Reddy and Tanner [40]. The steady state die swell ratios calculated using the Power law and Carreau–Yasuda rheology model are nearly identical and significantly lower than the swell ratio of B = 1.199 computed using the PTT model. Furthermore, the die swell profile obtained using the SV model converges to that computed with the PTT model as the profiles reach steady state. Note that simulation time using the SV model was 75 s which is much less than the 328 s when using the PTT model with the mesh given in Figure 5. Therefore, for a larger size of flow domain or a finer mesh quality, the SV model is a good candidate to qualitatively solve the flow problem with less computational cost.

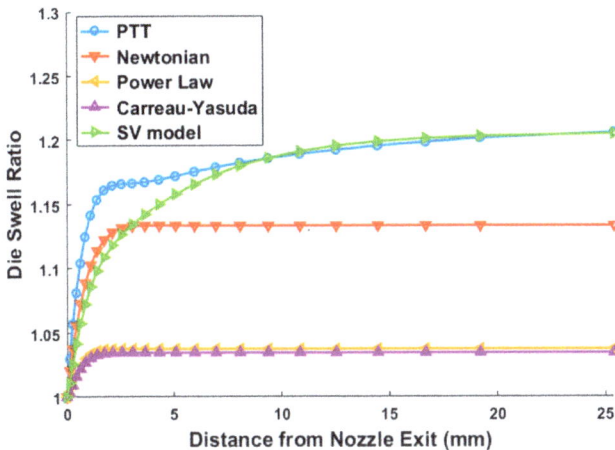

**Figure 8.** Die swell profiles predicted by using different rheology models.

**Table 3.** Apparent swell ratio values resulted by applied rheology models.

| Model Name | Apparent Swell Ratio |
|---|---|
| Newtonian model | 1.133 |
| Power law model | 1.037 |
| Carreau–Yasuda model | 1.035 |
| PTT model | 1.199 |
| SV model | 1.197 |

*4.2. Computed Fiber Orientation Distribution*

In this work, the fiber orientation is computed as described in Section 3.2 above. Our primary interest is in the direction of extrusion, i.e., in the direction of the positive $x_2$ axis in Figure 5. Therefore, our primary focus is on fiber alignment in $x_2$, which is best represented by the $A_{22}$ component of the second order orientation tensor **A**.

The solution of $A_{22}$ along various streamlines shown in Figure 9 is computed based on the flow kinematics solved using the PTT rheology model with Q = 100 mm$^3$/s. Values of $A_{22}$ computed for each of the rheology models showed similar trends as that appearing in Figure 9, so that these other results are omitted here for conciseness. ANSYS-CFD-Post [41] generates the 2D surface streamlines based on the mesh quality. Due to the mesh defined in Figure 5, we consider flow velocity fields along eight streamlines computed from the finite element results achieved by ANSYS-Polyflow. Note that $A_{22}$ values near unity indicate fibers are highly aligned along the $x_2$ direction. It can be seen that the fiber orientation starts from steady state at flow inlet (appearing as FI in Figure 9) as assumed. The $A_{22}$ components then starts to separate as the flow approaches the nozzle convergent zone (appearing as CZS in Figure 9). Orientation states along all streamlines increase before the flow reaches the exit of the convergent zone (appearing as CZE in Figure 9). The peak value of the $A_{22}$ component occurs at the convergent zone exit. Then, the orientation tensors at inner region streamlines decrease while those located at outer region increase as the flow propagating to the nozzle exit (appearing as NE in Figure 9). Once the polymer melt passes the nozzle exit at NE, values of $A_{22}$ in the outer region increase immediately and those more central begin to increase. This change occurs due to the shear rate limitation vanishing at the outer boundary just after nozzle exit. The velocity along the outer boundary accelerates first, causing fibers nearby to orientate in the flow direction. In addition, the elongational flow near the center of the nozzle accelerates so that the extrudate attains a uniform speed at some point not far from the nozzle exit. The final state of fiber orientation is set once variation across the bead ceases and a plug flow develops.

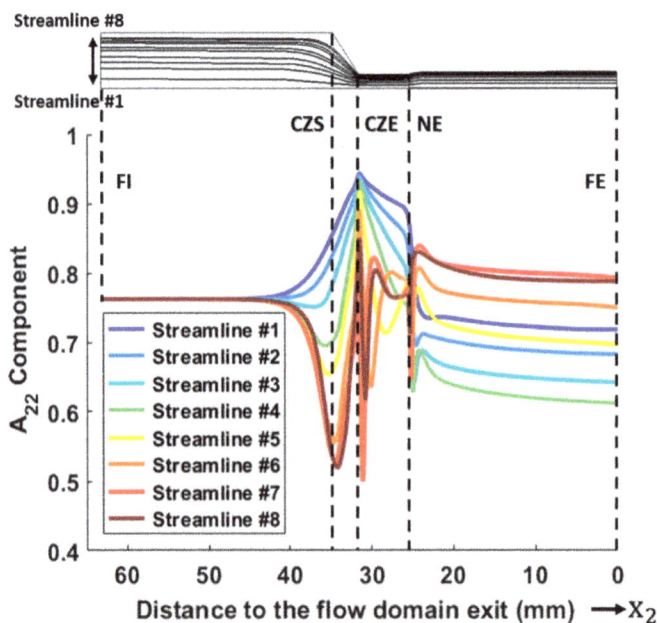

**Figure 9.** $A_{22}$ component of fiber orientation solution computed using the flow kinematics solved by the PTT model at Q equates 100 mm$^3$/s.

Similar to the results appearing in Figure 9, we also compute the fiber orientation tensor along streamlines in the flow domain using other rheology models. Values of $A_{22}$ at the end of the flow domain (i.e., across $\Gamma_6$, in Figure 5) for all simulations considered here appear in Figure 10. Results indicate that fibers are highly aligned near the edge of the flow where shear rates are high.

Fiber alignment then decreases just inside this outer band forming an intermediate slightly misaligned region. An increase in alignment then occurs within the core region towards the center of the extrudate. In addition, it can be seen that the PTT model yields the lowest alignment in $x_2$ among the applied rheology models. Alternatively, the Power law and Carreau–Yasuda law result in a similar steady state fiber orientation, which are the highest among these results. Moreover, the orientation result obtained using the SV model shows a good agreement with that of the PTT model in the shear dominant region but varies at other locations within the flow. Finally, the Newtonian model yields an intermediate value of the fiber orientation, somewhat positioned between the GNF laws and the viscoelastic model results.

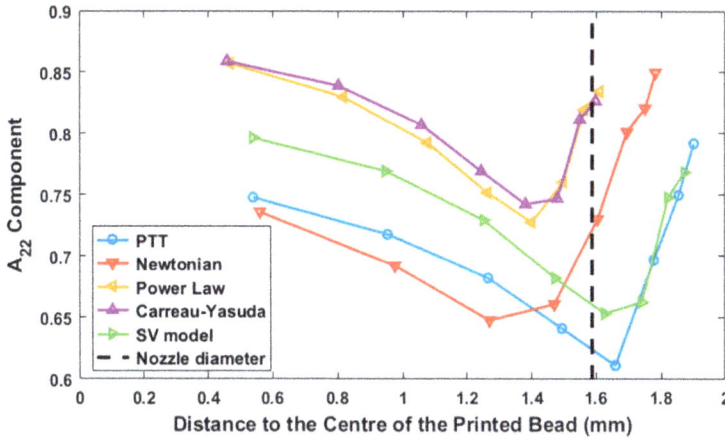

**Figure 10.** $A_{22}$ component at the flow domain exit solved using employed rheology models.

### 4.3. Elastic Properties across the Exrudate

Finally, we calculate the elastic properties for the fiber-reinforced polymer using the volume-averaged stiffness tensor computed from the steady state fiber orientation tensor (cf. Jack et al. [25]) and the Tandon–Wang analytical equation (cf. Tandon et al. [39]). Computed elastic properties across the extrudate at boundary $\Gamma_6$ (cf. Figure 5) for each rheology model considered here appear in Figure 11. For conciseness, we omit the results obtained using the Carreau–Yasuda rheology model since these results are very similar to that obtained in our Power law model simulations.

Our results show that the elastic properties of the composite extrudate are enhanced by the fiber reinforcement in comparison to the properties of neat ABS (dash lines appearing in Figure 11), particularly the $E_{22}$ component, which is elastic modulus along the extrusion direction. In addition, the elastic modulus along the extrusion direction ($E_{22}$) is not quite uniform while the moduli at other directions ($E_{11}$, $E_{33}$, $G_{12}$, $G_{23}$, $G_{13}$) are in small variance across the extrudate. In comparison, the Power law yielded relatively higher estimation in the principal modulus ($E_{22}$) and the PTT model results in the lowest prediction. In addition, the Power law model, the PTT model as well as the SV model show a more sharp variation in the $E_{22}$ component than that yielded by the Newtonian model, which shows trends that are similar to results calculated for a small FDM nozzle in Heller et al. [14].

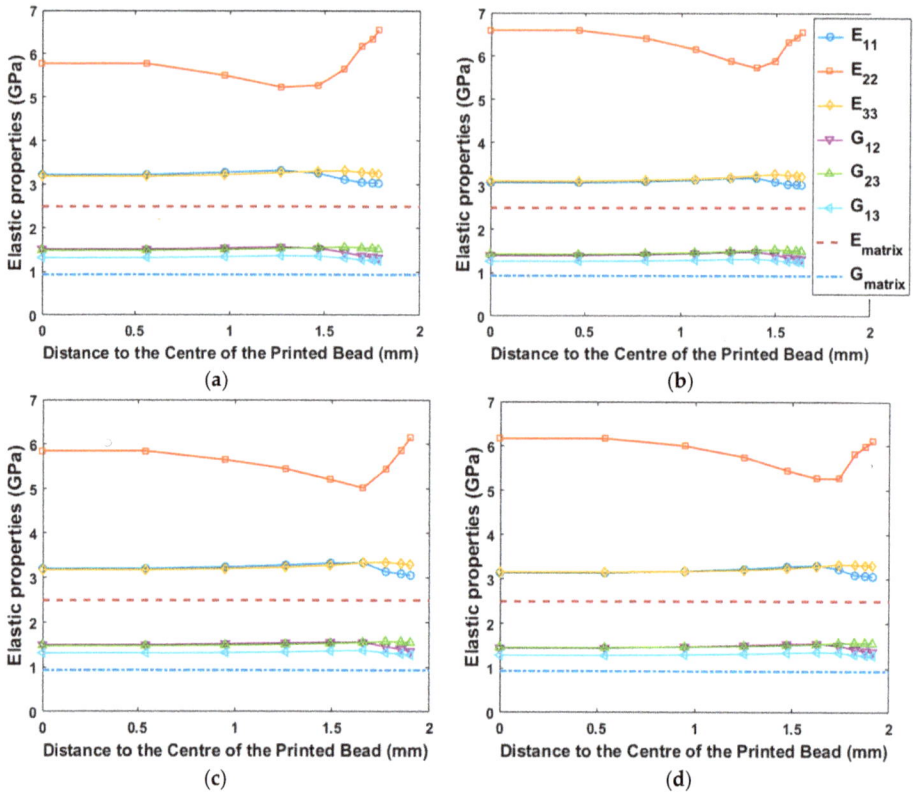

**Figure 11.** Elastic properties across the printed extrudate computed with flow kinematics obtained by (a) the Newtonian model; (b) the Power law model; (c) the PTT model; (d) the Simplified Viscoelastic (SV) model. Note that the legend shown in (b) works for all four sub-figures.

Furthermore, we average the elastic properties of the extrudate by numerically integrating the data appearing in Figure 11 as written in Equation (24) by the trapezoidal rule [42] and the results are given in Table 4:

$$X_{avg} = \frac{1}{\pi r_0^2} \int_0^{2\pi} \int_0^{r_0} X \cdot r \, dr d\Theta, \tag{24}$$

where X refers to any of the data points appearing in Figure 11, $X_{avg}$ is the area-averaged data, and $r_0$ is the outer radius of the extrudate. Calculated values of $X_{avg}$ corresponding to the elastic properties in Figure 11 appear in Table 4. Note that we assume that the centre of the extudate exhibits the same orientation behavior as the data in the streamline nearest to the centre-line, so as to achieve the fiber orientation over the entire cross-section of the extrudate (cf. Figure 11). In addition, due to the flow domain being built as a 2D axis-symmetric model, the integration is carried out in polar axis in order to include the effects over the entire cross-section area of the extrudate.

The data computed in Table 4 indicate that the printed composite exhibit quasi transverse isotropic mechanical behavior. In detail, the Carreau–Yasuda law (appearing as Carreau-Y. in Table 4) yields the highest averaged $E_{22}$ value, while the PTT model results in the lowest. In contrast, the shear moduli yielded by the PTT model is higher than the GNF models. Finally, as we consider more non-viscous effects of the flow (see data from the Power law (as well as the Carreau–Yasuda law), the Newtonian model, the SV model to the PTT model), it can be seen that the principal modulus ($E_{22}$) decreases while other moduli increase. In addition, the transverse isotropic behavior of results predicted by the PTT model is the most obvious one, in which the $E_{22}$ and $G_{13}$ are unique values and $E_{11}$, $E_{33}$ as well as $G_{12}$, $G_{23}$ show high agreement. Our material stiffness predictions agree remarkably well with test data appearing in the literature. Duty et al. [4] measured the Young's modulus of a 13 per cent carbon fiber reinforced ABS printed bead, reporting a mean value of 7.24 GPa and standard deviation of 0.59 GPa. Our predictions of the elastic modulus for the same material system shown as $E_{22}$ in Table 4 are in good agreement with these previously published experimental results. Note that all predicted values of $E_{22}$ in Table 4 are within one standard deviation of Duty's mean experimental value, regardless of rheology model used, except for the value of 6.57 GPa obtained using the PTT model, which is at 1.1 standard deviation from the experimental mean. In addition, Duty found the stiffness of a BAAM printed tensile test sample to be highly anisotropic, which is also seen in our results. Furthermore, the previously reported test sample transvers moduli show some anisotropy in the transverse plane (i.e., Ey = 2.26 GPa and Ez = 2.56 GPa). Our results obtained using GNF fluids (i.e., power law and Carreau–Yasuda fluids) show similar trends. While we are not able to show experimental data specific to each rheology model, the overall favorable comparison with previously published experimental work supports our computational approach.

**Table 4.** Averaged elastic properties of the printed extrudate.

| Model Name | $E_{11}$ (GPa) | $E_{22}$ (GPa) | $E_{33}$ (GPa) | $G_{12}$ (GPa) | $G_{23}$ (GPa) | $G_{13}$ (GPa) | $\nu_{12}$ | $\nu_{23}$ | $\nu_{13}$ |
|---|---|---|---|---|---|---|---|---|---|
| PTT | 3.48 | 6.57 | 3.50 | 1.68 | 1.68 | 1.43 | 0.22 | 0.12 | 0.21 |
| Power law | 3.32 | 7.45 | 3.40 | 1.56 | 1.63 | 1.37 | 0.20 | 0.10 | 0.22 |
| Carreau-Y. | 3.31 | 7.49 | 3.40 | 1.56 | 1.62 | 1.37 | 0.20 | 0.10 | 0.22 |
| Newtonian | 3.45 | 6.66 | 3.50 | 1.65 | 1.69 | 1.42 | 0.22 | 0.12 | 0.22 |
| SV model | 3.43 | 6.86 | 3.46 | 1.64 | 1.66 | 1.41 | 0.22 | 0.11 | 0.22 |

## 5. Conclusions

Polymer melt flow through a large-scale polymer deposition extrusion nozzle was simulated with the finite element method using ANSYS-Polyflow. These simulations included flow within the nozzle in addition to the free surface die swell flow just outside the nozzle exit. Several rheology models were compared in this work including a Newtonian model, a Power law model, a Carreau–Yasuda model, as well as a multi-mode PTT model and a Simplified Viscoelastic (SV) material model. Rheology data obtained experimentally using the HAAKE MARS 40 rheometer.

It was found that characterizing the melt flow by different rheology models yielded noticeable variation in predicted die swell, fiber orientation distribution and the ultimate elastic behavior of the extruded composites. The predicted die swell yielded by the PTT model was higher than those resulted by the Generalized Newtonian Fluid (GNF) models including the Newtonian model. The SV model yielded die swell results that agreed well with those from the PTT model by careful adjustments of the rheology model parameters.

Through the weakly couple formulation, the fiber orientation distribution within the extrudate was calculated from the melt flow velocity field. High fiber alignment in the direction of extrusion occurred near the high-shear flow edge region of the extrudate as well as the near-center region, which was due to the elongational effects of the free flow. Among the applied rheology models, the PTT model yielded the lowest principal fiber alignment while the Power law model resulted in the highest fiber orientation in polymer extrusion direction.

The elastic properties of a printed extrudate were evaluated based on the predicted fiber orientation distributions, in which the estimated elastic modulus along extrusion direction showed noticeable variance across the extrudate. The numerically-integrated averaged elastic moduli showed a good agreement with the published experimental work. The estimation indicated the composite extrudate exhibited quasi transverse isotropic behavior. In detail, the GNF models yielded higher Young's modulus along the principal direction while the PTT model resulted in a lower principal

Young's modulus but higher values of shear moduli. This indicates that, by considering the non-viscous rheology effects, the elastic properties of extrudate through Additive Manufacturing (AM) systems reduced at a longitudinal direction but increased at shear directions.

In addition, the SV model yielded relatively similar data of fiber orientation distribution as well as elastic properties in comparison with the PTT model, especially in the shear dominant flow boundary, yet cost less computational time than the PTT model. In the future study of 3D deposition modelling of large-scale AM, the computationally cost-effective SV model is a reasonable alternative for conventional viscoelastic fluid models (e.g., PTT model).

**Acknowledgments:** The authors would like to thank the Strangpresse Corporation for donating the Model-19 extruder as well as the financial support offered by Baylor University and Oak Ridge National Lab (RAMP-UP 40001455134).

**Author Contributions:** Douglas E. Smith. conceived and designed the experiments; Zhaogui Wang performed the experiments; Douglas E. Smith and Zhaogui Wang analyzed the data; All contributed materials are provided by Polyone Inc.; The extruder model is donated by Strangpresse Inc.; Douglas E. Smith and Zhaogui Wang. wrote the paper.

**Conflicts of Interest:** The founding sponsors had no role in the design of the study; in the collection, analyses, or interpretation of data; in the writing of the manuscript, and in the decision to publish the results.

## References

1.	Love, L.J. *Utility of Big Area Additive Manufacturing (BAAM) for the Rapid Manufacture of Customized Electric Vehicles*; Oak Ridge National Laboratory (ORNL); Manufacturing Demonstration Facility (MDF): Oak Ridge, TN, USA, 2015.

2.	Brenken, B.; Favaloro, A.; Barocio, E.; Denardo, N.; Kunc, V.; Pipes, R.B. Fused Deposition Modeling of Fiber-Reinforced Thermoplastic Polymers: Past Progress and Future Needs. In Proceedings of the American Society for Composites: Thirty-First Technical Conference, Williamsburg, VA, USA, 19–22 September 2016.

3.	Love, L.J.; Kunc, V.; Rios, O.; Duty, C.E.; Elliott, A.M.; Post, B.K.; Smith, R.J.; Blue, C.A. The importance of carbon fiber to polymer additive manufacturing. *J. Mater. Res.* **2014**, *29*, 1893–1898. [CrossRef]

4.	Duty, C.E.; Kunc, V.; Compton, B.; Post, B.; Erdman, D.; Smith, R.; Lind, R.; Lloyd, P.; Love, L. Structure and mechanical behavior of big area additive manufacturing (BAAM) materials. *Rapid Prototyp. J.* **2017**, *23*, 181–189. [CrossRef]

5.	Verweyst, B.E.; Tucker, C.L. Fiber suspensions in complex geometries: Flow/orientation coupling. *Can. J. Chem. Eng.* **2002**, *80*, 1093–1106. [CrossRef]

6.	Evans, J.G. The effect of non-Newtonian properties of a suspension of rod-like particles on flow fields. In *Theoretical Rheology*; Halstead Press: New York, NY, USA, 1975; pp. 224–232.

7.	Lipscomb, G.G.; Denn, M.M.; Hur, D.U.; Boger, D.V. The flow of fiber suspensions in complex geometries. *J. Non-Newton. Fluid Mech.* **1988**, *26*, 297–325. [CrossRef]

8.	Tucker, C.L. Flow regimes for fiber suspensions in narrow gaps. *J. Non-Newton. Fluid Mech.* **1991**, *39*, 239–268. [CrossRef]

9.	Bay, R.S.; Tucker, C.L. Fiber orientation in simple injection moldings. Part I: Theory and numerical methods. *Polym. Compos.* **1992**, *13*, 317–331. [CrossRef]

10.	Jackson, W.C.; Advani, S.G.; Tucker, C.L. Predicting the orientation of short fibers in thin compression moldings. *J. Compos. Mater.* **1986**, *20*, 539–557. [CrossRef]

11.	Nixon, J.; Dryer, B.; Lempert, I.; Bigio, D.I. Three parameter analysis of fiber orientation in fused deposition modeling geometries. In Proceedings of the PPS Conference, Cleveland, OH, USA, 8–12 June 2014.

12.	Phelps, J.H.; Tucker, C.L. An anisotropic rotary diffusion model for fiber orientation in short-and long-fiber thermoplastics. *J. Non-Newton. Fluid Mech.* **2009**, *156*, 165–176. [CrossRef]

13.	Kunc, V. Advances and challenges in large-scale polymer additive manufacturing. In Proceedings of the 15th SPE Automotive Composites Conference, Novi, MI, USA, 9–11 September 2015.

14.	Heller, B.; Smith, D.E.; Jack, D.A. The Effects of Extrudate Swell, Nozzle Shape, and the Nozzle Convergence Zone on Fiber Orientation in Fused Deposition Modeling Nozzle Flow. In Proceedings of the American Society of Composites—30th Technical Conference, Lansing, MI, USA, 29–30 September 2015.

15. Advani, S.G.; Tucker, C.L., III. The use of tensors to describe and predict fiber orientation in short fiber composites. *J. Rheol.* **1987**, *31*, 751–784. [CrossRef]
16. Crochet, M.J.; Keunings, R. Die swell of a Maxwell fluid: Numerical prediction. *J. Non-Newton. Fluid Mech.* **1980**, *7*, 199–212. [CrossRef]
17. Luo, X.-L.; Tanner, R.I. A streamline element scheme for solving viscoelastic flowproblems part II: Integral constitutive models. *J. Non-Newton. Fluid Mech.* **1986**, *22*, 61–89. [CrossRef]
18. Luo, X.-L.; Mitsoulis, E. An efficient algorithm for strain history tracking in finite element computations of non-Newtonian fluids with integral constitutive equations. *Int. J. Numer. Methods Fluids* **1990**, *11*, 1015–1031. [CrossRef]
19. Béraudo, C.; Fortin, A.; Coupez, T.; Demay, Y.; Vergnes, B.; Agassant, J.F. A finite element method for computing the flow of multi-mode viscoelastic fluids: Comparison with experiments. *J. Non-Newton. Fluid Mech.* **1998**, *75*, 1–23. [CrossRef]
20. Thien, N.P.; Tanner, R.I. A new constitutive equation derived from network theory. *J. Non-Newton. Fluid Mech.* **1977**, *2*, 353–365. [CrossRef]
21. Ganvir, V.; Lele, A.; Thaokar, R.; Gautham, B.P. Prediction of extrudate swell in polymer melt extrusion using an arbitrary Lagrangian Eulerian (ALE) based finite element method. *J. Non-Newton. Fluid Mech.* **2009**, *156*, 21–28. [CrossRef]
22. Limtrakarn, W.; Pratumwal, Y.; Krunate, J.; Prahsarn, C.; Phompan, W.; Sooksomsong, T.; Klinsukhon, W. Circular die swell evaluation of LDPE using simplified viscoelastic model. *King Mongkut's Univ. Technol. North Bangk. Int. J. Appl. Sci. Technol.* **2013**, *6*, 59–68.
23. Clemeur, N.; Rutgers, R.P.G.; Debbaut, B. Numerical simulation of abrupt contraction flows using the Double Convected Pom–Pom model. *J. Non-Newton. Fluid Mech.* **2004**, *117*, 193–209. [CrossRef]
24. *ANSYS Polyflow User's Guide*; Ansys Inc.: Canonsburg, PA, USA, 2013.
25. Jack, D.A.; Smith, D.E. The effect of fibre orientation closure approximations on mechanical property predictions. *Compos. Part Appl. Sci. Manuf.* **2007**, *38*, 975–982. [CrossRef]
26. Jack, D.A. Advanced Analysis of Short-Fiber Polymer Composite Material Behavior. Ph.D. Dissertation, University of Missouri, Columbia, MO, USA, 2006.
27. *ANSYS Polymat User's Guide*; Ansys Inc.: Canonsburg, PA, USA, 2013.
28. Tadmor, Z.; Gogos, C.G. *Principles of Polymer Processing*; John Wiley & Sons: New York, NY, USA, 2013.
29. Ajinjeru, C.; Kishore, V.; Liu, P.; Hassen, A.A.; Lindahl, J.; Kunc, V.; Duty, C. Rheological evaluation of high temperature polymer. In Proceedings of the 28th Annual International Solid Freeform Fabrication Symposium, Austin, TX, USA, 7–9 August 2017.
30. Michaeli, W. *Extrusion Dies for Plastics and Rubber: Design and Engineering Computations*; Hanser Verlag: München, Germany, 2003.
31. Reddy, J.N.; Gartling, D.K. *The Finite Element Method in Heat Transfer and Fluid Dynamics*; CRC Press: Boca Raton, FL, USA, 2010; Volume 32.
32. Fu, S.-Y.; Yue, C.-Y.; Hu, X.; Mai, Y.-W. Characterization of fiber length distribution of short-fiber reinforced thermoplastics. *J. Mater. Sci. Lett.* **2001**, *20*, 31–33. [CrossRef]
33. Verleye, V.; Couniot, A.; Dupret, F. Prediction of fiber orientation in complex injection molded parts. In Proceedings of the Ninth International Conference on Composite Materials (ICCM/9), Madrid, Spain, 12–16 July 1993; Volume 3, pp. 642–650.
34. Dupret, F.; Verleye, V.; Languillier, B. Numerical prediction of the molding of composite parts. *ASME-Publ.-FED* **1997**, *243*, 79–90.
35. Cintra, J.S., Jr.; Tucker, C.L., III. Orthotropic closure approximations for flow-induced fiber orientation. *J. Rheol.* **1995**, *39*, 1095–1122. [CrossRef]
36. Chung, D.H.; Kwon, T.H. Improved model of orthotropic closure approximation for flow induced fiber orientation. *Polym. Compos.* **2001**, *22*, 636–649. [CrossRef]
37. Meyer, K.J.; Hofmann, J.T.; Baird, D.G. Initial conditions for simulating glass fiber orientation in the filling of center-gated disks. *Compos. Part Appl. Sci. Manuf.* **2013**, *49*, 192–202. [CrossRef]
38. Mase, G.T.; Smelser, R.E.; Mase, G.E. *Continuum Mechanics for Engineers*; CRC Press: Boca Raton, FL, USA, 2009.
39. Tandon, G.P.; Weng, G.J. The effect of aspect ratio of inclusions on the elastic properties of unidirectionally aligned composites. *Polym. Compos.* **1984**, *5*, 327–333. [CrossRef]

40. Reddy, K.R.; Tanner, R.I. Finite element solution of viscous jet flows with surface tension. *Comput. Fluids* **1978**, *6*, 83–91. [CrossRef]
41. *ANSYS CFD-Post User's Guide*; Ansys Inc.: Canonsburg, PA, USA, 2012.
42. Chapra, S.C. *Applied Numerical Methods with MATLAB for Engineers and Scientists*; McGraw Hill Publications: New York City, NY, USA, 2012.

*Journal of*
*composites science*

MDPI

*Article*

# Simulation of Reinforced Reactive Injection Molding with the Finite Volume Method

Florian Wittemann [1,*], Robert Maertens [2], Alexander Bernath [1], Martin Hohberg [1], Luise Kärger [1] and Frank Henning [1,2]

[1] Karlsruhe Institute of Technology (KIT), 76131 Karlsruhe, Germany; alexander.bernath@kit.edu (A.B.); martin.hohberg@kit.edu (M.H.); luise.kaerger@kit.edu (L.K.); frank.henning@ict.fraunhofer.de (F.H.)

[2] Fraunhofer Institute of Chemical Technology (ICT), 76327 Pfinztal, Germany; robert.maertens@ict.fraunhofer.de

\* Correspondence: florian.wittemann@kit.edu; Tel.: +49-721-60845379

Received: 19 December 2017; Accepted: 22 January 2018; Published: 31 January 2018

**Abstract:** The reactive process of reinforced thermoset injection molding significantly influences the mechanical properties of the final composite structure. Therefore, reliable process simulation is crucial to predict the process behavior and relevant process effects. Virtual process design is thus highly important for the composite manufacturing industry for creating high quality parts. Although thermoset injection molding shows a more complex flow behavior, state of the art molding simulation software typically focusses on thermoplastic injection molding. To overcome this gap in virtual process prediction, the present work proposes a finite volume (FV) based simulation method, which models the multiphase flow with phase-dependent boundary conditions. Compared to state-of-the-art Finite-Element-based approaches, Finite-Volume-Method (FVM) provides more adequate multiphase flow modeling by calculating the flow at the cell surfaces with an Eulerian approach. The new method also enables the description of a flow region with partial wall contact. Furthermore, fiber orientation, curing and viscosity models are used to simulate the reinforced reactive injection molding process. The open source Computational-Fluid-Dynamics (CFD) toolbox OpenFOAM is used for implementation. The solver is validated with experimental pressure data recorded during mold filling. Additionally, the simulation results are compared to commercial Finite-Element-Method software. The simulation results of the new FV-based CFD method fit well with the experimental data, showing that FVM has a high potential for modeling reinforced reactive injection molding.

**Keywords:** reinforced reactive injection molding; thermoset processing; process simulation; FVM; OpenFOAM

## 1. Introduction

Reinforced reactive injection molding (RRIM) is one of the most important manufacturing processes for short fiber reinforced composites with thermoset matrices. The process significantly affects the mechanical and optical properties of the final part. Hence, the process control is crucial for manufacturing high performance parts. Furthermore, an adequate process simulation is needed to achieve this process control and fulfill the high standards of the automotive and polymer industry [1–5].

Despite the importance of the RRIM process, well-known simulation software is often focused on thermoplastic injection molding and uses the same models for RRIM. However, thermosets show a more complex flow behavior, triggered by a low-viscosity surface layer resulting from the hotter mold [1,2,5]. Furthermore, the flow is not shear dominated, like a plug flow, which is a typical adoption for thermoplastics [1,2]. Additionally, commercial software does not simulate the air in the mold as a separate flow phase. This simplification is made to reduce the calculation time, but consequently neglects phenomena like air traps with significant influence on the process. These are only taken into account with empirical models in the post processing. For good simulation of RRIM, many material and process aspects need to be considered. As the resin is a non-Newtonian fluid, the viscosity must be modeled as function of temperature, shear rate and degree of cure to accurately simulate the pressure field [6]. For that purpose, and to have a reliable prediction of cycle time, curing kinetics have to be modeled in addition to rate-dependent viscosity models. Furthermore, fiber orientations should be predicted, since they have a significant influence on the mechanical and thermal properties of the final part [3,4,7].

On the one hand, commercial software is often based on the Finite-Element-Method (FEM) to simulate the RRIM process [8–11]. On the other hand, Finite-Volume-based solvers, representing the state of research, focus on thermoplastic injection molding using incompressible and isothermal models [12]. In this study, the Finite-Volume-Method (FVM) is used, resolving the flux at the cell faces and using an Eulerian approach [13–15]. Due to this flow modeling, FVM provides a more realistic multiphase flow with physical significance on the fluxes, which is not possible in FEM by solving a Lagrangian mesh at the nodes. For implementation and simulation, the open source Computational-Fluid-Dynamics (CFD) toolbox OpenFOAM 4.1 (OpenCFD Ltd., Bracknell, UK) [7,12–15] is used and well-known viscosity, curing and fiber orientation models are implemented to model the reinforced reactive injection molding process. A solver for compressible, non-isothermal multiphase flow is extended, using a phase depending boundary condition, defined to enable mold-filling simulation, by separating and interpolating boundary conditions for polymer and air. Additionally, the solver predicts fiber orientation, resulting from the flow during mold filling.

The simulation results are compared to a commercial FEM software (Moldflow 2018.1, Autodesk, San Rafael, CA, USA) and to experimental injection molding trials of a glass fiber reinforced phenolic resin.

## 2. Models and Implementations

The following subsections present the models and methods used for phase-dependent boundary conditions, curing kinetics, viscosity and fiber orientation. The models are implemented in OpenFOAM 4.1, using C++ as the program language and finite volume theory for the numerical solution [13–15]. In principal, the Navier-Stokes equations can be solved on structured and unstructured meshes. Due to the open source code of the CFD toolbox OpenFOAM, several solvers for scalar transport, transient problems, multiphase problems etc. are available and can be extended in a user-defined way.

In the present work, a new solver is created for mold filling simulation of fiber reinforced thermoset materials. The new solver is based on the compressible, non-isothermal multiphase solver compressibleInterFoam. This solver uses the Volume-of–Fluid-Method (VoF) for modeling multiphase flows. VoF separates the different phases using the dimensionless scalar $\alpha$, which is equal to one in cells completely filled by phase 1, zero in cells filled with phase 2 and $0 < \alpha < 1$ in the interface regions [13,15]. In this work, $\alpha = 1$ means that a control volume is completely filled with polymer resin, and $\alpha = 0$ if it is filled with air. The boundary conditions of the injection flow are accordingly extended to phase-dependent formulations. Additionally, material models for curing kinetics, viscosity and fiber orientation are implemented to suitably model the reinforced reactive flow behavior.

### 2.1. Governing Equations

The governing equations are the continuity equation, momentum and energy balance to form the Navier-Stokes equations.

The continuity equation describes the change of the density $\rho$ as follows:

$$\frac{d\rho}{dt} + \frac{\partial(\rho U_i)}{\partial x_i} = 0, \tag{1}$$

where incompressibility is not assumed, so $\frac{d\rho}{dt} = 0$ is not generally applicable and $U_i$ is the velocity vector.

For momentum balance the general formulation

$$\frac{d\rho U_i}{dt} + \frac{\partial(\rho U_i U_j)}{\partial x_j} = -\frac{\partial p}{\partial x_i} + \eta \frac{\partial U_i}{\partial x_j \partial x_j} \tag{2}$$

is used with $p$ being the pressure and $\eta$ being the viscosity (see Section 2.4).

The energy balance describes the distribution of the temperature $T$ with:

$$\frac{d\rho T}{dt} + \frac{\partial(\rho U_i T)}{\partial x_i} = \frac{\lambda}{c_p} \frac{\partial T}{\partial x_i \partial x_i}, \tag{3}$$

according to Fourier's law. The parameters $\lambda$ and $c_p$ are thermal conductivity and specific heat capacity in this case. The wall temperature of the mold is assumed to be constant, hence no heat transfer between the material and the mold is needed. Radiation is neglected.

## 2.2. Phase-Dependent Boundary Condition

Three types of boundary faces can be distinguished in OpenFOAM: An inlet for flow into the system, an outlet for flow out of the system and a wall, which cannot be passed by any media. To enable mold-filling simulation, a special boundary condition (BC) for the outlet is developed, which allows the air phase, but not the polymer phase, to leave the mold through the outlet face. Hence, the outlet BC for velocity vector $U$ is implemented as an interpolation between a Dirichlet boundary condition, defining the absolute value of $U$ to be zero, and a Neumann boundary condition, defining the gradient of $U$ to be zero. The interpolation depends on the VoF factor $\alpha$, in such a way that a cell completely filled with polymer has a pure Dirichlet BC and a cell completely filled with air has a pure Neumann BC. Consequently, the boundary face changes from an outlet to a wall during filling, represented by:

$$BC = \alpha \cdot DirichletBC + (1 - \alpha) \cdot NeumannBC, \tag{4}$$

with

$$DirichletBC \stackrel{\text{def}}{=} U = \begin{pmatrix} 0 \\ 0 \\ 0 \end{pmatrix} \text{ and } NeumannBC \stackrel{\text{def}}{=} \frac{dU}{dt} = \begin{pmatrix} 0 \\ 0 \\ 0 \end{pmatrix}. \tag{5}$$

For implementation of the phase-dependent BC, a GroovyBC is used, which is part of the third party tool SWAK4FOAM 0.4.1 (Free Software Foundation, Boston, MA, USA). The program code is described in Appendix A.

## 2.3. Model for Curing Kinetics

Modeling of curing kinetics is an important aspect since cross-linking significantly influences the viscosity and consequently the process pressure and cycle time [16,17]. In this study, the Kamal-Malkin model is used to model curing kinetics, since it is a well-known and often used model for curing kinetics of thermoset polymers [11,16–21]. The model describes the cure rate by

$$\frac{dc}{dt} = (K_1 + K_2 c^m)(1 - c)^n \tag{6}$$

with

$$K_1 = A_1 exp\left(-\frac{E_1}{R \cdot T}\right), \tag{7}$$

$$K_2 = A_2 exp\left(-\frac{E_2}{R \cdot T}\right), \tag{8}$$

where $c$ is the degree of cure between 0 and 1, R is the ideal gas constant and $m, n, A_1, E_1, A_2, E_2$ are material specific constants which have to be identified by experimental measurements like Differential Scanning Calorimetry (DSC) and curve fitting [17]. The curing heat is neglected because of the thin part geometry, hence the curing kinetics have no influence on the temperature distribution. The reaction kinetics model is implemented as a thermo-chemical model within the OpenFOAM structure.

### 2.4. Model for Viscosity

The Castro-Macosko viscosity model [6] is an often used and well-established model to simulate the viscosity of thermoset materials [11,20,21]. The model describes the viscosity as a function of temperature $T$, shear rate $\dot{\gamma}$ and degree of cure $c$:

$$\eta(T,\dot{\gamma},c) = \frac{\eta_0(T)}{1 + \left(\frac{\eta_0(T)\dot{\gamma}}{\tau^*}\right)^{(1-n)}} \left(\frac{c_g}{c_g - c}\right)^{(c_1 + c_2 \cdot c)} \tag{9}$$

and

$$\eta_0(T) = B \cdot exp\left(\frac{T_b}{T}\right), \tag{10}$$

with $\tau^*, n, c_1, c_2, B$ and $T_b$ being material specific constants. $c_g$ denotes the material specific degree of cure at gelation. The Castro-Macosko model is implemented as a transport model in the thermo-physical model library of OpenFOAM.

### 2.5. Model for Fiber Orientation

For calculation of the fiber orientations, the Folgar Tucker model is used [22–25]. A well-described implementation of this model has already been published by Kerstin Heinen [7]. Heinen's code is extended in the present work to model multiphase flows. Hence, fiber orientation is calculated only in cells with $\alpha \geq 0.5$. In all other cells the initial conditions remain. The result is a symmetric tensor [23], with the diagonal entries being the probability of fibers oriented in this coordinate direction with a value between zero and one, where zero means no fibers are aligned in this direction and one, that all fibers are aligned in this direction. The summation of the diagonal entries must be equal to one.

The fourth order fiber orientation tensor $A_{klij}$ is described by second order tensors using the hybrid closure approximation, as described in [7,23,24]. The differential equation for the second order fiber orientation tensor in a control volume is given by:

$$\frac{dA_{ij}}{dt} + \frac{\partial v_i A_{jk}}{\partial x_k} = -\left(W_{ik}A_{kj} - A_{ik}W_{kj}\right) + \lambda_F\left(D_{ik}A_{kj} - A_{ik}D_{kj} - 2D_{kl}A_{klij}\right) + 2C_I\dot{\gamma}(\delta_{ij} - 3A_{ij}), \tag{11}$$

with

$$D_{ij} = \frac{1}{2}\frac{\partial U_i}{\partial x_j} + \frac{1}{2}\left(\frac{\partial U_i}{\partial x_j}\right)^T \tag{12}$$

$$W_{ij} = \frac{1}{2}\frac{\partial U_i}{\partial x_j} - \frac{1}{2}\left(\frac{\partial U_i}{\partial x_j}\right)^T \tag{13}$$

$$\lambda_F = \frac{r^2 - 1}{r^2 + 1} \tag{14}$$

where $C_I$ is the interaction coefficient calculated with the Bay's equation [7] and $r$ is the aspect ratio of the fibers defined as quotient of fiber length and diameter. The vector $v$ is the flux of the velocity vector.

### 3. Test Setup for Experimental Validation

*3.1. Test Structure and Process Conditions*

The thermoset molding compound used for the experimental studies is a phenolic resin of the novolac type, reinforced with glass fibers. The material was provided for the trials by SBHPP Vyncolit (Gent, Belgium).

For the experimental study, an electrically heated plate mold equipped with pressure and temperature sensors is used. It is a symmetrical two-cavity mold with a variable plate thickness between two and five millimeters. For the experimental studies described here, a constant wall thickness of 2 mm is chosen. Both plates have a square shape with an edge length of 190 mm. The sprue has a start diameter of 9 mm (inlet), expands to 15.5 mm and is 185 mm high. Figure 1 shows the complete molded part with sprue system. The position of the pressure and temperature sensors in one of the mold cavities is shown in Figure 2.

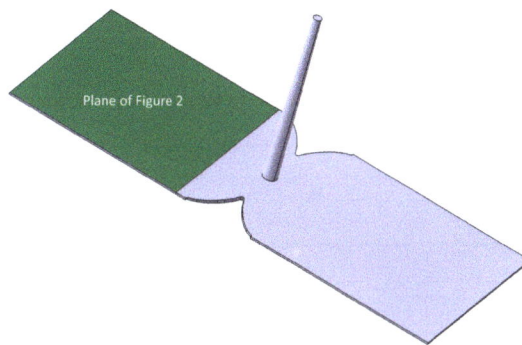

**Figure 1.** Molded part with sprue system.

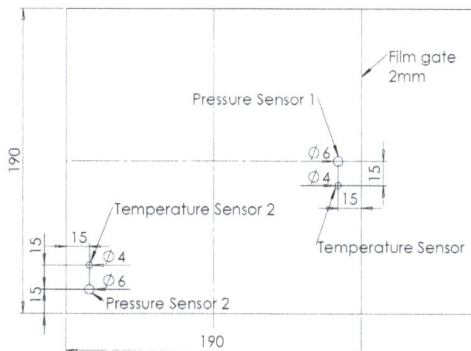

**Figure 2.** Position of pressure and temperature sensors in mold.

The sensor positions are in the region where the first material should appear after entering the plate mold (position 1) and the last region where material should appear before complete filling (position 2). The positions are chosen to see when the plate filling starts and ends. The sensors are nearly in plane with the mold, so the influence on material flow behavior can be neglected. For pressure measurement, sensors of the type 6163 manufactured by Kistler Instrumente GmbH (Winterhur, Switzerland) are used. These sensors are equipped with a diaphragm in order to resist the low viscosity resin. The temperature sensors are Kistler type 6192 NiCr-Ni thermocouples.

The injection molding machine is a KraussMaffei 550/2000 GX (Munich, Germany) equipped with a standard 60 mm thermoset screw without a non-return valve. The temperature control of the plasticizing unit is realized using four individually controlled, oil-tempered zones. The clamping unit has a maximum clamping force of 5500 kN.

The process parameters for obtaining the pressure and temperature signals are summarized in Tables 1 and 2.

**Table 1.** Constant process parameters used in the RRIM study.

| Parameter | Value | Unit |
|---|---|---|
| Plasticizing unit temperature profile from inlet to nozzle | 60–70, 70–80, 80–90 | °C |
| Screw speed | 40 | 1/min |
| Back pressure | 40 | bar |
| Switchover point | 5 | cm$^3$ |
| Hold pressure stage 1 | 800 | bar |
| Time period of stage 1 | 30 | s |
| Hold pressure stage 2 | 800–15 (linear) | bar |
| Time period of stage 2 | 10 | s |
| Curing time period | 40 | s |

**Table 2.** Varied parameters used in the RRIM study.

| Parameter | Value Min | Value Max | Unit |
|---|---|---|---|
| Mold temperature | 170 | 190 | °C |
| Injection speed | 137.5 | 250 | cm$^3$/s |
| Dosage volume | 230 | 232 | cm$^3$ |

The filling is injection speed controlled until the switchover point, meaning that there is a constant material inflow until 5 cm$^3$ of the dosage volume remains in the plasticizing unit. After the switchover, a constant pressure (hold pressure) acts on the inlet. After the hold pressure stages, the mold is separated from the plasticizing unit and no material flow in the mold is possible. The curing time ensures the shape stability of the composite part.

The aim of this study is to investigate the flow behavior with dependence on varying mold temperature and injection speed and, thus, to validate the proposed FVM-based CFD method. In order to compensate the difference in material backflow at higher injection speeds, the dosage volume is adapted accordingly. The purpose of this adaption is to keep the maximum cavity pressure constant within the study.

### 3.2. Numerical Model, Boundary Conditions and Model Parameters

Since the mold is symmetric in two directions (cf. Figure 1), only a quarter of the mold is simulated to reduce the computation time. The model with predefined boundary face types is shown in Figure 3. To replicate the experiments, four simulations with different process settings are conducted. Thus, two different injection speeds and two different mold temperatures are modeled. For pressure validation, the FVM-based solver is compared to the experimental results and also to FEM results. The FEM simulations are conducted with Autodesk Moldflow Insight 2018.1, also using Castro-Macosko viscosity and the Kamal-Malkin kinetic model [8–10]. The FEM model is meshed in Moldflow with a global edge length of 3.06 mm and eight elements over thickness. The FVM model is meshed with the OpenFOAM meshing tools BlockMesh and snappyHexMesh. Therefore, a base mesh with an edge length 2 mm is used, the refinement of all surfaces is set to level 2 and one additional surface layer is created.

The initial values of the internal field and the different boundary types of faces are given in Tables 3–6.

The hydrostatic pressure is given by the scalar $p_{rgh}$. In this study it is equal to the pressure $p$, because gravity is neglected.

At the symmetry planes the symmetry condition is set for every field variable. The flow rate at the inlet in Table 4 equals one quarter of 137.5 cm$^3$/s and 250 cm$^3$/s, because only one quarter of the mold is simulated. According to the experiments, the filling simulation is initially volume-flux controlled (Table 4), and pressure-controlled after the first switchover point. Two switchover points are also used in the FEM and FVM simulations, where the FVM settings are chosen according to the FEM simulations: After the first switchover point, the pressure is set constant at the actual values, calculated for this time step. After the second one, the pressure is set according to the holding pressure of the experiments, see Table 1. At the first switchover point, the inlet boundary condition for the velocity is changed to pressureInletOutletVelocity, so that the velocity is calculated relating to the fixed pressure. The switchover points are calculated from the experimental parameters and the mold volume. They are given in Table 7.

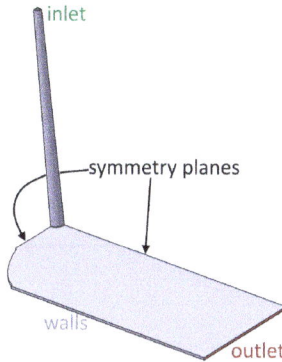

**Figure 3.** Finite-Volume-Method (FVM) model for OpenFOAM with boundary faces.

**Table 3.** Initial conditions and values for the internal field.

| Field Variable | Value | Unit |
|---|---|---|
| A (fiber tensor) | $\begin{pmatrix} 0.33 & 0 & 0 \\ 0 & 0.33 & 0 \\ 0 & 0 & 0.33 \end{pmatrix}$ | - |
| $\alpha$ | 0 | - |
| $c$ | 0.001 | - |
| cure rate | 0 | 1/s |
| $\dot{\gamma}$ | 0 | 1/s |
| $p$ and $p_{rgh}$ | $10^5$ | Pa |
| $T$ | 443.15 or 463.15 | K |
| $u$ | (0 0 0) | m/s |

**Table 4.** Boundary conditions and values for the inlet.

| Field Variable | Type | Value | Unit |
|---|---|---|---|
| A (fiber tensor) | fixedValue | $\begin{pmatrix} 0.33 & 0 & 0 \\ 0 & 0.33 & 0 \\ 0 & 0 & 0.33 \end{pmatrix}$ | - |
| $\alpha$ | fixedValue | 1 | - |
| $c$ | zeroGradient | - | - |

Table 4. *Cont.*

| Field Variable | Type | Value | Unit |
|---|---|---|---|
| cure rate | zeroGradient | - | 1/s |
| $\dot\gamma$ | zeroGradient | - | 1/s |
| $p$ and $p_{rgh}$ | fixedfluxPressure | $10^5$ | Pa |
| $T$ | fixedValue | 423.15 | K |
| $u$ | flowRateInletVelocity | $34.375 \times 10^{-6}$ or $62.5 \times 10^{-6}$ | $m^3/s$ |

Table 5. Boundary conditions and values for the outlet.

| Field Variable | Type | Value | Unit |
|---|---|---|---|
| A (fiber tensor) | zeroGradient | - | - |
| $\alpha$ | zeroGradient | - | - |
| cure | zeroGradient | - | - |
| cure rate | zeroGradient | - | 1/s |
| gammadot | zeroGradient | - | 1/s |
| $p$ and $p_{rgh}$ | zeroGradient | - | Pa |
| $T$ | fixedValue | 443.15 or 463.15 | K |
| $u$ | BC (see Section 2.2) | - | m/s |

Table 6. Boundary conditions and values for the wall.

| Field Variable | Type | Value | Unit |
|---|---|---|---|
| A (fiber tensor) | zeroGradient | - | - |
| $\alpha$ | zeroGradient | - | - |
| cure | zeroGradient | - | - |
| cureRate | zeroGradient | - | 1/s |
| gammadot | zeroGradient | - | 1/s |
| $p$ and $p_{rgh}$ | fixedfluxPressure | $10^5$ | Pa |
| $T$ | fixedValue | 443.15 or 463.15 | K |
| $u$ | noSlip | - | m/s |

For fiber orientation prediction, a fiber volume fraction of 0.35 and an aspect ratio of 25 are assumed.

Table 7. Switchover points of the simulations.

| Filling Rate | First Point | Second Point |
|---|---|---|
| 137.5 $cm^3/s$ | 1.46 s | 1.5 s |
| 250 $cm^3/s$ | 0.8 s | 0.82 s |

Table 8 summarizes the parameter values of reaction kinetics of the material system used for experimental validation.

Table 8. Parameters for the Kamal-Malkin kinetic model.

| Parameter | Value | Unit |
|---|---|---|
| R | 8.3144598 | J/(K·mol) |
| A1 | $1.9454 \times 10^{12}$ | 1/s |
| A2 | 3041.4 | 1/s |
| E1 | 2,878,805.64 | J/mol |
| E2 | 38,425.6452 | J/mol |
| m | 1.643 | - |
| n | 0.4893 | - |

The parameter values of the viscosity model of the material system used for experimental validation are given in Table 9. Air is assumed to be a perfect gas.

**Table 9.** Parameters for the Castro-Macosko viscosity model.

| Parameter | Value | Unit |
|-----------|-------|------|
| $\tau^*$ | 0.79 | Pa |
| n | 0.5 | - |
| $c_1$ | 17 | - |
| $c_2$ | 17 | - |
| B | $1.123 \times 10^{-7}$ | Pa·s |
| $T_b$ | 13.750 | K |
| $c_g$ | 0.4 | - |

Values received from manufacturer.

### 3.3. Numerical Model for Verification of Fiber Orientations

For comparison of fiber orientation results of FVM and FEM a different geometry is regarded. To verify all room directions, a mold with dominant flow in every direction is created. The boundary types are the same as mentioned in Tables 5–8. Geometry and boundary faces are shown in Figure 4.

For the velocity, an injection speed of 3.8 cm$^3$/s is chosen, leading to a filling time of 1 s. The inlet and outlet are squares with an edge length of 5 mm. The FEM mesh is built up in Moldflow with tetrahedral elements, while hexahedral volumes created with BlockMesh are used for the FVM mesh. In both meshes a global edge length of 1 mm is set.

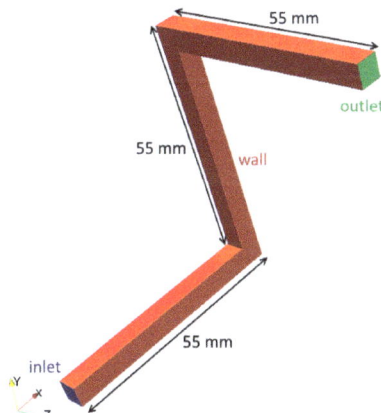

**Figure 4.** Geometry for fiber orientation validation.

## 4. Results

In this section, the simulation results are presented and compared to experimental and FEM results.

### 4.1. Results of the Filling Simulation

Figure 5 shows the FVM simulation results for a filling rate of 137.5 cm$^3$/s and a mold temperature of 170 °C. Displayed are (from top to bottom) the filling level (via VoF), the velocity in m/s and the pressure in MPa at 1 s (Figure 5a), where the filling is velocity-controlled, and at 1.5 s (Figure 5b), where it is pressure-controlled and should be just filled.

At 1.5 s the mold is not filled completely, visible at the corners, where air is still left in the system (blue). For a sum of $\alpha$ over all cells, divided by the total number of cells, the value should be equal to one for a complete filled system. In this case, the value is 0.998, which matches a filling state of 99.8%. The 0.2% air can be neglected and the mold is regarded as completely filled. The mold gets completely filled during holding pressure stage 1. There is no possibility to make a statement when exactly the mold in the trials is filled completely, because the material can only be detected at the sensor points. Based on the pressure data, the pressure rise and difference between sensor 1 and 2 in the trials, the filling also takes approximately 1.5 s for an injection speed of 137.5 cm$^3$/s in the experimental studies (see Section 4.2).

A detailed look at the flow front (Figure 5a top) displays that there is no sharp interface between polymer and air, but an interface region with $0 < \alpha \leq 1$ over a few cells. This area indicates the region with partial wall contact, which is a typical phenomenon of RRIM, although a clear line between partial and full wall contact is not detectable as described in [1,2,5]. This phenomenon cannot be observed in the FEM simulation, where a sharp interface is predicted and an element is always either polymer or empty, see Figure 6.

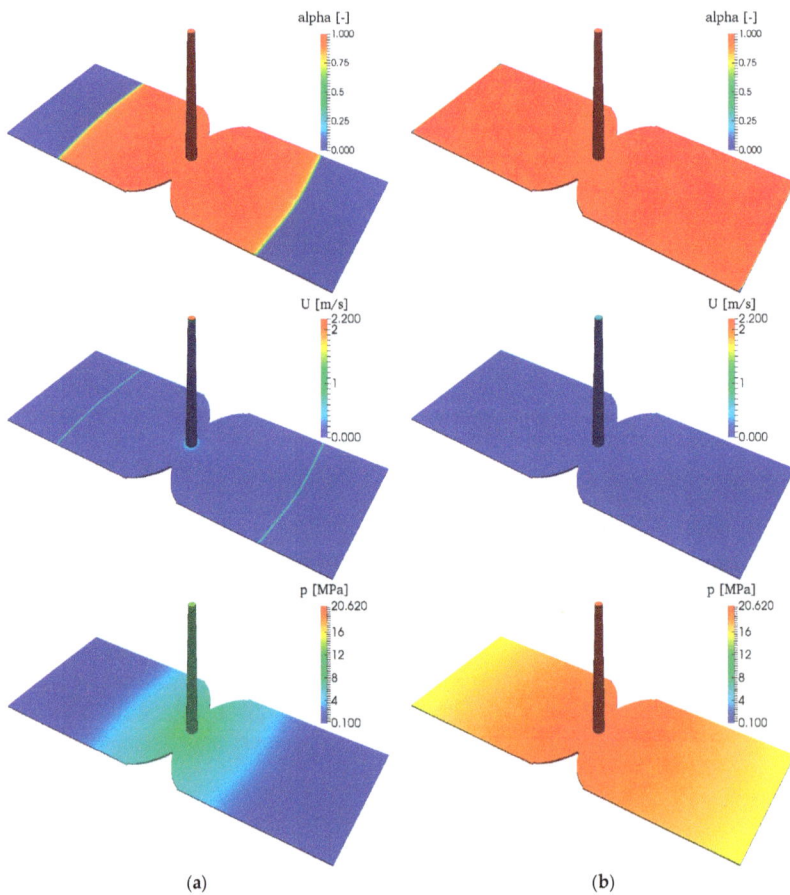

**Figure 5.** FVM-simulation results of polymer filling state alpha (top), velocity $U$ in m/s (middle) and pressure $p$ in MPa (bottom) after 1 s (a) and 1.5 s (b) for a filling rate of 137.5 cm$^3$/s and a mold temperature of 170 °C.

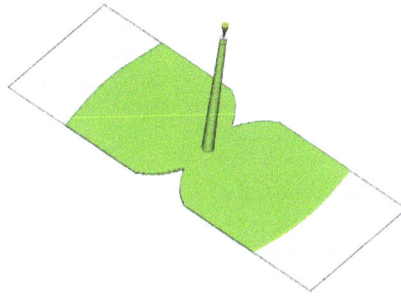

**Figure 6.** Filling state of the FEM-simulation at 1 s.

The velocity (Figure 5 middle) ranges in a reasonable spectrum. The polymer enters the mold with an injection speed of 2.2 m/s at the inlet, which corresponds to a volumetric flow rate of 137.5 cm$^3$/s based on the inlet area. Furthermore, it can be seen that the velocity is nearly zero after 1.5 s (Figure 5b middle), although there is still a pressure gradient at that moment (Figure 5 bottom). This aspect approves the phase-dependent boundary condition, which stops the flow when the mold is filled.

### 4.2. Comparisson of Pressure

The pressure during filling with an injection speed of 137.5 cm$^3$/s is shown in Figure 7a for a mold temperature of 170 °C and Figure 7b for 190 °C. The experiments at 170 °C are more reproducible than the ones at 190 °C, as visualized by the scatter beams in Figure 7. The large scatter at 190 °C might be caused by the fact that 190 °C is the maximum temperature for injection molding according to the manufacturer. Every curve of sensor position 1 shows a continuous pressure growth during filling (0–1.46 s). After filling, the experimental curves of sensor 1 and 2 are nearly identical, which shows good process control and filling behavior. The simulation results of FEM and FVM considerably differ from each other. Compared to measurements at sensor 1, FEM predicts a higher pressure, while the FVM results fit better to the experiments and are just slightly higher for 170 °C. At sensor 2, a significant pressure rise after switchover (1.46 s) is detectable in experiments as well as in FEM and FVM simulations. However, both simulations show a too fast pressure rise at sensor 2 compared to the measurements.

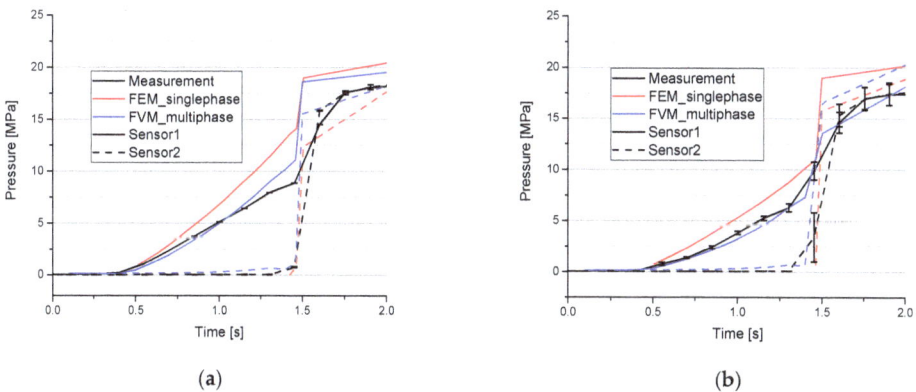

(a)

(b)

**Figure 7.** Pressure over time for a filling speed of 137.5 cm$^3$/s and a mold temperature of 170 °C (a) and 190 °C (b). Comparing measurement (black), FEM (red) and FVM (blue) at sensor position 1 (solid line) and 2 (dotted line).

The pressure difference between sensor 1 and 2 after switchover is too high in both simulations for both temperatures, where it is even higher in the FEM than in the FVM calculations. In general, the pressure during filling is lower at 190 °C than for 170 °C in experiments and simulations. This is caused by the lower viscosity, resulting from the higher temperature.

Although material backflow is not simulated, the overall pressure rise at both sensor positions is quite similar in experiments and simulations. Hence, material backflow could be neglected, although no non-return valve is used in the experimental trials.

For a filling rate of 250 cm$^3$/s, there is also a greater scatter in pressure measurement for 190 °C (Figure 8b) than for 170 °C (Figure 8a) and the pressure is again lower at 190 °C because of the lower viscosity. Up to the first switchover, the experimental data of a filling rate of 250 cm$^3$/s show a higher pressure growth during filling than the data of the filling rate of 137.5 cm$^3$/s (Figure 7), which is a consequence of the higher velocity. Hence, there is no visible pressure rise at switchover (0.8 s) for sensor 1. After switchover, the experimental curves of sensor 1 and 2 do not immediately fit to each other as well as for 137.5 cm$^3$/s. For 190 °C the measured pressure at sensor 2 is even higher than the one at sensor 1 between 0.85 and 1 s.

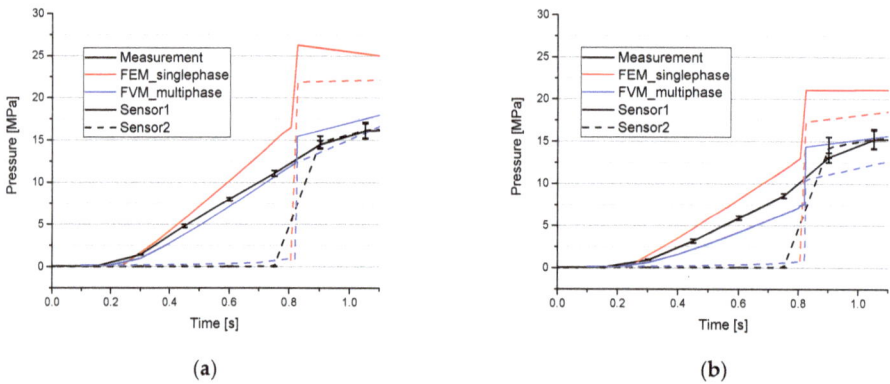

**Figure 8.** Pressure as a function of time for a filling speed of 250 cm$^3$/s and a mold temperature of 170 °C (**a**) and 190 °C (**b**). Comparing measurement (black), FEM (red) and FVM (blue) at sensor Position 1 (solid line) and 2 (dotted line).

Both simulation methods overestimate the pressure rise at switchover, which is distinctly higher in the FEM simulations. As before, the FEM calculated pressure is too high for sensor 1. The pressure distribution of the FVM simulation fits well during filling for 170 °C and is too low for 190 °C. Regarding the curve course of sensor 2, the pressure rise of both methods appears too fast. However, the real pressure curve in this area is not entirely known, because there are only measure points at 0.75 and 0.9 s (switchover at 0.8 s), making the distribution look flatter.

In summary, the results of the FVM simulation are satisfying, although no further material characterization has been done except for the material data provided by the manufacturer. By ranging in a correct spectrum and rising at the right time, the FVM simulation results show a good agreement with the experimental data. The proper pressure modeling indicates a suitable viscosity modeling and hence a suitable simulation of curing kinetics, as further evaluated in the following section.

*4.3. Comparison of Curing Kinetics*

The correct implementation of the curing kinetics modeling of the new FVM-solver is verified by comparison with commercial FEM software, which also uses the Kamal-Malkin model. The increase of the curing degree at sensor position 1 is illustrated in Figure 9. The results are quite similar for FEM and FVM, confirming the correct implementation in the curing kinetics modeling in the FVM-solver.

At the end of filling (1.5 s) the degree of cure is about 0.85% (in the FVM simulation). This raises the viscosity by 44.5% (Equation (9)) related to the same temperature and shear rate but with a degree of curing equal to zero ($c = 0$).

The kinetic model (see Section 2.3) only depends on the state variables time and temperature, which are not as much influenced by flow as pressure and velocity. That might be the reason for the good fitting results of FEM and FVM.

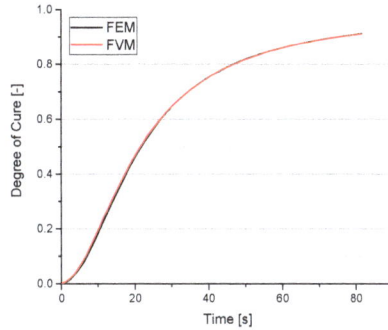

**Figure 9.** Degree of cure as a function of time for FEM (black) and FVM (red) for a filling speed of 137.5 cm$^3$/s and a mold temperature of 170 °C.

### 4.4. Comparison of Fiber Orientation

The fiber orientation computed by FVM simulation is compared to those of commercial FEM software. For the FEM simulation, the Moldflow standard model (Moldflow-rotation-diffusion) is chosen. The results are based on the geometry of Figure 4. The geometries in Figure 10 might seem to be different for FEM and FVM, which is only caused by the perspective.

Regarding the fiber alignment in $x$-direction (Figure 10a), a deviation especially in the edges can be detected. In the FEM model, a high orientation right at the beginning can be detected, while the FVM result aligns only after a few millimeters. The results in $y$-direction fit well, while in $z$-direction the FVM-solver predicts a slightly higher alignment of about 10%. The FVM solution has a higher scattering in the transition area.

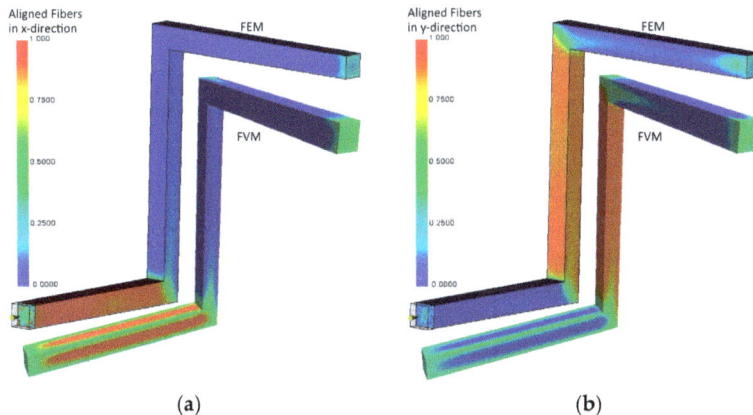

(a)                              (b)

**Figure 10.** *Cont.*

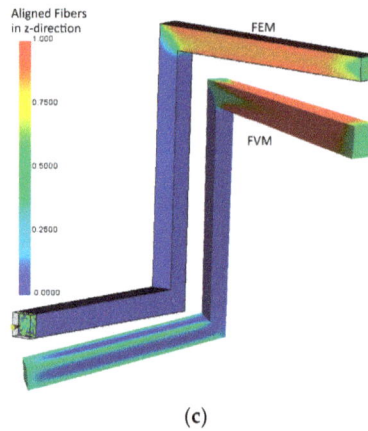

(c)

**Figure 10.** Comparison of predicted fiber orientation between FEM and FVM in *x*-direction (**a**), *y*-direction (**b**) and *z*-direction (**c**).

These slight differences may be caused by the different flow modeling. The plug flow, modeled in FEM, allows a faster orientation especially in regions near the walls (like edges and corners) because of the full wall contact, whereas the fountain flow with partial wall contact modeled in FVM does not lead to such a rapid orientation, because of the different velocity gradient. In summary the results fit well, which confirms a correct implementation.

## 5. Discussion and Conclusions

A new method for simulation of reinforced reactive injection molding is presented. The solver is implemented in the open source CFD toolbox OpenFOAM, using the Finite-Volume-Theory. Compressible, non-isothermal multiphase flows are simulated, modeling air and polymer in the mold at every iteration. Comparison with experimental data proves that the FVM-solver is able to predict pressure distribution during mold filling for several filling speeds and mold temperatures. This validates a good viscosity and flow modeling. Regions with partial wall contact are detectable at the flow front, which is a typical phenomenon of RRIM and cannot be seen in the FEM simulations. Furthermore, the new FVM-based solver simulates fiber orientation and curing kinetics at the level of commercial FEM software. Hence, this study shows the encouraging opportunity of FVM for simulation of reaction injection molding with a realistic two-phase flow modeling for thermoset materials.

The open source structure of OpenFOAM in combination with the good quality of the results achieved so far, reveal a high potential for additional features, applications and process phenomena, which can be regarded and simulated in the future. For example, diffusion and transport models can be used for analyzing the arising weld lines. The energy balance can be extended with a source term for curing heat, to have a better temperature and hence better curing and viscosity modeling for thick parts. Moreover, anisotropic viscosity models and viscoelastic behavior for modeling extensional flows can be implemented. This could lead to a better simulation of the region with partial wall contact, which is fundamental for accurate and detailed flow modeling of RRIM, having an impact on the filling, fiber orientation and hence the mechanical and thermal properties of the final part.

**Acknowledgments:** We would like to thank the Vector Foundation, Stuttgart, Germany for the financial support and SBHPP Vyncolit, Gent, Belgium for providing the material and the material data.

**Author Contributions:** Florian Wittemann performed the major part of the work in terms of method development and implementation. He also wrote the first draft of the paper. The curing kinetic and viscosity model was implemented by Alexander Bernath. All experimental work was conducted by Robert Maertens. Martin Hohberg initiated the work, supported the FE-based simulations and revised the paper. Luise Kärger supervised the method development in general, supported the discussion of simulation results and thoroughly revised the paper. Frank Henning supervised the work in terms of composite process knowledge and relevance of the addressed subjects.

**Conflicts of Interest:** The authors declare no conflict of interest.

## Appendix A

For implementation of the phase-dependent boundary condition, the third party tool SWAK4FOAM must be installed. The following code for the boundary condition must be placed in the *U* file of the 0 folder (or other corresponding time folders) in the boundary field block. In this case it is specified for a patch called "outlet".

```
outlet
    {
        type            groovyBC;
        aliases         {alpha1 alpha.polymer;}
        valueExpression "vector(0,0,0)";
        gradientExpression "vector(0,0,0)";
        fractionExpression "alpha1";
        value uniform (0 0 0);
    }
```

Furthermore, the relevant library must be read at the end of the ControlDict file:

```
libs ("libgroovyBC.so");
```

## References

1. Michaeli, W.; Hunold, D.; Kloubert, T. Formfüllung beim Spritzgießen von Duroplasten. *Plastverarbeiter* **1992**, *43*, 42–46.
2. Thienel, P.; Hoster, B. Ermittlung der Füllbildkonstruktion. *Plastverarbeiter* **1992**, *43*, 36–41.
3. Höer, M. *Einfluss der Material- und Verarbeitungseigenschaften von Phenolharzformmassen auf die Qualität Spritzgegossener Bauteile*; Universität Chemnitz: Chemnitz, Germany, 2014.
4. Englich, S. *Strukturbildung bei der Verarbeitung*; Universität Chemnitz: Chemnitz, Germany, 2015.
5. Ohta, T.; Yokoi, H. Visual analysis of cavity filling and packing process in injection molding of thermoset phenolic resin by the gate-magnetization method. *Polym. Eng. Sci.* **2001**, *41*, 806–819. [CrossRef]
6. Castro, J.M.; Macosko, C.W. Studies of mold filling and curing in the reaction injection molding process. *AIChE J.* **1982**, *28*, 250–260. [CrossRef]
7. Heinen, K. *Mikrostrukturelle Orientierungszustände Strömender Polymerlösungen und Fasersuspensionen*; Universität Dortmund: Dortmund, Germany, 2007.
8. Autodesk. *Autodesk Simulation Moldlfow Insight, Fundamentals 2014, Student Guide—Theory and Concepts Manual*; Ascent: Charlottesville, VA, USA, 2013.
9. Autodesk. *Simulation Moldflow Insight, Fundamentals 2014, Student Guide—Practice Manual*; Ascent: Charlottesville, VA, USA, 2013.
10. Autodesk. *Autodesk Simulation Moldflow Insight, Advanced Flow 2014, Student Guide—Therory and Concepts Manual*; Ascent: Charlottesville, VA, USA, 2013.
11. Tamil, J.; Ore, S.H.; Gan, K.Y. Molding Flow Modeling and Experimental Study on Void Control for Flip Chip Package Panel Molding with Molded Underfill Technology. *J. Microelectron. Electron. Packag.* **2012**, *9*, 19–30. [CrossRef]
12. Ospald, F. Numerical Simulation of Injection Molding using OpenFOAM. *Proc. Appl. Math. Mech.* **2014**, *14*, 673–674. [CrossRef]
13. Deshpande, S.S.; Anumolu, L.; Trujillo, M.F. Evaluating the performance of the two-phase flow solver interFoam. *Comput. Sci. Disc.* **2012**, *5*, 14–16. [CrossRef]

14. Marić, T.; Höpken, J.; Mooney, K. *The OpenFOAM Technology Primer*; Sourceflux: Duisburg, Germany, 2014; ISBN 978-3-00-046757-8.

15. Damián, S.M. *An Extended Mixture Model for the Simultaneous Treatment of Short and Long Scale Interfaces*; Universidad Nacional del Litoral: Santa Fe, Argentina, 2013.

16. Ruiz, E.; Waffo, F.; Owens, J. Modeling of resing cure kinetics for molding cycle optimization. In Proceedings of the 8th International Conference on Flow Process in Composite Materials, Douai, France, 11–13 July 2006.

17. Bernath, A.; Kärger, L.; Henning, F. Accurate Cure Modeling for Isothermal Processing of Fast Curing Epoxy Resins. *Polymers* **2016**, *8*, 390. [CrossRef]

18. Kamal, M.R.; Sourour, S. Kinetics and thermal characterization of thermoset cure. *Polym. Eng. Sci.* **1973**, *13*, 59–64. [CrossRef]

19. Liang, G.; Chandrashekhara, K. Cure kinetics and rheology characterization of soy-based epoxy resin system. *J. Appl. Polym. Sci.* **2006**, *102*, 3168–3180. [CrossRef]

20. Khalil Abdullah, M.; Abdullah, M.Z.; Mujeebu, M.A.; Kamaruddin, S. A Study on the Effect of Epoxy Molding Compound (EMC) Rheology during Encapsulation of Stacked-CHIP Scale Packages (S-CSP). *J. Reinf. Plast. Compos.* **2009**, *28*, 2527–2538. [CrossRef]

21. Khor, C.Y.; Abdullah, M.Z.; Lau, C.-S.; Azid, I.A. Recent fluid–structure interaction modeling challenges in IC encapsulation—A review. *Microelectron. Reliab.* **2014**, 1511–1526. [CrossRef]

22. Folgar, F.; Tucker, C.L. Orientation Behavior of Fibers in Concentrated Suspensions. *J. Reinf. Plast. Compos.* **1984**, *3*, 98–119. [CrossRef]

23. Advani, S.G.; Tucker, C.L. The Use of Tensors to Describe and Predict Fiber Orientation in Short Fiber Composites. *J. Rheol.* **1987**, 751–784. [CrossRef]

24. Advani, S.G.; Tucker, C.L. Closure approximations for three-dimensional structure tensors. *J. Rheol.* **1990**, 367–386. [CrossRef]

25. Skrabala, O.; Bonten, C. Orientierungsverhalten plättchenförmiger Zusatzstoffe in Kunststoffsuspensionen. *J. Plast. Technol.* **2016**, 157–183. [CrossRef]

*Journal of*
*composites science*

MDPI

Article

# Simulative Prediction of Fiber-Matrix Separation in Rib Filling During Compression Molding Using a Direct Fiber Simulation

**Christoph Kuhn [1],\*, Ian Walter [2], Olaf Täger [1] and Tim Osswald [2]**

[1]   Volkswagen Group Research, Postbox 011/14990, 38440 Wolfsburg, Germany; olaf.taeger@volkswagen.de
[2]   Polymer Engineering Center (PEC), University of Wisconsin-Madison, 1513 University Ave,
      Madison, WI 53706, USA; iwalter@wisc.edu (I.W.); tosswald@wisc.edu (T.O.)
\*    Correspondence: christoph.kuhn@volkswagen.de; Tel.: +49-5361-9-14631

Received: 22 November 2017; Accepted: 13 December 2017; Published: 28 December 2017

**Abstract:** Compression molding of long fiber reinforced composites offers specific advantages in automotive applications due to the high strength to weight ratio, the comparably low tooling costs and short cycle times. However, the manufacturing process of long fiber composite parts presents a range of challenges. The phenomenon of fiber matrix separation (FMS) is causing severe deviations in fiber content, especially in complex ribbed structures. Currently, there is no commercial software that is capable to accurately predict FMS. This work uses a particle level mechanistic model to study FMS in a rib filling application. The direct fiber simulation (DFS) is uniquely suited to this application due to its ability to model individual fibers and their bending, as well as the interaction amongst fibers that leads to agglomeration. The effects of mold geometry, fiber length, viscosity, and initial fiber orientation are studied. It is shown that fiber length and initial fiber orientation have the most pronounced effects on fiber volume percentage in the ribs, with viscosity and part geometry playing a smaller role.

**Keywords:** fiber reinforced plastics; long fiber reinforced thermoplastics (LFT); process simulation; compression molding; fiber content; direct fiber simulation; mechanistic model

---

## 1. Introduction

Fiber reinforced composites offer high mechanical strength to weight ratios. This makes them interesting for a number of industries, including the automotive industry, where tightening emissions restrictions have forced designers to find suitable materials and processes to reduce component weight in large scale production [1]. Compression molding using glass mat thermoplastics (GMT) or long fiber reinforced thermoplastics (LFT) is widely used in high volume production [2,3]. The process can be fully automated with a high degree of quality consistency in production, which reduces costs and cycle time [2,4]. Simulation tools are included at early stages of the product design process to determine processing conditions and predict part performance [5–7], however there are some aspects of processing that are not yet implemented in commercially available tools. One aspect which is particularly difficult to predict is fiber matrix separation (FMS), which occurs in systems that are highly filled with long fibers [8–10]. Due to increasing fiber interaction, fibers are restrained in their movement and accumulate inside complex flow geometries. This effect of FMS leads to significant variations in fiber content throughout the part. Current mold-filling software is not able to account for this behavior as it uses continuum models that are based on the motion and interactions observed in short, rigid fiber experiments [6,11]. This presents an issue, since the mechanical properties of fiber reinforced parts are highly dependent on fiber content, orientation and length distribution [5,12].

In order to predict FMS and the occurring variations in fiber content, a different approach is needed that accounts for complex fiber movement and interactions during processing. To this end, the direct fiber simulation (DFS) was developed to study complex interactions occurring during processing. The DFS utilizes a particle level mechanistic model, in which fibers are simulated as rigid beams connected by ball and socket joints [13,14]. During simulation, fiber-fiber interaction, hydrodynamic, fiber-mold interaction, and intra-fiber forces are computed. With this approach, complex phenomena leading to FMS can be predicted. However, using a particle level approach leads to far greater computational times than conventional models.

In this work, the DFS is applied to study FMS inside ribbed components in long fiber compression molding. Earlier simulations with the instigated component geometry (Figure 1) were conducted in [15] and showed limitations regarding mesh exiting and fiber freezing due to boundary conditions and model simplifications unsuitable for full scale DFS. In this paper, improvements are presented to accurately display the fiber behavior during processing. With the implemented improvements, further simulations are conducted to evaluate the effects of fiber length, initial orientation, rib geometry, and viscosity on rib filling. The simulation results are then compared to experimental studies, which were obtained earlier in compression molding experiments [15]. The research of this paper focuses on the prediction of fiber content distribution in compression molding. Results regarding the fiber orientation prediction will be published consecutively.

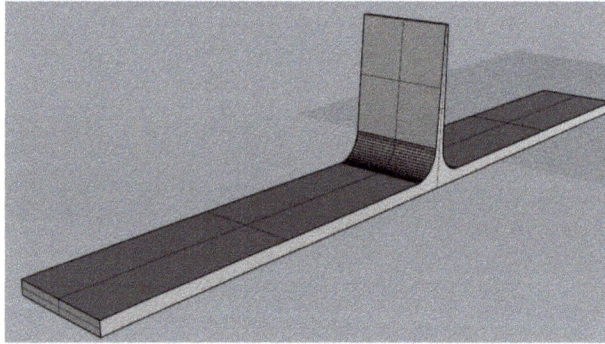

**Figure 1.** Simplified ribbed plate geometry.

## 2. Direct Fiber Simulation

The DFS simulation chain is initially started with a traditional mold filling simulation, using the commercial software Moldex3D™ by CoreTech System Co., Ltd., Chupei City, Taiwan. The conventional filling simulation is conducted with SABIC STAMAX 30YM240 material data, a fiber reinforced polypropylene with an initial fiber content of 30 wt %. After the conventional simulation has been completed, the flow field is extracted and utilized in the mechanistic model. Since the flow-field velocity is output at discreet steps, it is necessary to use a high number of output files to improve simulative accuracy. Once the Moldex3D™ simulation is completed, the DFS is implemented by placing fibers within the volume of the mold. These fibers are randomly generated in the specified volume, according to the desired fiber volume fraction and initial fiber orientation. For a more complete description of the fiber generation, please refer to [16].

Once the bundle is placed in the cavity, fiber motion is determined by interaction with the flow field. First, a force balance is calculated for each fiber including all the effects shown in Figure 2. The "excluded volume effect" listed below is a penalty function used to model fiber-fiber interaction and fiber-mold interaction and keep fibers from occupying the same volume. As fibers approach one another, a repulsive force is applied along the shortest path between the approaching fibers, or fiber and mold wall. Fibers are also able to bend at the joints between segments, unlike in the Folgar-Tucker model, which assume short, rigid fibers [6,17]. This is important for calculating fiber behavior in parts whose dimensions are smaller than the fiber length.

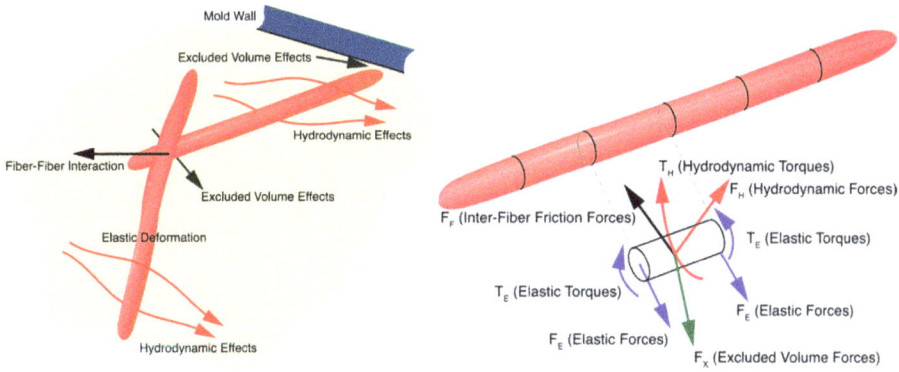

**Figure 2.** Depiction of forces and interactions calculated by direct fiber simulation.

The DFS applies the approach of Schmid [18] and Lindstroem [17], and the code used in this work was developed by Ramírez and Pérez [13,14]. Fibers are modeled as rigid rods connected by ball and socket joints [13]. Positions, velocities and angular velocities are stored at each node and bending may occur at these nodes. Each segment *i* experiences drag forces ($F_H$), inter-fiber interactions ($F_F$), and intra-fiber forces from neighboring segments ($F_X$) as seen in Figure 2. Due to the low Reynolds numbers seen in polymer flows, inertial effects are ignored, as are gravitational effects. Long range hydrodynamic forces are also not calculated. It was found that calculation of these effects is prohibitively expensive in systems with large numbers of fibers and less important than contact forces [8]. Additionally, no extensional deformation of the fibers is taken into account. It is possible to make this assumption due to the high stiffness of the fibers and the relatively small effect stretching has compared to bending. More details on the fiber modelling in the mechanistic model are presented by Ramírez and Pérez [13,14]. Once the force balance is calculated, a linear system of equations is solved to find the velocity of the nodes, and the fibers are advanced one time step. One important note is that fiber-fiber and fiber-mold interactions are not calculated every integration, rather it is up to the user to specify how many integrations may pass in between this calculation. This is done to decrease computation time, as fiber-fiber interactions are the most expensive effect to calculate. In the simulations discussed in this paper, interactions were calculated once every 50 integrations which showed optimized simulative performance in earlier simulations.

With the described method, fiber movement inside the compression charge is calculated. Figure 3 shows the fiber movement during compression molding for the ribbed-plate component. It is observed how the reinforcement fibers are dragged inside the rib by the polymer flow, gradually filling the part. The fibers display complex fiber movement, bending and interaction during flow. The governing mechanisms on FMS are investigated in detail after the simulation if improved regarding the simulative boundary conditions in the following chapter.

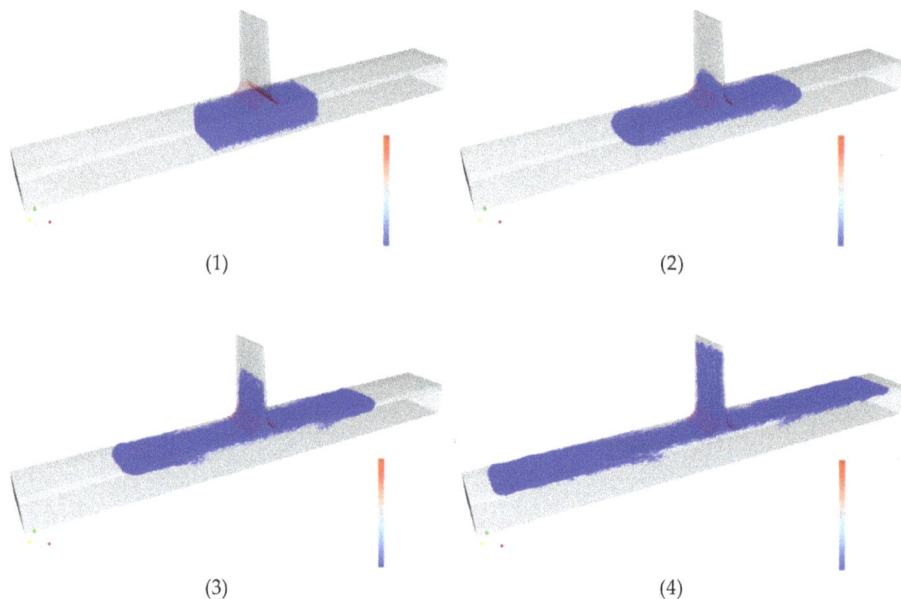

**Figure 3.** Fiber movement during compression molding simulations with the ribbed plate at consecutive time steps 1–4.

## 3. Simulation Improvements

In previous simulations [15], fibers were often observed exiting the part cavity mesh. This occurs because there is no Boolean variable that stops fibers from exiting. During fiber movement, the excluded volume force applies a force away from the wall that deters fibers from exiting the mesh. This force increases with smaller distance between fibers or to the cavity walls. The issue of fibers exiting the mesh arises due to the increasing excluded volume forces within the fiber bed. Due to fiber interactions and bundling, especially as observed in the rib base, fiber interaction forces are higher than excluded volume forces from the wall. At these locations, fibers are pushed outside the wall during numerical calculations. When fibers exit the mesh, they become frozen in place. This presents an issue because the part of the fiber inside of the cavity also freezes, leaving frozen fibers hanging out into the flow field. These fibers act as hooks and artificially prevent other fibers from flowing past their location, as presented in Figure 4. In an effort to mitigate this, four identical simulations using 20 mm fibers were run and two new methods were implemented in this work.

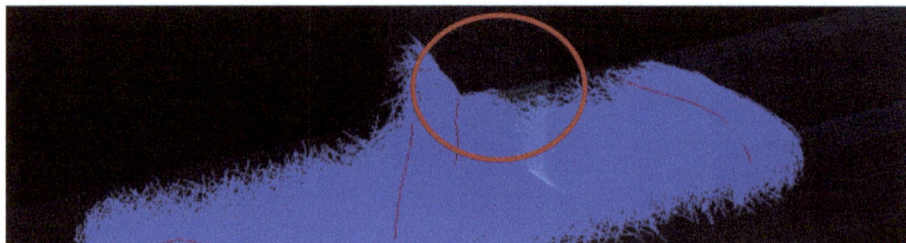

**Figure 4.** The circled frozen fiber in the center can be seen to stop other fibers from flowing past it.

In order to stop frozen fibers hindering the movement of other fibers, a function was added that removes fibers from the simulation once they exit the mesh and freeze. Their positions are saved before deletion and are used in the analysis of fiber position, bending, etc. When fibers are removed after freezing, a dramatic increase in fiber volume fraction at the top of the rib is seen. Enhancing boundary conditions was also used in an attempt to prevent fibers from exiting the mesh in the first place, as shown in Figure 5. Mesh boundary conditions are enhanced at the edge of the mesh in problematic areas such as corners or ribs. These extra barriers increase the excluded volume force applied away from the wall and help prevent fibers from exiting the mesh. As can be seen in Figure 5, this leads to more realistic results. All full charge simulations done within this work used both a fiber barrier, and removed fibers when they were frozen to avoid artificial bridging.

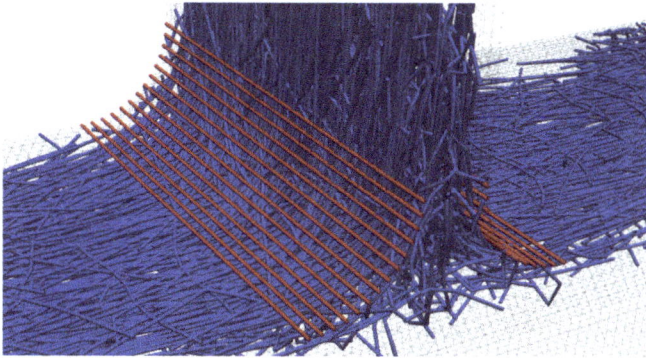

**Figure 5.** Increased boundary conditions shown in red. Note: Fiber radius is shown as larger for illustration purposes.

## 4. Simulative Study

Simple rib simulations are conducted using the entire bottom of the plate as the compression surface as shown in Figure 6. Initially, a fractional factorial design of experiments (DOE) is completed to give an overview of the different variables and assess their relative effect on fiber volume fraction throughout the rib. For analysis, the 35 mm rib is divided into three sections, including the top (11 mm), middle (12 mm), and bottom (12 mm). Additionally, the fiber orientation in the plate extending 10 mm in either direction from the center of the rib is evaluated.

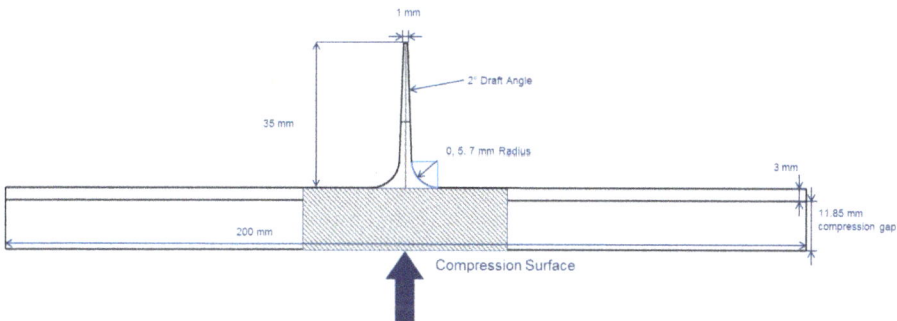

**Figure 6.** Cross section of rib. Width (into page) was adjusted based on fiber length. The striped sections show charge placement.

For the following analysis, each variable was used at two different levels, −/+, as shown in Table 1. The average fiber volume fraction values at each section are measured and used to determine the relative magnitude of the effects of each parameter.

**Table 1.** Variables for initial design of experiments (DOE).

| Parameter | − | + |
|---|---|---|
| Fiber Length (mm) | 10 | 20 |
| Rib Radius (mm) | 5 | 7 |
| Viscosity (Pa·s) | 0.1 | 150 |
| Initial Orientation (-) | Cross Rib (CR.) | Random (Rand.) |
| Fiber Diameter (m) | $1.9 \times 10^{-5}$ | |
| Fiber Young's Modulus (MPa) | 70 | |
| Initial Fiber Content in Charge (vol %) | 0.5 | |

As observed in Figure 7, fiber length displays the largest effect on filling percentage in the top third of the rib. Initial fiber orientation had the next largest effect, with fibers oriented across the rib showing excessive fiber bridging and FMS. Longer fibers and fibers oriented across the rib opening create more fiber bridging which made it more difficult for other fibers to flow past and into the rest of the rib. This led to lower volume fractions in the rib due to the velocity gradient between the fiber and matrix.

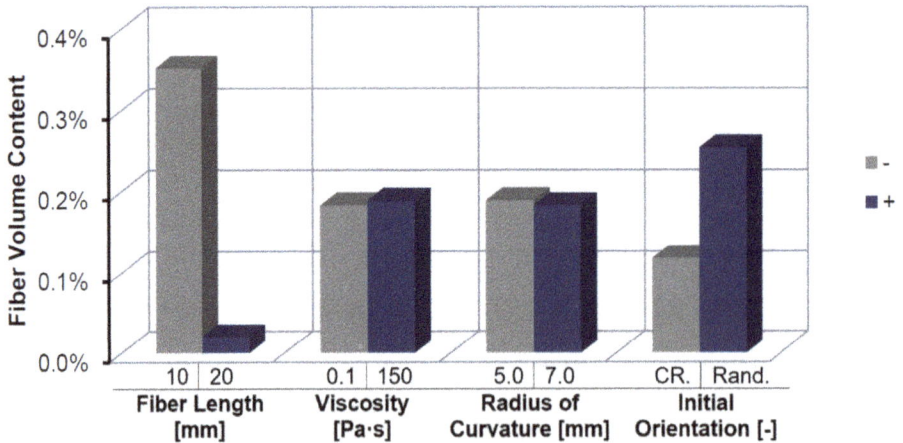

**Figure 7.** Average effect of each variable on volume fraction at top of rib.

Figure 8 illustrates the difference seen between the most favorable conditions and the least favorable. A large variation in fiber content is seen, and a large difference in orientation can also be observed at the rib base. On the left-hand side, random short fibers can be seen to easily make it into the rib as they are not oriented across the opening. On the right-hand side, longer fibers oriented across the rib tend to get stuck in this area and experience great difficulty moving past the bottom of the rib.

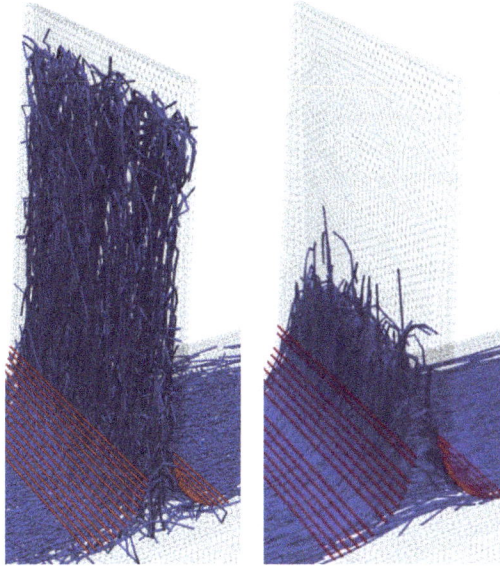

**Figure 8.** Ten-millimeter fibers' random orientation (**left**) vs. 20 mm fibers with cross rib orientation (**right**) after filling completed. Fiber diameter is exaggerated for visibility purposes.

Inside the midsection of the rib (Figure 9), it is observed that varying fiber length and orientation creates the largest content variations, although the viscosity and geometry begin to have a larger effect as well. Higher viscosity and a larger rib entrance both have favorable effects on the volume content. These results are in agreement with previous work by Londoño et al. [10].

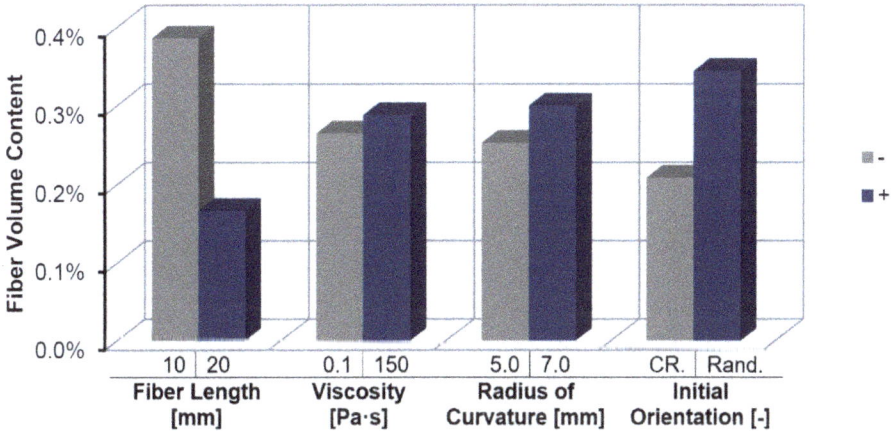

**Figure 9.** Average effect on volume fraction in middle section of rib.

Figure 10 displays the fiber content at the bottom of the rib. Here we see that viscosity and fiber length do not play as important of a role, but rather the size of the rib entrance and the orientation have a larger, but yet comparably low effect. This implies that a sufficient length for fiber jamming and bridging has been reached and the orientation begins to play a larger role on fiber bridging and jamming.

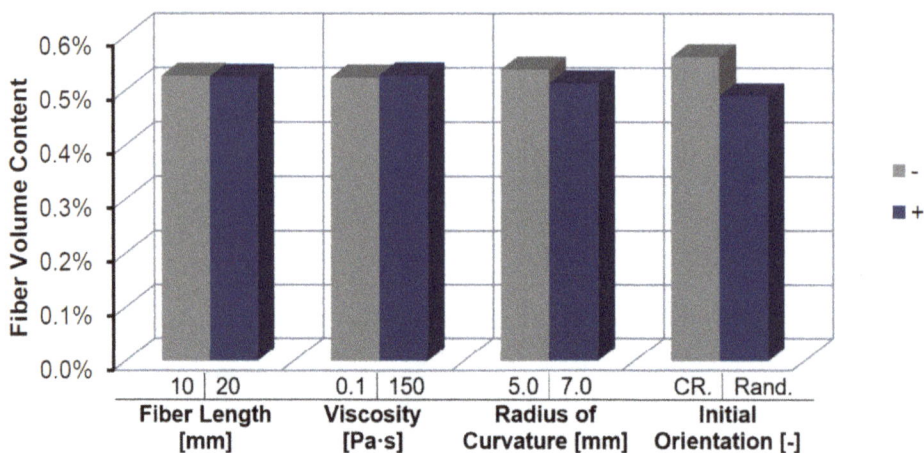

**Figure 10.** Average effect on volume fraction in bottom section of rib.

The effect of fiber bridging can be seen in both the DFS results as well as physical experiments as shown by Kuhn et al. [15] in Figure 11. Several fiber bundles that appear to be flowing around an obstruction are seen, and on the right, the same behavior is observed during simulation. This implies that the DFS is able to model fiber-fiber interaction effects that lead to FMS.

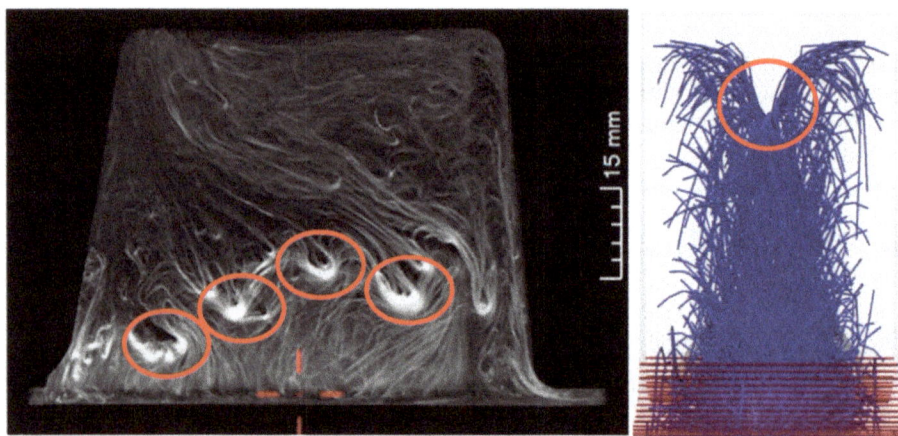

**Figure 11.** Examples of fiber bridging seen in real parts (**left**, circles) and in the direct fiber simulation (DFS) (**right**).

To further investigate the influencing parameters on FMS in ribs, additional simulations are conducted. Figure 12 displays the change of fiber content inside the rib sections with different initial fiber lengths.

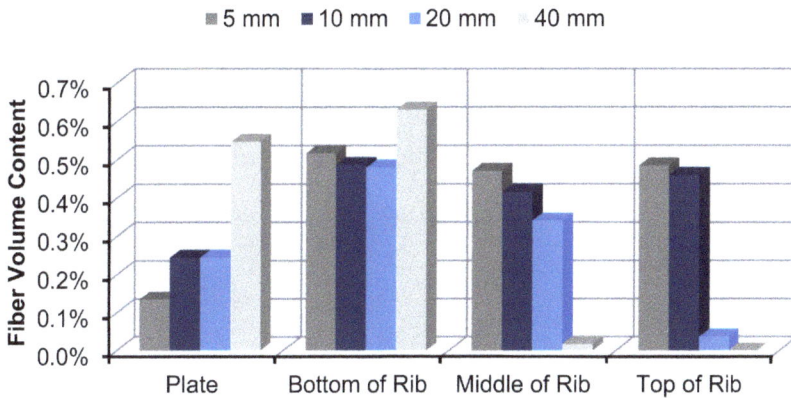

**Figure 12.** Fiber volume fraction throughout rib for different fiber lengths. All simulations used a center charge, random initial fiber orientation and a rib entrance radius of 7 mm.

It is observed that 5 and 10 mm fibers exhibit less deviations in fiber content than 20 and 40 mm. With fibers which are shorter than, or approximately the same size as the rib entrance, significantly less fiber bridging is observed. With 20 and 40 mm fibers, intensive fiber bridging occurs which is reflected in the high fiber content at the base of the rib, and smaller fiber content in the upper two sections of the rib. Simulations are conducted to further assess the effect of initial fiber orientation on fiber volume content deviations inside the rib. Figure 13 shows that there is a dramatic difference as the initial orientation moved from cross rib aligned to aligned in the direction of the rib. This agrees with the hypothesis that bridging plays a large role in preventing fibers from entering the rib.

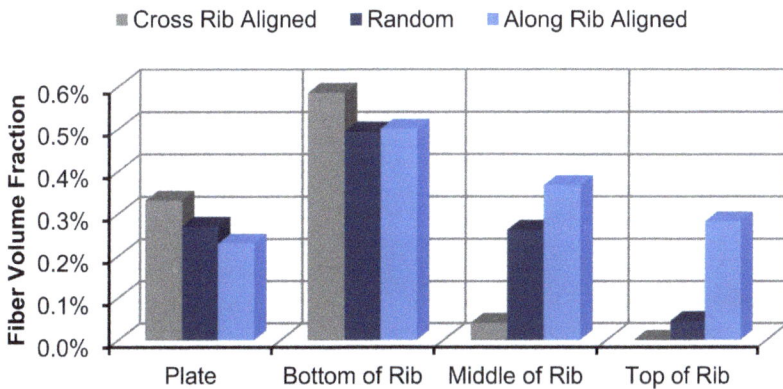

**Figure 13.** Fiber volume fraction throughout the rib for different initial fiber orientations. All simulations used a center charge with 20 mm fibers and a rib entrance radius of 5 mm.

Fewer fibers are oriented across the rib, which leads to fewer fibers as candidates for bridging. This reduces potential flow obstacles and allows more fibers to enter the rib. Further evidence is observed in the change of volume fraction at the bottom of the rib, where there are more fibers in the cross rib bundle. The same set of simulations was run at a slightly higher fiber volume fraction but, this did not prove to be a sufficiently high volume fraction to change the filling behavior appreciably. The effect of the radius of the rib entrance was also studied, and simulations using a rib that had no radius at the bottom were run to compare against the 5 mm and 7 mm radii, as shown in Figure 14. Adding a radius to the bottom of the rib improved the fiber volume fraction in the middle portion of the rib. Fibers had difficulties getting to the top of the rib regardless of the rib entrance, with no significant difference seen amongst the three different geometries. The absence of a radius also led to a higher volume fraction immediately below the rib entrance and a slightly higher volume fraction in the bottom-most portion of the rib. This higher volume fraction shows that there are more fibers that span the rib opening and become stuck in this area. With a smaller opening, a much wider range of orientations can facilitate bridging.

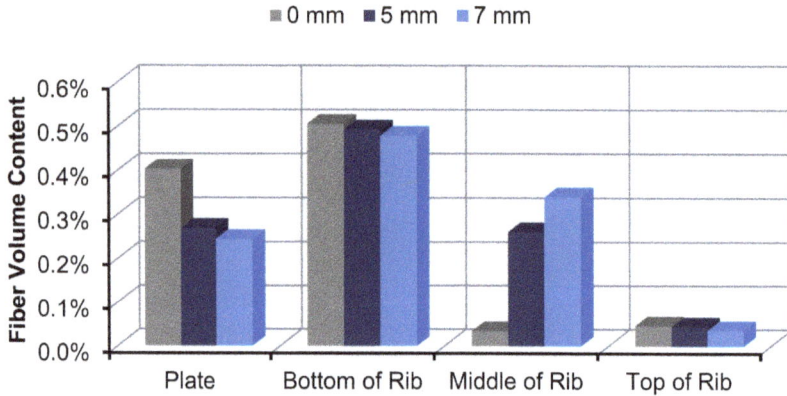

**Figure 14.** Volume fraction throughout rib for different rib entrance radii. All simulations used a center charge with 20 mm fibers and a random orientation.

Furthermore, the influence of polymer viscosity is investigated in Figure 15. With increasing viscosity, the volume fraction increases from the plate to the rib, and then decreases. The highest level of accumulation in the rib base is seen at the lowest viscosity level, where the hydrodynamic forces are unable to overcome the jamming that occurs. This also leads to a lower level in the midsection, with higher viscosities showing higher content here. Finally, in the top of the rib, none of the viscosity levels show a high volume fraction.

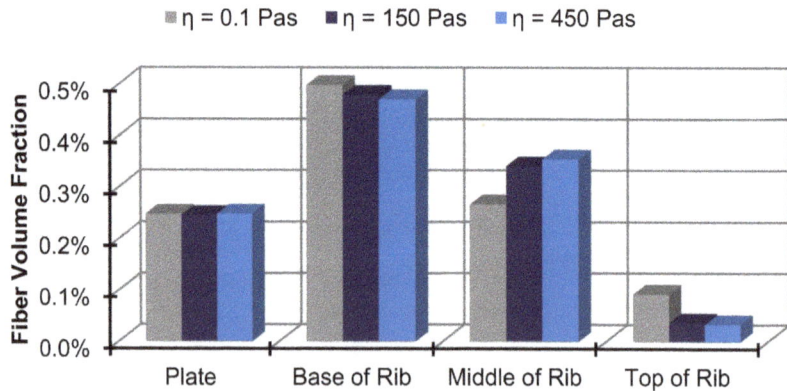

**Figure 15.** Influence of polymer viscosity on fiber content distribution inside the rib.

## 5. Experimental Results Comparison

The results of the DFS were compared to fiber content values taken from micro-computed tomography scans (µCT) of compression molded samples of the star rib geometries as shown in

Figure 11 [15]. As shown in Figure 16, it was found that the DFS was able to capture the general trend seen in molded parts. There is an initial increase in the fiber content from the plate to the base of the rib, followed by a decrease in the upper portions of the rib. The DFS generally shows a more drastic variation in fiber content than observed in the molded parts. For future comparisons, experimental data with the applied ribbed mold is necessary.

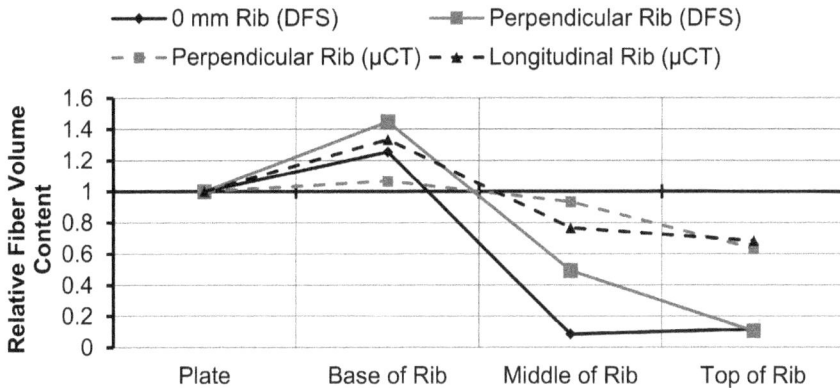

**Figure 16.** Comparison between fiber volume content taken from molded parts [15], and predicted values from the direct fiber simulation. Measured values are shown with dashed lines and simulation results with solid.

## 6. Conclusions and Outlook

Using the mechanistic model for DFS, it was found that initial fiber orientation and fiber length has the largest impact on the fiber volume fraction in ribs. Using simplified rib geometries, it was shown that decreasing the radius at the bottom of the rib has a negative impact on fiber volume fraction. Viscosity has not shown to play a large role in fiber content distributions, but it is believed that the initial fiber volume fractions used in this work are not high enough to capture the differences that have been found in other works. In general, the behavior observed in the mechanistic model generally complies with earlier experiments and shows great advantage over the predictions of traditional process simulation tools.

Future simulative work will focus on increasing fiber volume fractions, as well as investigating other parameters such as charge placement location and fiber bending stiffness and their effect on fiber volume fraction in ribs. Simulations using more complex geometry such as the star mold used for the preliminary simulations will be investigated. Analysis of other fiber effects including orientation and bending will also be performed. Regarding future experiments, the geometry presented in the simulation, the ribbed-plate, is applied to further study the fiber behavior during molding and to compare the simulative results directly.

**Acknowledgments:** The Volkswagen Group Research would like to thank the University of Wisconsin-Madison for their ongoing support in joint research projects.

**Author Contributions:** Christoph Kuhn and Ian Walter conceived and designed the experiments; Ian Walter performed the simulations; Christoph Kuhn and Ian Walter. analyzed the data and conceived further simulations; All contributed materials are provided by the Volkswagen Group Research under Olaf Taeger The Mechanistic Model is provided by the Polymer Engineering Center under Tim Osswald; Christoph Kuhn wrote the paper. All presented work was conducted at the Volkswagen Group under supervision of Christoph Kuhn.

**Conflicts of Interest:** The authors declare no conflict of interest.

## References

1.  Teixeira, D.; Giovanela, M.; Gonella, L.; Crespo, J. Influence of flow restriction on the microstructure and mechanical properties of long glass fiber-reinforced polyamide 6.6 composites. *Mater. Des.* **2015**, *85*, 695–706.
2.  Stauber, R.; Vollrath, L. *Plastics in Automotive Engineering*; Carl Hanser Verlag: Munich, Germany, 2007.
3.  Osswald, T.; Menges, G. *Material Science of Polymers for Engineers*; Carl Hanser Verlag: Munich, Germany, 2012.
4.  Davis, B.; Gramann, P.T.O.; Rios, A. *Compression Molding*; Carl Hanser Verlag: Munich, Germany, 2003.
5.  Wang, J.; O'gara, J.; Tucker, C. An Objective model for slowing orientation kinetics in concentrated fiber suspensions: Theory and rheological evidence. *J. Rheol.* **2008**, *52*, 1179–1200. [CrossRef]
6.  Folgar, F.; Tucker, C. Orientation behavior of fibers in concentrated suspensions. *J. Reinf. Plast. Compos.* **1984**, *3*, 98–119. [CrossRef]
7.  Phelps, J.; Tucker, C. An anisotropic rotary diffusion model for fiber orientation in short and long-fibre thrmoplastics. *J. Non-Newton. Fluid Mech.* **2009**, *156*, 165–176. [CrossRef]
8.  Londoño-Hurtado, A. *A Mechanistic Model for Fiber Flow*; University of Wisconsin-Madison: Madison, WI, USA, 2009.
9.  Goris, S.; Osswald, T.A. Process-Induced Fiber Matrix Separation in Long Fiber-Reinforced Thermoplastics. *Compos Part A Appl. Sci. Manuf.* **2018**, *105*, 321–333. [CrossRef]
10. Londoño, A.; Osswald, T.A. Fiber Jamming and Fiber Matrix Separation during Compression Molding. *J. Plast. Technol.* **2006**, *15*, 109.
11. Jeffery, G.B. The Motion of Ellipsoidal Particles Immersed in a Viscous Fluid. *R. Soc.* **1922**, *102*, 161–179. [CrossRef]
12. Thomason, J.; Vlug, M. Influence of fibre length and concentration on the properties of glass fibre-reinforced polypropylene: 1. Tensile and flexural modulus. *Compos. Part A Appl. Sci. Manuf.* **1996**, *27*, 419–503.
13. Pèrez, C. *The Use of a Direct Particle Simulation to Predict Fiber Motion in Polymer Processing*; University of Wisconsin-Madison: Madison, WI, USA, 2016.
14. Ramìrez, D. *Study of Fiber Motion in Molding Processes by Means of a Mechanistic Model*; University of Wisconsin-Madison: Madison, WI, USA, 2014.
15. Kuhn, C.; Walter, I.; Taeger, O.; Osswald, T. Experimental and Numerical Analysis of Fiber Matrix Separation During Compression Molding of Reinforced Thermoplastics. *J. Compos. Sci.* **2017**, *1*, 2. [CrossRef]
16. Pèrez, C.; Ramìrez, D.; Osswald, T. Mechanistic model simulation of a compression molding process: Fiber orientation and fiber-matrix separation. In Proceedings of the Technical Conference & Exhibition, Orlando, FL, USA, 23–25 March 2015.
17. Lindstroem, S.; Uesaka, T. Simulation of the motion of flexible fibers in viscous fluid flow. *Phys. Fluids* **2007**, *19*, 113307. [CrossRef]
18. Schmid, C.; Switzer, L.; Klingenberg, D. Simulations of fiber flocculation: Effects of fiber properties and interfiber friction. *J. Rheol.* **2000**, *44*, 781–809. [CrossRef]

MDPI

St. Alban-Anlage 66

4052 Basel

Switzerland

Tel. +41 61 683 77 34

Fax +41 61 302 89 18

www.mdpi.com

*Journal of Composites Science* Editorial Office

E-mail: jcs@mdpi.com

www.mdpi.com/journal/jcs

MDPI
St. Alban-Anlage 66
4052 Basel
Switzerland

Tel: +41 61 683 77 34
Fax: +41 61 302 89 18

www.mdpi.com

MDPI

ISBN 978-3-03897-492-5

www.ingramcontent.com/pod-product-compliance
Lightning Source LLC
Chambersburg PA
CBHW051846210326
41597CB00033B/5790